W9-AZC-498

Global Physical Climatology

This is Volume 56 in the
INTERNATIONAL GEOPHYSICS SERIES
A series of monographs and textbooks
Edited by RENATA DMOWSKA and JAMES R. HOLTON

A complete list of the books in this series appears at the end of this volume.

Global Physical Climatology

Dennis L. Hartmann
DEPARTMENT OF ATMOSPHERIC SCIENCES
UNIVERSITY OF WASHINGTON
SEATTLE, WASHINGTON

ACADEMIC PRESS
An Imprint of Elsevier
San Diego New York Boston
London Sydney Tokyo Toronto

Front cover photograph: Apollo 17 view of Earth courtesy of the National Aeronautics and Space Administration. The view extends from the Mediterranean Sea at the top to the Antarctic ice cap at the bottom. The coastline of Africa, the Arabian Peninsula, and the island of Madagascar are clearly visible, and the Asian mainland is on the horizon toward the northeast. Note the bright clouds over equatorial Africa, with a band of dark vegetated land just to the north of it in the Sahel region, and to the north of the dark band the much brighter surfaces of the Sahara Desert. Cover design assistance was provided by K. M. Dewar and G. C. Gudmundson, Department of Atmospheric Sciences, University of Washington.

ISBN-13: 978-0-12-328530-0
ISBN-10: 0-12-328530-5

This book is printed on acid-free paper. 

Copyright © 1994 by ACADEMIC PRESS

All Rights Reserved.
No part of this publication may be reproduced or transmitted in any form or by any means, electronic or mechanical, including photocopy, recording, or any information storage and retrieval system, without permission in writing from the publisher.

Permissions may be sought directly from Elsevier's Science and Technology Rights Department in Oxford, UK. Phone: (44) 1865 843830, Fax: (44) 1865 853333, e-mail: permissions@elsevier.co.uk. You may also complete your request on-line via the Elsevier homepage: http://www.elsevier.com by selecting "Customer Support" and then "Obtaining Permissions".

Explicit permission from Academic Press is not required to reproduce a maximum of two figures or tables from an Academic Press chapter in another scientific or research publication provided that the material has not been credited to another source and that full credit to the Academic Press chapter is given.

Academic Press
An Imprint of Elsevier
525 B Street, Suite 1900, San Diego, California 92101-4495, USA
http://www.academicpress.com

Academic Press
Harcourt Place, 32 Jamestown Road, London NW1 7BY, UK
http://www.academicpress.com

Library of Congress Cataloging-in-Publication Data

Hartmann, Dennis L.
Global physical climatology / Dennis L. Hartmann.
p. cm. – (International geophysics : v. 56)
Includes index.
ISBN-13: 978-0-12-328530-0 ISBN-10: 0-12-328530-5
1. Climatology. 2. Atmospheric physics. I. Title. II. Series.
QC981.H32 1994
551.6—dc20 93-39578
CIP

PRINTED IN THE UNITED STATES OF AMERICA

09 10 11 MM 16 15 14

Contents

Preface

The science of climatology began to evolve rapidly in the last third of the twentieth century. This rapid development arose from several causes. During this period the view of Earth from its moon made people more aware of the exceptional nature of their planetary home at about the same time that it became widely understood that humans could alter our global environment. Scientific and technological developments gave us new and quantitative information on past climate variations, global observations of climate parameters from space, and computer models with which we could simulate the global climate system. These new tools together with concern about global environmental change and its consequences for humanity caused an increase in the intensity of scientific research about climate.

Modern study of the Earth's climate system has become an interdisciplinary science incorporating the atmosphere, the ocean, and the land surface, which interact through physical, chemical, and biological processes. A fully general treatment of this system is as yet impossible, because the understanding of it is just beginning to develop. This textbook provides an introduction to the physical interactions in the climate system, viewed from a global perspective. Even this endeavor is a difficult one, since many earth science subdisciplines must be incorporated, such as dynamic meteorology, physical oceanography, radiative transfer, glaciology, hydrology, boundary-layer meteorology, and paleoclimatology. To make a book of manageable size about such a complex topic requires many difficult choices. I have endeavored to provide a sense of the complexity and interconnectedness of the climate problem without going into excessive detail in any one area. Although the modern approach to climatology has arisen out of diverse disciplines, a coherent collection of concepts is emerging that defines a starting point for a distinct science. This textbook is my attempt to present the physical elements of that beginning, with occasional references to where the chemical and biological elements are connected.

This book is intended as a text for upper-division undergraduate physical science majors and, especially in the later chapters, graduate students. I have used the first seven chapters as the basis for a 10-week undergraduate course for atmospheric sciences majors. A graduate course can be fashioned by supplementing the text with readings from the current literature. Most climatology textbooks are descriptive and written from the perspective of geographers, but this one is written from the perspective of a physicist. I have attempted to convey an intuition for the workings of the climate system that is based on physical principles. When faced with a choice between providing easy access to an important concept and providing a rigorous and comprehensive treatment, I have chosen easy access. This approach should allow students to acquire the main ideas without great pain. Instructors may choose to elaborate on the presentation where their personal interests and experience make it desirable to do so.

This book could not have been produced without the assistance of many people. It evolved from 15 years of teaching undergraduate and graduate students, and I thank the ATMS 321 and ATMS 571 students at the University of Washington who have endured my experimentation and provided comments on early drafts of this book. Professor Steve Esbensen and his AtS 630 class at Oregon State University provided commentary on a near final draft of Chapters 1–7 in the spring of 1993. Valuable comments and suggestions on specific chapters were also provided by David S. Battisti, Robert J. Charlson, James R. Holton, Conway B. Leovy, Gary A. Maykut, Stephen G. Porter, Edward S. Sarachik, J. Michael Wallace, and Stephen G. Warren. The encouragement and advice given by James R. Holton were critical to the completion of this book. Many people contributed graphics, and I am particularly grateful for the special efforts given by Otis Brown, Frank Carsey, Jim Coakley, Joey Comiso, Scott Katz, Gary Maykut, Pat McCormick, Robert Pincus, Norbert Untersteiner, and Stephen Warren.

Grace C. Gudmundson applied her professional editorial skills to this project with patience, dedication, and good humor. Her efforts greatly improved the quality of the end product. Similarly, Kay M. Dewar's artistic and computer skills produced some of the more appealing figures. Marc L. Michelsen's genius with the computer extracted data from many digital archives and converted them into attractive and informative computer graphics. Luanna Huynh and Christine Rice were especially helpful with the appendices and tables.

My efforts to understand the climate system have been generously supported over the years by research grants and contracts from the United States government. I am particularly happy to acknowledge support from the Climate Dynamics Program in the Atmospheric Sciences Division of the National Science Foundation, and the Earth Radiation Budget Experiment and Earth Observing System programs of the National Aeronautics and Space Administration. I also thank all of my colleagues from whom I have learned, who have shared their ideas with me, and who have given me the respect of serious argument.

This book is dedicated to my family, especially my wife, Lorraine, and my children, Alan and Jennifer, whose love and sacrifice were essential to its completion. I hope this book will help to explain why I spend so many evenings and weekends in my study. I thank my parents, Alfred and Angeline, for a good start in life and support along the way toward happy employment.

Dennis L. Hartmann

Global Physical Climatology

Chapter 1 Introduction to the Climate System

1.1 Atmosphere, Ocean, and Land Surface

Climate is the synthesis of the weather in a particular region. It can be defined quantitatively using the expected values of the meteorological elements at a location during a certain month or season. The expected values of the meteorological elements can be called the climatic elements and include such variables as the average temperature, precipitation, wind, pressure, cloudiness, and humidity. In defining the climate we usually employ the values of these elements at the surface of Earth. Thus one can characterize the climate of Seattle by stating that the average annual mean precipitation is 36 in. and the annual mean temperature is 52°F. One might need a great deal more information than the annual means, however. For example, a farmer would also like to know how the precipitation is distributed through the year and how much rain would fall during the critical summer months. A hydroelectric plant engineer needs to know how much interannual variability in rainfall and snow accumulation to expect. A homebuilder should know how much insulation to install and the size of the heating or cooling unit needed to provide for the weather in the region.

The importance of climate is so basic that we sometimes overlook it. If the climate were not more or less as it is, life and civilization on this planet would not have developed as they have. The distribution of vegetation and soil type over the land areas is determined primarily by the local climate. Climate affects human lives in many ways; for example, climate influences the type of clothing and housing that people have developed. In the modern world, with the great technological advances of the last century, one might think that climate no longer constitutes a force capable of changing the course of human history. It is apparent, on the contrary, that we are as sensitive now as we have ever been to climate fluctuations and climate change.

Because food, water, and energy supply systems are strained to meet demand and are optimized to the current average climatic conditions, fluctuations or trends in climate can cause serious difficulties for humanity. Moreover, since the population has grown to absorb the maximum agricultural productivity in much of the world, the absolute number of human lives at risk of starvation during climatic anomalies has never been greater. In addition to natural year-to-year fluctuations in the weather, which are an important aspect of climate, we must be concerned with the effects of human activities in producing long-term trends in the climate. It is now clear that humans are affecting local climate and our influence on global climate will increase in the future. The actions of humankind that can influence the global climate include altering the nature of Earth's surface and the composition of Earth's atmosphere.

The surface climate of Earth is varied, ranging from the heat of the tropics to the cold of the polar regions, and from the drought of a desert to the moisture of a rain forest. Nonetheless, the climate of Earth is favorable for life, and living creatures exist in every climatic extreme. The climate of a region depends on latitude, altitude, and orientation in relation to water bodies, mountains, and the prevailing wind direction. In this book we are concerned primarily with the global climate and its geographic variation on scales of hundreds to thousands of kilometers. In order to focus on these global issues, climate variations on horizontal spatial scales smaller than several tens of kilometers are given only minimal discussion.

The climate of Earth is defined in terms of measurable weather elements. The weather elements of most interest are temperature and precipitation. These two factors together largely determine the species of plants and animals that survive and prosper in a particular location. Other variables are also important, of course. The *humidity,* the amount of water vapor in the air, is a critical climate factor that is related closely to the temperature and precipitation. Condensation of water in the atmosphere produces clouds of water droplets or ice particles that greatly change the radiative properties of the atmosphere. Cloudiness influences the amount of solar radiation that reaches the surface and the transmission of terrestrial radiation through the atmosphere. The occurrence of clouds is important in itself for aviation and other activities, but clouds also play a role in determining both precipitation and surface temperature. The mean wind speed and direction are important considerations for local climate, air-pollution dispersion, aviation, navigation, wind energy, and many other purposes. The climate system of Earth determines the distribution of energy and water near the surface and consists primarily of the atmosphere, the oceans, and the land surface. The workings of this global system are the topic of this book (Fig. 1.1).

1.2 Atmospheric Temperature

Temperature is the most widely recognized climatic variable. The global average temperature at the surface of Earth is 288 K, 15°C, or 59°F. The range of temperatures encountered at the surface is favorable for the life forms that have developed on Earth. The extremes of recorded surface temperature range from the coldest temperature of −89°C (−128.6°F) at Vostok, Antarctica to the warmest temperature of 58°C (136.4°F) at Al Aziziyah, Libya. These temperature extremes reflect the well-known decrease of temperature from the tropics, where the warmest temperatures occur, to the polar regions, which are much colder. The cold temperature at Vostok is a result of both the high latitude and altitude at this location, which is 3450 m above sea level.

Fig. 1.1 View of Earth from space (Apollo Saturn, AS10, NASA, May 18–26, 1969).

An important feature of the temperature distribution is the decline of temperature with height above the surface in the lowest 10–15 km of the atmosphere (Fig. 1.2). This rate of decline, called the *lapse rate,* is defined by

$$\Gamma \equiv -\frac{\partial T}{\partial z} \tag{1.1}$$

where T is the temperature and z is altitude. The global mean tropospheric lapse rate is about 6.5 K km^{-1}, but the lapse rate varies with altitude, season, and latitude. In the upper *stratosphere* the temperature increases with height up to about 50 km. The

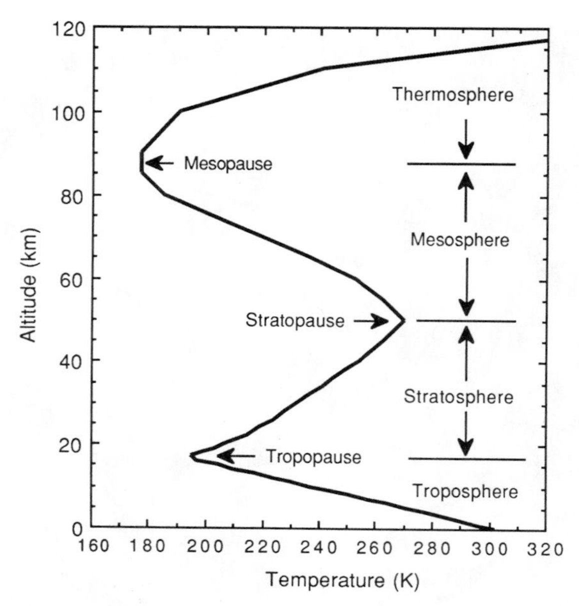

Fig. 1.2 The main zones of the atmosphere defined according to the temperature profile of the standard atmosphere profile at 15°N for annual-mean conditions. [Data from U.S. Standard Atmosphere Supplements (1966).]

increase of temperature with height that characterizes the stratosphere is caused by the absorption of solar radiation by ozone. Above the stratopause at about 50 km the temperature begins to decrease with height in the *mesosphere*. The temperature of the atmosphere increases rapidly above about 100 km because of heating produced by absorption of ultraviolet radiation from the sun, which dissociates oxygen and nitrogen molecules and ionizes atmospheric gases in the *thermosphere*.

The decrease of temperature with altitude in the *troposphere* is crucial to many of the mechanisms whereby the warmth of the surface temperature of Earth is maintained. The lapse rate in the troposphere and the mechanisms that maintain it are also central to the determination of climate sensitivity, as discussed in Chapter 9. The lapse rate and temperature in the troposphere are determined primarily by a balance between radiative cooling and convection of heat from the surface. The vertical distribution of temperature varies with latitude and season. At the equator the temperature decreases with altitude up to about 17 km (Fig. 1.3). The tropical *tropopause* is the coldest part of the lowest 20 km of the atmosphere in the annual mean. In middle and high latitudes the temperature of the lower stratosphere is almost independent of height. The tropospheric lapse rate in polar latitudes is less than it is nearer the equator. In the winter and spring at high latitudes the temperature actually increases with altitude in the lower troposphere (Fig. 1.4). A region of negative lapse rate is called a *temperature inversion*. The polar temperature inversion has important implications

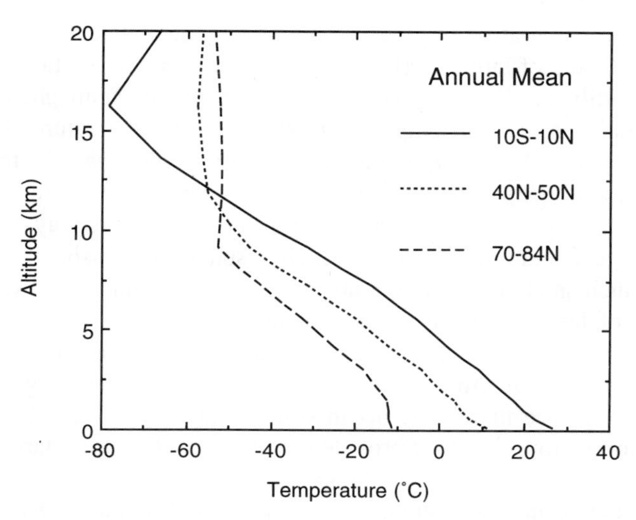

Fig. 1.3 Annual-mean temperature profiles for the lowest 20 km of the atmosphere in three latitude bands. [Data from Oort (1983).]

for the climate of the polar regions. It arises because the surface cools very efficiently through emission of infrared radiation in the absence of insolation during the winter darkness. The air does not emit radiation as efficiently as the surface, and heat transported poleward in the atmosphere keeps the air in the lower troposphere warmer than the surface.

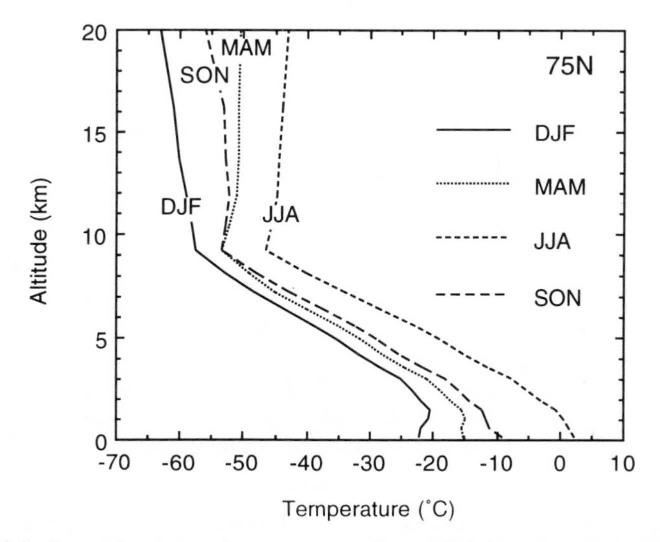

Fig. 1.4 Seasonal variation of temperature profiles at 75°N. [Data from Oort (1983).]

The surface temperature is greatest near the equator, where it exceeds 26°C across a broad band of latitudes (Fig. 1.5). Outside this belt, surface temperature decreases steadily toward both poles. In the Northern Hemisphere we see a strong seasonal variation of temperature. The coldest temperatures in the polar regions occur during February, and are about 26°C colder than the temperatures in July. The amplitude of the seasonal cycle in temperature decreases from the North Pole to the equator, where the zonal-mean temperature stays within a degree of 27°C year round. In the Southern Hemisphere the seasonal cycle of temperature is much smaller than in the north. The largest contrast between the seasonal cycles of temperature in the two hemispheres occurs in the latitude belt between 45 and 60 degrees. The smaller seasonal variation of air temperature in midlatitudes of the Southern Hemisphere is associated with the larger fraction of ocean-covered surface there. The ocean stores heat very effectively. During the summer season it stores the heat provided by the sun. Because a large amount of heat is required to raise the surface temperature of the oceans, the summer insolation raises the surface temperature by only a small amount. During winter a large amount of heat is released to the atmosphere with a relatively small change in sea surface temperature. Land areas heat up and cool down much more quickly than oceans.

The geographic distributions of January and July surface temperature are shown in Fig. 1.6. The interiors of the northern continents become very cold during winter, but during summer they are warmer than ocean areas at the same latitude. Seasonal variations of surface temperature in the interiors of North America and Asia are very large (Fig. 1.7). Seasonal variation in the Southern Hemisphere is much smaller because of the greater fraction of the surface covered by ocean.

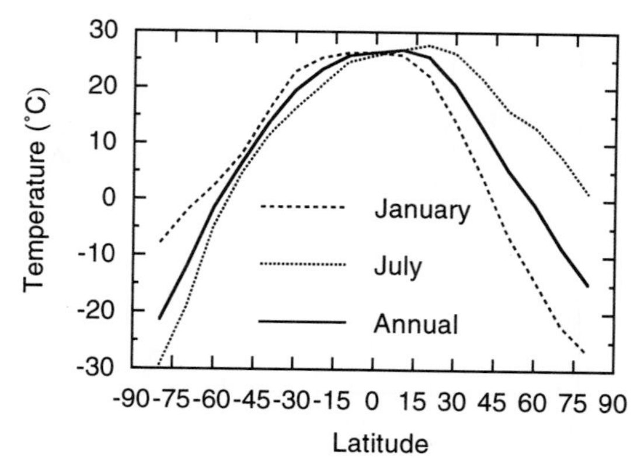

Fig. 1.5 Near-surface air temperature as a function of latitude for January, July, and annual-average conditions. [Data from Oort (1983).]

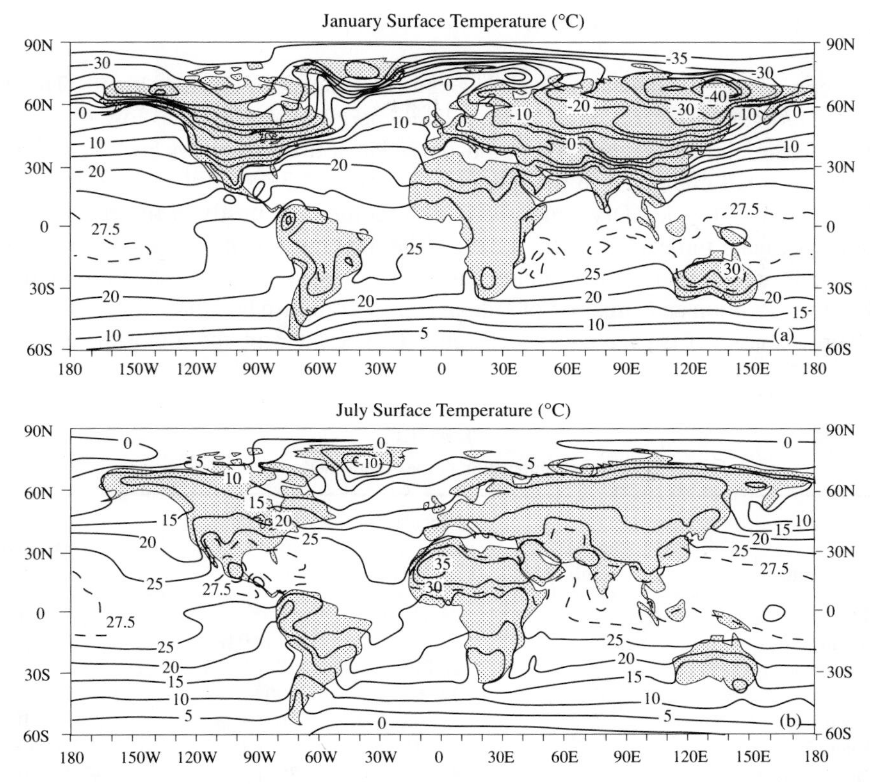

Fig. 1.6 Global map of the (a) January and (b) July surface temperature. [From Shea (1986). Reproduced with permission from the National Center for Atmospheric Research.]

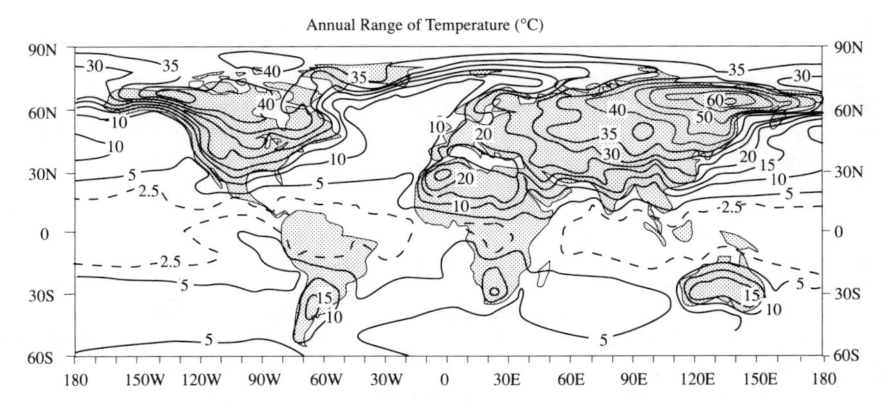

Fig. 1.7 Map of the amplitude of the annual cycle of surface temperature. [From Shea (1986). Reproduced with permission from the National Center for Atmospheric Research.]

1.3 Atmospheric Composition

The composition of the atmosphere is a key determinant of Earth's climate. The interaction of atmospheric gases with radiant energy modulates the flow of energy through the climate system. The atmosphere has a mass of about 5.14×10^{18} kg, which is small compared to the mass of the ocean, 1.39×10^{21} kg, and the solid Earth, 5.98×10^{24} kg. Dry atmospheric air is composed mostly of molecular nitrogen (78%) and molecular oxygen (21%). The next most abundant gas in the atmosphere is argon (1%), an inert noble gas. The atmospheric gases that are important for the absorption and emission of radiant energy comprise less than 1% of the atmosphere's mass. These include water vapor (3.3×10^{-3} of the atmosphere's total

Table 1.1

Composition of the Atmosphere

Constituent	Chemical formula	Molecular weight ($^{12}C = 12$)	Fraction by volume in dry air	Total mass (g)
Total atmosphere		28.97		5.136×10^{21}
Dry air		28.964	100.0 %	5.119×10^{21}
Nitrogen	N_2	28.013	78.08 %	3.87×10^{21}
Oxygen	O_2	31.999	20.95 %	1.185×10^{21}
Argon	Ar	39.948	0.934 %	6.59×10^{19}
Water vapor	H_2O	18.015	Variable	1.7×10^{19}
Carbon dioxide	CO_2	44.01	353 ppmv[a]	$\sim 2.76 \times 10^{18}$
Neon	Ne	20.183	18.18 ppmv	6.48×10^{16}
Krypton	Kr	83.80	1.14 ppmv	1.69×10^{16}
Helium	He	4.003	5.24 ppmv	3.71×10^{15}
Methane	CH_4	16.043	1.72 ppmv[a]	$\sim 4.9 \times 10^{15}$
Xenon	Xe	131.30	87 ppbv	2.02×10^{15}
Ozone	O_3	47.998	Variable	$\sim 3.3 \times 10^{15}$
Nitrous oxide	N_2O	44.013	310 ppbv[a]	$\sim 2.3 \times 10^{15}$
Carbon monoxide	CO	28.01	120 ppbv	$\sim 5.9 \times 10^{14}$
Hydrogen	H_2	2.016	500 ppbv	$\sim 1.8 \times 10^{14}$
Ammonia	NH_3	17.03	100 ppbv	$\sim 3.0 \times 10^{13}$
Nitrogen dioxide	NO_2	46.00	1 ppbv	$\sim 8.1 \times 10^{12}$
Sulfur dioxide	SO_2	64.06	200 pptv	$\sim 2.3 \times 10^{12}$
Hydrogen sulfide	H_2S	34.08	200 pptv	$\sim 1.2 \times 10^{12}$
CFC-12	CCl_2F_2	120.91	480 pptv[a]	$\sim 1.0 \times 10^{13}$
CFC-11	CCl_3F	137.37	280 pptv[a]	$\sim 6.8 \times 10^{12}$

[Data excerpted with the permission of the Macmillan Company from *Evolution of the Atmosphere* by J. C. G. Walker, © 1977 by Macmillan Publishing Company; Verniani, 1966 © American Geophysical Union; and Williamson (1973).]

[a] Values of trace constituents valid in 1990 (ppmv = 10^{-6}, ppbv = 10^{-9}, pptv = 10^{-12}) (ppmv, ppbv, pptv = parts per million, billion, trillion by volume).

mass), carbon dioxide (5.3×10^{-7}), and ozone (6.42×10^{-7}), in order of importance for surface temperature, followed by methane, nitrous oxide, and a host of other minor species (Table 1.1).

1.4 Hydrostatic Balance

The atmosphere is composed of gases held close to the surface of the planet by gravity. The vertical forces acting on the atmosphere at rest are gravity, which pulls the air molecules toward the center of the planet, and the pressure force, which tries to push the atmosphere out into space. These forces are in balance to a very good approximation, and by equating the pressure gradient force and the gravity force one obtains the *hydrostatic balance*. Since force is mass times acceleration, we may express the vertical force balance per unit mass as an equation between the downward acceleration of gravity, g, and the upward acceleration that would be caused by the increase of pressure toward the ground, if gravity were not present to oppose it.

$$g = -\frac{1}{\rho}\frac{dp}{dz} \tag{1.2}$$

For an ideal gas, pressure (p), density (ρ), and temperature (T) are related by the formula

$$p = \rho RT \tag{1.3}$$

where R is the gas constant. After some rearrangement, (1.2) and (1.3) yield

$$\frac{dp}{p} = -\frac{dz}{H} \tag{1.4}$$

where

$$H = \frac{RT}{g} = \text{scale height} \tag{1.5}$$

If the atmosphere is *isothermal*, then the temperature and scale height are constant and the hydrostatic equation may be integrated from the surface, where $p = p_s = 1.01325 \times 10^5$ Pa, to an arbitrary height, z, yielding an expression for the distribution of pressure with height.

$$p = p_s\, e^{-z/H} \tag{1.6}$$

The pressure thus decreases exponentially away from the surface, declining by a factor of $e = 2.71828$ every scale height. The scale height for the mean temperature

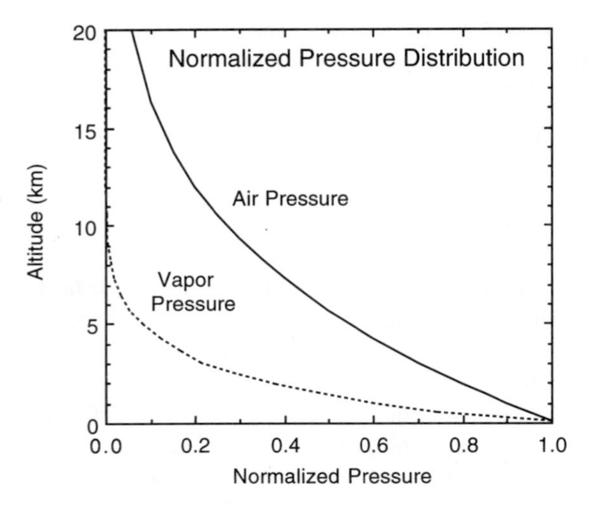

Fig. 1.8 Vertical distributions of air pressure and partial pressure of water vapor as functions of altitude for globally and annually averaged conditions. Values have been normalized by dividing by the surface values of 1013.25 and 17.5 mb (millibars), respectively.

of Earth's atmosphere is about 7.6 km. Figure 1.8 shows the distribution of atmospheric pressure with altitude. The pressure is largest at the surface and decreases rapidly with altitude in accord with the exponential decline given by (1.6). We can rearrange (1.2) to read

$$dm \equiv \rho\, dz = \frac{-dp}{g} \tag{1.7}$$

The mass between two altitudes, *dm,* is related to the pressure change between those two levels. Because of hydrostatic balance, the total mass of the atmosphere may be related to the global mean surface pressure.

$$\text{Atmospheric mass} = \frac{p_s}{g} = 1.03 \times 10^4 \ \text{kg m}^{-2} \tag{1.8}$$

The vertical column above every square meter of Earth's surface contains about 10,000 kg of air.

Because the surface climate is of primary interest, and because the mass of the atmosphere is confined to within a few scale heights of the surface, or several tens of kilometers, it is the lower atmosphere that is of most importance for climate. For this reason most of this book will be devoted to processes taking place in the troposphere, at the surface, or in the ocean. The stratosphere has some important effects on climate, however, and these will be described where appropriate.

1.5 Atmospheric Humidity

Atmospheric humidity is the amount of water vapor carried in the air. The atmosphere must carry away the water evaporated from the surface and supply water to areas of rainfall. Water that flows from the land to the oceans in rivers was brought to the land areas by transport in the atmosphere as vapor. Atmospheric water vapor is also the most important greenhouse gas in the atmosphere. Water vapor condenses to form clouds, which can release rainfall and are also extremely important in both reflecting solar radiation and reducing the infrared radiation emitted by Earth.

The mass mixing ratio of water vapor in the atmosphere decreases very rapidly with altitude (Fig. 1.9). When one considers that much of the mass of the atmosphere is also within 5 km of the surface it is clear that most of the atmospheric water vapor is within a few kilometers of the surface. The partial pressure of water vapor decreases to half of its surface value by 2 km above the surface and to less than 10% of its surface value at 5 km (Fig. 1.8). Atmospheric water vapor also decreases rapidly with latitude. The amount of water vapor in the atmosphere at the equator is nearly 10 times that at the poles. The rapid upward and poleward decline in water vapor abundance in the atmosphere is associated with the strong temperature dependence of the saturation vapor pressure. Warmer air can contain a much larger fraction of water vapor.

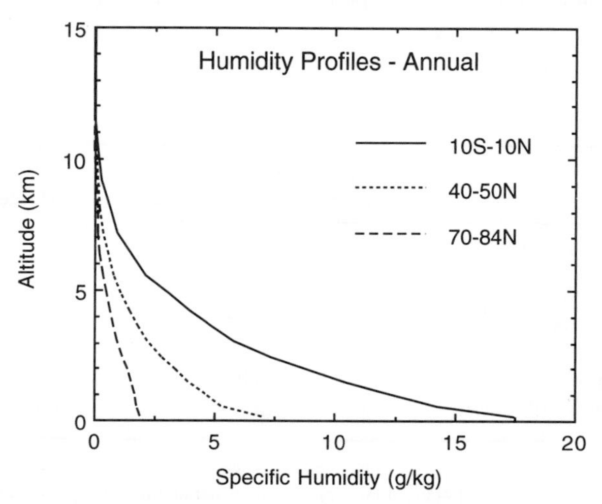

Fig. 1.9 Specific humidity or mass mixing ratio of water vapor for annual-mean conditions at latitude belts centered on the equator, 45°N and 75°N. [Data from Oort (1983).]

1.6 The World Ocean

The atmosphere contains a minute fraction of the total water of the climate system—about one part in 10^5. Most of the surface water of Earth is contained in the oceans and in ice sheets (Table 1.2). Earth contains about 1.35×10^9 km^3 of water, of which about 97% is seawater. Since all the oceans are connected to some degree, we can think of them collectively as the world ocean. The world ocean is a key element of the physical climate system. Ocean covers about 71% of Earth's surface to an average depth of 3729 m. The ocean has tremendous capability to store and release heat on time scales of seasons to centuries. About half of the equator-to-pole energy transport that works to warm the poles and cool the tropics is provided by the oceans. The world ocean is the reservoir of water that supplies atmospheric water vapor for rain and snowfall over land. The ocean plays a key role in determining the composition of the atmosphere through the exchange of gases and particles across the air–sea interface. The ocean removes carbon dioxide from the atmosphere and produces molecular oxygen, and participates in other key geochemical cycles that regulate the surface environment of Earth.

Temperature in the ocean generally decreases with depth from a temperature very near that of the surface air temperature to a value near the freezing point of water in the deep ocean (Fig. 1.10). A thin, mixed surface layer is stirred by winds and waves so efficiently that its temperature and salinity are almost independent of depth. Most temperature change occurs in the *thermocline,* the first kilometer or so of the ocean. Below the thermocline is a deep layer of almost uniform temperature.

Salinity of seawater is defined as the number of grams of dissolved salts in a kilogram of seawater. Salinity in the open ocean ranges from about 33–38 g kg^{-1}. In

Table 1.2

Water on Earth

Water reservoir	Depth if spread over the entire surface of Earth (m)	Percent of total
Oceans	2650	97
Icecaps and glaciers	60	2.2
Groundwater[a]	20	0.7
Lakes and streams[a]	0.35	0.013
Soil moisture[a]	0.12	0.013
Atmosphere	0.025	0.0009
Total	2730	100

[Data are a composite of various sources: Nace (1964), used with permission from the American Museum of Natural History; Baumgartner and Reichel (1975), and Korzun et al. (1978).]

[a]Numbers uncertain.

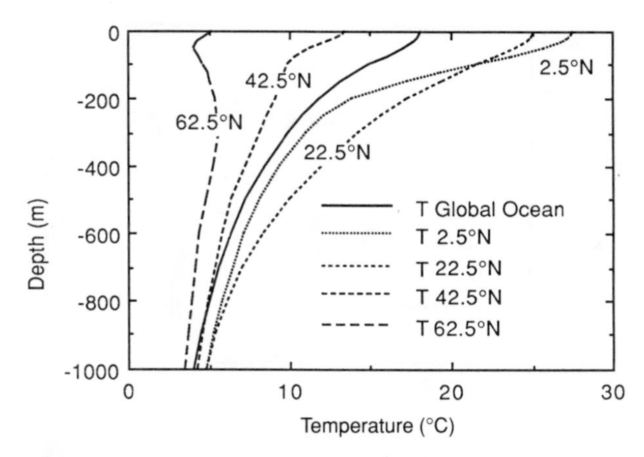

Fig. 1.10 Annual-mean ocean temperature profiles for various latitudes. [Data from Levitus (1982).]

seawater with a salinity of 35, about 30 g kg^{-1} are composed of sodium and chloride (Table 1.3). Salinity is an important contributor to variations in the density of seawater at all latitudes and is the most important factor in high latitudes, where the temperature is close to the freezing point of water. Variations in the density of seawater drive the deep-ocean circulation, which is critical for heat transport and the recirculation of nutrients necessary for sea life. Salinity of the global ocean varies

Table 1.3

Concentrations of the Major Components of Seawater with a Salinity of 35‰

Component	Grams per kilogram
Chloride	19.353
Sodium	10.76
Sulfate	2.712
Magnesium	1.294
Calcium	0.413
Potassium	0.387
Bicarbonate	0.142
Bromide	0.067
Strontium	0.008
Boron	0.004
Fluoride	0.001

[From Turekian (1968) © by Prentice-Hall, Inc. Englewood Cliffs, NJ. Used with permission from the publisher.]

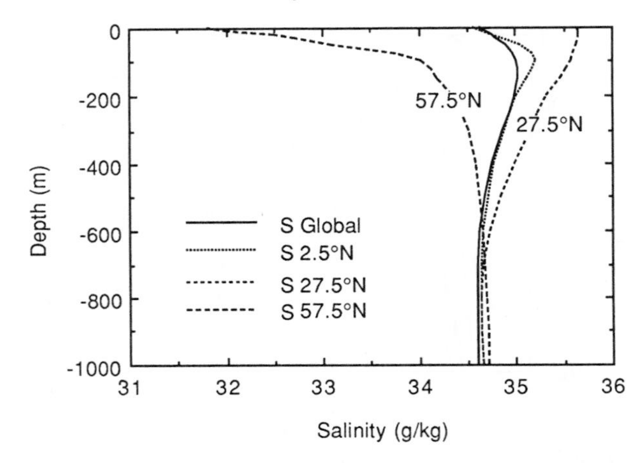

Fig. 1.11 Profiles of annual-mean salinity for the global mean and for various latitudes. [Data from Levitus (1982).]

systematically with latitude in the upper layers of the ocean (Fig. 1.11). In subtropical latitudes the surface salinity is large because evaporation exceeds precipitation and leaves the seawater enriched in salt. In middle and high latitudes precipitation of freshwater exceeds evaporation and surface salinities are quite low. In the deep ocean salinity variations are much smaller than near the surface, because the sources and sinks of freshwater are at the surface.

1.7 The Cryosphere

About 2% of the water on Earth is frozen, and this frozen water constitutes about 80% of the freshwater. Most of the mass of ice is contained in the great ice sheets of Antarctica (89%) and Greenland (8.6%) (Table 1.4). All of the ice near the surface of Earth is called the *cryosphere*. For climate, it is often not the mass of ice that is of primary importance, but rather the surface area that is covered by ice of any depth. This is because surface ice of any depth generally is a much more effective reflector of solar radiation than the underlying surface. Also, sea ice is a good insulator and allows air temperature to be very different from that of the seawater under the ice. Currently, year-round (perennial) ice covers about 11% of the land area and 7% of the world ocean. During some seasons the amount of land covered by seasonal snowcover exceeds the surface area covered by perennial icecover. The surface areas covered by ice sheets, seasonal snowcover, and sea ice are comparable. Ice sheets cover about 16×10^6 km^2, seasonal snow about 50×10^6 km^2, and sea ice up to 23×10^6 km^2.

Table 1.4
Estimated Global Inventory of Land and Sea Ice[a]

			Area (km^2)	Volume (km^3)	Percent of total ice mass
Land ice	Antarctic ice sheet		13.9×10^6	30.1×10^6	89.3
	Greenland ice sheet		1.7×10^6	2.6×10^6	8.6
	Mountain glaciers		0.5×10^6	0.3×10^6	0.76
	Permafrost	Continuous	8×10^6	(Ice content) $0.2–0.5 \times 10^6$	0.95
		Discontinuous	17×10^6		
	Seasonal snow (average maximum)	Eurasia	30×10^6	$2\text{-}3 \times 10^3$	
		America	17×10^6		
Sea ice	Southern Ocean	Max	18×10^6	2×10^4	
		Min	3×10^6	6×10^3	
	Arctic Ocean	Max	15×10^6	4×10^4	
		Min	8×10^6	2×10^4	

[a]Not included in this table is the volume of water in the ground that annually freezes and thaws at the surface of permafrost ("active layer"), and in regions without permafrost but with subfreezing winter temperatures. [After Untersteiner (1984). Printed with permission from Cambridge University Press.]

1.8 The Land Surface

Although the land surface plays a lesser role in the global climate system than the atmosphere or the ocean, the climate over the land surface is extremely important to us because humans are land-dwelling creatures. Cereal grains are the world's most important food source, and supply about half of the world's calories and much of the protein. About 80% of the animal protein consumed by humans comes from meat, eggs, and dairy products, and only 20% from seafood. In addition, most of our construction materials grow on land or are mined from the continents.

Over the land surface, temperature and soil moisture are key determinants of natural vegetation and the agricultural potential of a given area. Vegetation, snowcover, and soil conditions also affect the local and global climate, so that local climate and land surface conditions participate in a two-way relationship.

Land covers only about 30% of the surface area of Earth. The arrangement of land and ocean areas on Earth plays a role in determining global climate. The arrangement of land and ocean varies on time scales of millions of years as the continents drift about. At the present time about 70% of Earth's land area is in the Northern Hemisphere (Fig. 1.12), and this asymmetry causes significant differences in the climates of the Northern and Southern Hemispheres. The topography of the land sur-

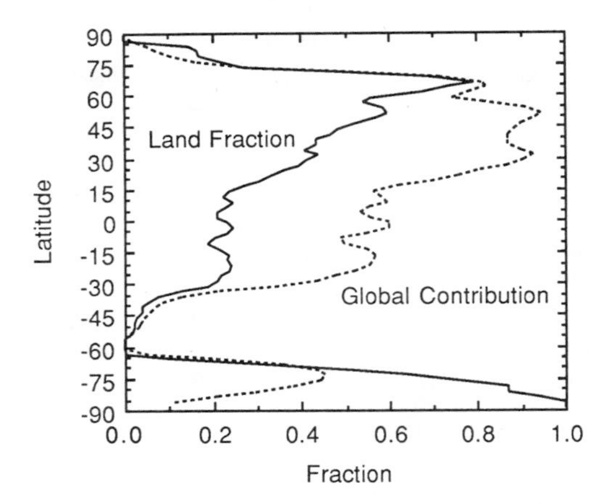

Fig. 1.12 Fraction of surface area covered by land as a function of latitude (solid line) and contribution of each latitude belt to the global land surface area (dashed line).

face and the arrangement and orientation of mountain ranges are key determinants of climate in land areas. The Northern Hemisphere has much more dramatic east-west variations in continental elevation, especially in middle latitudes where the Himalaya and Rocky Mountains are prominent features (Fig. 1.13).

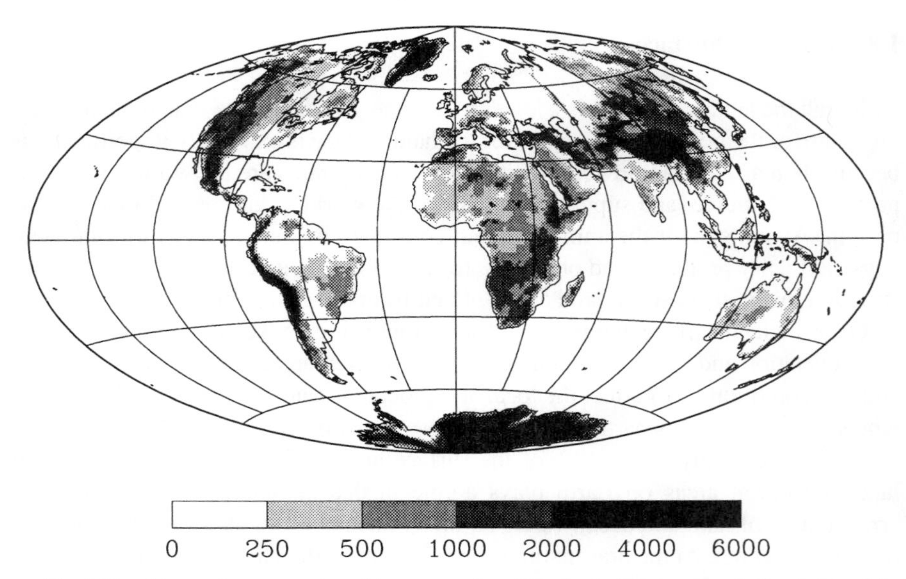

Fig. 1.13 Gray-scale contour plot of the topography of Earth relative to sea level. Resolution of the basic data is one degree of latitude and longitude. Scale is in meters.

Table 1.5

Land Use as a Percentage of Total Land Area

Land use	Percent
Arable mixed farming and human areas	10–13
Grazing land	20–25
Extratropical forests (mostly conifer)	10–15
Tropical forests and woodlands	13–18
Deserts	25–30
Tundra, high latitude	6–9
Swamp and marshes, lakes and streams	2–3

[From Dickinson (1983).]

Climate places limits on how land may be used. Table 1.5 shows the fraction of Earth's land areas that are currently in use for various purposes. About one-third is used for growing crops or grazing animals, one-third is forested, and one-third is desert or tundra. Historically, humans have converted forested lands to farming and living areas and grazing land. Overgrazing combined with drought can convert grazing land to desert in some climatic conditions. Humans have greatly altered the nature of the land surface, and continue to do so at an increasing rate.

Exercises

1. Give several reasons why the amplitude of the annual variation of surface temperature is greatest in Siberia (Fig. 1.7).
2. If you are standing atop Mount Everest at 8848 m, about what fraction of mass of the atmosphere is below you?
3. If the atmosphere warmed up by 5°C, would the atmospheric pressure at 5 km above sea level increase or decrease, and by approximately how much?
4. Explain why the North Polar temperature inversion is present in winter but not in summer?
5. Why do you think the salinity at 57.5°N is so much less than the salinity at 27.5°N (Fig. 1.11)?
6. What fraction of the land surface area that could be used for farming or grazing is currently being used for farming or grazing (Table 1.5)?

Chapter 2 The Global Energy Balance

2.1 Warmth and Energy

Temperature, a key climate variable, is a measure of the energy contained in the movement of molecules. To understand how the temperature is maintained, one must therefore consider the *energy balance* that is formally stated in the First Law of Thermodynamics. The basic global energy balance of Earth is between energy coming from the sun and energy returned to space by Earth's radiative emission. The generation of energy in the interior of Earth has a negligible influence on its energy budget. The absorption of solar radiation takes place mostly at the surface of Earth, whereas most of the emission to space originates in its atmosphere. Because its atmosphere efficiently absorbs and emits infrared radiation, the surface of Earth is much warmer than it would be in the absence of its atmosphere. When averaged over a year, more solar energy is absorbed near the equator than near the poles. The atmosphere and the ocean transport energy poleward to reduce the effect of this heating gradient on surface temperature. Much of the character of Earth's evolution and climate has been determined by its position within the solar system.

2.2 The Solar System

The source for the energy to sustain life on Earth comes from the sun. Earth orbits the sun once a year while keeping a relatively constant distance from it, so that our sun provides a stable, comfortable source of heat and light. Our sun is one of about 10^{11} stars in our galaxy, the Milky Way (Table 2.1). It is a single star, whereas about two-thirds of the stars we can see are in multiple star systems.

The luminosity is the total rate at which energy is released by the sun. We know of stars that are 10^{-4} times less luminous and 10^{5} times brighter than the sun. Their emission temperatures range from 2000 to 30,000 K, whereas the temperature of the photosphere of the sun is about 6000 K. The photosphere is the region of the sun from which most of its energy emission is released to space. Stellar radii range from 0.1 to 200 solar radii. Energy is produced in the core of the sun by nuclear fusion, whereby lighter elements are made into heavier ones, releasing energy in the process. For a smallish star such as the sun, the projected lifetime on the main

Table 2.1
Characteristics of the Sun

Mass	1.99×10^{30} kg
Radius	6.96×10^{8} m
Luminosity	3.9×10^{26} J s^{-1}
Mean distance from Earth	1.496×10^{11} m

sequence is about 11 billion years, of which about half has passed. The sun is thus a solitary, middle-aged, medium-bright star. Theories of stellar evolution predict that the luminosity of the sun has increased by about 30% during the lifetime of Earth, about 5 billion years.

The solar system includes nine planets. They may be divided into the *terrestrial,* or inner, planets and the *Jovian,* or outer, planets (Table 2.2). The terrestrial planets include Mercury, Venus, Mars, and Earth. The Jovian planets include Jupiter, Saturn, Uranus, and Neptune. Pluto doesn't fit neatly into either category (Table 2.3).

2.2.1 Planetary Motion

The planets orbit about the sun in ellipses, which have three characteristics: the mean planet–sun distance, the eccentricity, and the orientation of the orbital plane. The mean distance from the sun controls the amount of solar *energy flux density* (energy delivered per unit time per unit area) arriving at the planet. The mean distance from the sun also controls the length of the planetary year, the time it takes the planet to complete one orbit. The planetary year increases with increasing distance from the sun.

The *eccentricity* of an orbit is a measure of how much the orbit deviates from being perfectly circular. It controls the amount of variation of the solar flux density at the planet as it moves through its orbit during the planetary year. If the eccentricity is not zero (circular orbit) then the distance of the planet from the sun varies

Table 2.2
Characteristics of Inner and Outer Planets

Characteristic	Jovian (outer)	Terrestrial (inner)
Density	Small	Large
Mass	Large	Small
Sun distance	Large	Small
Atmosphere	Extensive	Thin or none
Satellites	Many	Few or none
Composition	H, He, CH_4, NH_3	Mostly silicates, rocks

Table 2.3

Physical Data for the Planets

Planet[a]	Mass (10^{26} g)	Mean radius (km)	Mean density (g cm^{-3})	Average distance from sun (10^6 km)	Length of year (days)	Obliquity (degrees)	Orbital eccentricity	Period of rotation (days)	Albedo
Mercury	3.35	2439	5.51	58	88	(0)[b]	0.206	58.7	0.058
Venus	48.7	6049	5.26	108	225	<3	0.007	–243[c]	0.71
Earth	59.8	6371	5.52	150	365	23.45	0.017	1.00	0.30
Mars	6.43	3390	3.94	228	687	24.0	0.093	1.03	0.16
Jupiter	19,100	69,500	1.35	778	4330	3.1	0.048	0.41	0.34
Saturn	5690	58,100	0.69	1430	10,800	26.8	0.056	0.43	0.34
Uranus	877	24,500	1.44	2870	30,700	98.0	0.047	–0.72[c]	0.34
Neptune	1030	25,100	1.65	4500	60,200	28.8	0.009	0.76	0.29
Pluto	(0.16)	(1500)	(1.10)	5900	90,700	(57.5)	0.247	(6.75)[b]	(0.4)

[Data from Allen (1973), de Vaucouleurs (1964), Ash *et al.* (1967), Shapiro (1967) © by the AAAS; and Dole (1970). Chamberlain and Hunten (1987).]

[a]The first four planets are similar in size, mass, density, and probably chemical composition. They are the *inner planets*. The remaining five are very different from Earth, but apart from Pluto, are similar to one another. They are the *outer planets*.

[b]Data in parentheses are uncertain.

[c]Venus and Uranus rotate in the opposite sense to the other planets.

through the year. The orientation of the orbital plane doesn't have too much direct bearing on the climate. Most planets except Pluto are in more or less the same orbital plane.

In addition to the orbit parameters, the parameters of the planets' rotation and their relationship to the orbit are very important. The rotation rate controls the temporal behavior of the insolation at a point (diurnal cycle) and is also an important control on the response of the atmosphere and ocean to solar heating and thereby on the patterns of winds and currents that develop.

The *obliquity* or tilt is the angle between the axis of rotation and the normal to the plane of the orbit. It influences the seasonal variation of insolation, particularly in high latitudes. It also strongly affects the annual mean insolation that reaches the polar regions. Currently the obliquity of Earth's axis of rotation is about 23.45°.

The longitude of perihelion measures the phase of the seasons relative to the planet's position in the orbit. For example, at present Earth passes closest to the sun (*perihelion*) during Southern Hemisphere summer, on about January 5. As a result the Southern Hemisphere receives more top-of-atmosphere insolation than the Northern Hemisphere.

The effect of these orbital parameters on climate will be discussed in more detail in Chapter 11, where the orbital parameter theory of climate change is described. For the time being we consider only the distance from the sun and the obliquity or declination angle.

2.3 Energy Balance of Earth

2.3.1 First Law of Thermodynamics

The First Law of Thermodynamics states that energy is conserved. The first law for a closed system may be stated as, "The heat added to a system is equal to the change in internal energy minus the work extracted." This law may be expressed symbolically as

$$dQ = dU - dW \tag{2.1}$$

Here dQ is the amount of heat added, dU is the change in the internal energy of the system, and dW is the work extracted from the system.

Heat can be transported to and from a system in three ways.

1. *Radiation:* No mass is exchanged, and no medium is required. Pure radiant energy moves at the speed of light.
2. *Conduction:* No mass is exchanged, but a medium is required to transfer heat by collisions between atoms or molecules.
3. *Convection:* Mass is exchanged. A net movement of mass may occur, but more commonly parcels with different energy amounts change places, so that energy is exchanged without a net movement of mass (Fig. 2.1).

Fig. 2.1 Cumulonimbus clouds over Zaire photographed from Shuttle 6, NASA, April 1983. Note the three-dimensional structure of the clouds, their shadows, and their various shapes and sizes. Clouds convect heat and moisture vertically and also influence radiative transfer in the atmosphere. The distance across the photo is approximately 80 km. The large convective cloud is approximately 30 km across.

The transmission of energy from the sun to Earth is almost entirely radiative. Some mass flux is associated with the particles of the solar wind, but the amount of energy is too small to have a measurable effect on Earth's surface temperature. Moreover, the work done by Earth on its environment is also negligible. To calculate the approximate energy balance of Earth, we need only consider radiative energy exchanges. The amount of matter in space that could affect the flow of energy between the sun and Earth is small, and we may thus consider the space between the photosphere of the sun and the top of Earth's atmosphere to be a vacuum. In a vacuum, only radiation can transport energy.

2.3.2 Energy Flux, Flux Density, and Solar Constant

The sun puts out a nearly constant flux of energy that we call the solar *luminosity*, $L_0 = 3.9 \times 10^{26}$ W. We can calculate the average flux density at the photosphere by dividing this energy flux by the area of the photosphere.

$$\begin{aligned} \text{Flux density}_{\text{photo}} &= \frac{\text{flux}}{\text{area}_{\text{photo}}} = \frac{L_0}{4\pi r_{\text{photo}}^2} \\ &= \frac{3.9\times10^{26}\ \text{W}}{4\pi\left[6.96\times10^8\,\text{m}\right]^2} = 6.4\times10^7\ \text{W m}^{-2} \end{aligned} \quad (2.2)$$

Since space is effectively a vacuum and energy is conserved, the amount of energy passing outward through any sphere with the sun at its center should be equal to the luminosity, or total energy flux from the sun. If we assume that the flux density is uniform over the sphere, and write the flux density at any distance d from the sun as S_d, then conservation of energy requires

$$\text{Flux} = L_0 = S_d 4\pi d^2 \quad (2.3)$$

From this we deduce that the flux density S_d is inversely proportional to the square of the distance to the sun. We define the solar constant as the energy flux density of the solar emission at a particular distance.

$$\text{Solar constant} = \text{flux density at distance } d = S_d = \frac{L_0}{4\pi d^2} \quad (2.4)$$

The solar constant is only a constant on a sphere with a fixed radius and the sun at its center. At the mean distance of Earth from the sun (1.5×10^{11} m), the solar constant is $S_0 = 1367$ W m^{-2}.

2.3.3 Cavity Radiation

The radiation field within a closed cavity in thermodynamic equilibrium has a value that is uniquely related to the temperature of the cavity walls, regardless of the material of which the cavity is made. This cavity radiant intensity, which is uniquely related to the wall temperature, is also called the *blackbody radiation,* since it corresponds to the emission from a surface with unit emissivity. Perfect blackbodies may not be easily found, but the radiation inside a cavity in equilibrium will always equal the blackbody radiation. The dependence of the blackbody emission on temperature follows the Stefan–Boltzmann Law.

$$E_{BB} = \sigma T^4; \quad \sigma = 5.67\times10^{-8}\ \text{W m}^{-2}\ \text{K}^{-4} \quad (2.5)$$

Example: Emission Temperature of the Sun

We calculated earlier that the solar flux density at the photosphere is about 6.4×10^7 W m^{-2}. We can equate this to the Stefan–Boltzmann formula and derive an effective emission temperature for the photosphere.

$$\sigma T_{\text{photo}}^4 = 6.4 \times 10^7 \text{ W m}^{-2}$$

$$T_{\text{photo}} = \sqrt[4]{\frac{6.4 \times 10^7 \text{ W m}^{-2}}{\sigma}} = 5796 \text{ K} \sim 6000 \text{ K}$$

2.3.4 Emissivity

In equilibrium, the radiant intensity inside a cavity at temperature T is $E_{BB} = \sigma T^4$. We may define the emissivity, ε, as the ratio of the actual emission of a body or volume of gas to the blackbody emission at the same temperature.

$$E_R = \varepsilon \sigma T^4 \rightarrow \varepsilon = \frac{E_R}{\sigma T^4} \tag{2.6}$$

2.4 Emission Temperature of a Planet

The emission temperature of a planet is the blackbody temperature with which it needs to emit in order to achieve energy balance. The basic idea is to equate the solar energy absorbed by a planet with the energy emitted by a blackbody. This defines the emission temperature of the planet.

Solar radiation absorbed = planetary radiation emitted

To calculate the solar radiation absorbed we begin with the solar constant, which measures the energy flux density of solar radiation arriving at the mean distance of the planet from the sun. The flux density is defined relative to a flat surface perpendicular to the direction of radiation. Solar radiation is essentially a parallel and uniform beam for a planetary body in the solar system, because the planets all have diameters that are small compared to their distance from the sun. The amount of energy incident on a planet is equal to the solar constant times the area that the planet sweeps out of the beam of parallel energy flux. We call this the *shadow area* (Fig. 2.2). Since the atmosphere of Earth is very thin, we can ignore its effect on the shadow area and use the radius of the solid planet, r_p, to calculate the shadow area.

We must also take account of the fact that not all of the solar energy incident on a planet is absorbed. Some fraction is reflected back to space without being absorbed

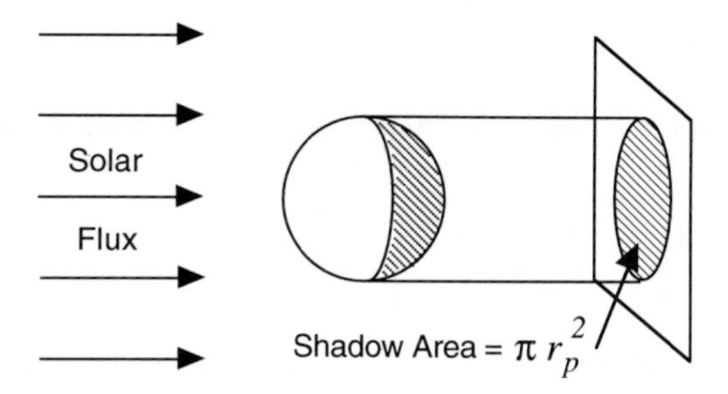

Fig. 2.2 Diagram showing the shadow area of a spherical planet.

and so does not enter into the planetary energy balance. We call this planetary reflectivity the *albedo*[1] and give it the symbol α_p. Thus we have

$$\text{Absorbed solar radiation} = S_0\left(1-\alpha_p\right)\pi r_p^2 \tag{2.7}$$

The globally averaged insolation at the top of the atmosphere is about 342 W m^{-2}. Since the *planetary albedo* for Earth is 30%, only 70% of this is absorbed by the climate system, about 240 W m^{-2}. This amount of energy must be returned to space by terrestrial emission. We assume that the terrestrial emission is like that of a blackbody. The area from which the emission occurs is the surface area of a sphere, rather than the area of a circle. The terrestrial emission flux is thus written

$$\text{Emitted terrestrial radiation} = \sigma T_e^4 \, 4\pi r_p^2 \tag{2.8}$$

If we equate the absorbed solar flux with the emitted terrestrial flux, we obtain the planetary energy balance, which will define the emission temperature.

$$\frac{S_0}{4}\left(1-\alpha_p\right) = \sigma T_e^4 \tag{2.9}$$

or

$$T_e = \sqrt[4]{\frac{(S_0/4)\left(1-\alpha_p\right)}{\sigma}} \tag{2.10}$$

The factor of 4 dividing the solar constant is the ratio of the global surface area of a sphere to its shadow area, which is the area of a circle with the same radius. The emission temperature may not be the actual surface or atmospheric temperature of the planet; it is merely the blackbody emission temperature a planet requires to balance the solar energy it absorbs.

[1]Albedo comes from a Latin word for "whiteness."

Example: Emission Temperature of Earth

Earth has an albedo of about 0.30. The emission temperature of Earth from (2.10) is therefore

$$T_e = \sqrt[4]{\frac{(1367 \text{ W m}^{-2}/4)(1-0.3)}{5.67\times10^{-8}\text{ W m}^{-2}\text{ K}^{-4}}} = 255 \text{ K} \cong -18^{\circ}\text{C}, 0^{\circ}\text{F}$$

The emission temperature of 255 K is much less than the observed global mean surface temperature of 288 K ≅ +15°C. To understand the difference we need to consider the greenhouse effect.

2.5 Greenhouse Effect

One may illustrate the greenhouse effect with a very simple elaboration of the energy balance model used to define the emission temperature. An atmosphere that is assumed to be a blackbody for terrestrial radiation, but is transparent to solar radiation, is incorporated into the global energy balance (Fig. 2.3). Since solar radiation is mostly visible and near-infrared, and Earth emits primarily thermal infrared radiation, the atmosphere may affect solar and terrestrial radiation very differently. The energy balance at the top of the atmosphere in this model is the same as in the basic energy balance model that defined the emission temperature (2.9). Since the atmospheric layer absorbs all of the energy emitted by the surface below it and emits like a blackbody, the only radiation emitted to space is from the atmosphere in this model. The energy balance at the top of the atmosphere is thus

$$\frac{S_0}{4}\left(1-\alpha_p\right) = \sigma T_A^4 = \sigma T_e^4 \tag{2.11}$$

Therefore we see that the temperature of the atmosphere in equilibrium must be the emission temperature in order to achieve energy balance. The surface tempera-

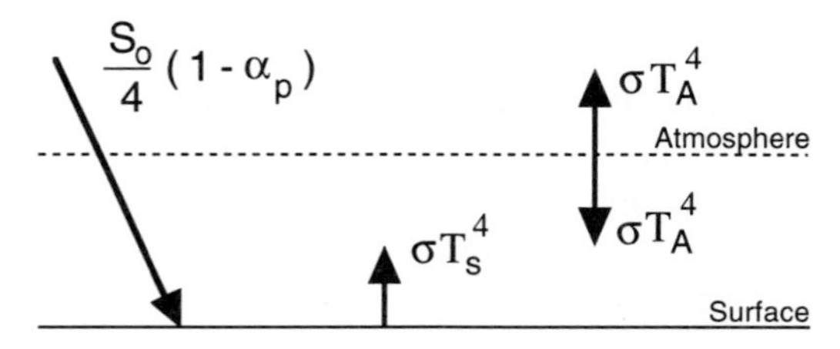

Fig. 2.3 Diagram of the energy fluxes for a planet with an atmosphere that is transparent for solar radiation but opaque to terrestrial radiation.

ture is much warmer, however, as we can see by deriving the energy balance for the atmosphere and the surface. The atmospheric energy balance gives

$$\sigma T_s^4 = 2\sigma T_A^4 \quad \Rightarrow \quad \sigma T_s^4 = 2\sigma T_e^4 \tag{2.12}$$

and the surface energy balance is consistent:

$$\frac{S_0}{4}\left(1-\alpha_p\right)+\sigma T_A^4 = \sigma T_s^4 \quad \Rightarrow \quad \sigma T_s^4 = 2\sigma T_e^4 \tag{2.13}$$

We can see from the diagram in Fig. 2.3 and the surface energy balance (2.13), that the surface temperature is increased because the atmosphere does not inhibit the flow of solar energy to the surface, but augments the solar heating of the surface with its own downward emission of longwave radiation, which in this case is equal to the solar heating. The atmospheric greenhouse effect warms the surface because the atmosphere is relatively transparent to solar radiation and yet absorbs and emits terrestrial radiation very effectively.

2.6 Global Radiative Flux Energy Balance

The vertical flux of energy in the atmosphere is one of the most important climate processes. The radiative and nonradiative fluxes between the surface, the atmosphere, and space are key determinants of climate. The ease with which solar radiation penetrates the atmosphere and the difficulty with which terrestrial radiation is transmitted through the atmosphere determine the strength of the greenhouse effect.

The contributions of radiative processes to the energy balance of the surface, troposphere, and stratosphere are shown schematically in Fig. 2.4. The values are given in percentages of the globally averaged available solar radiation at the top of the atmosphere, which is about 342 W m^{-2}. The planet absorbs about 70% of the incident solar radiation and reflects 30%. Fully 50% of the insolation available at the top of the atmosphere reaches the surface and is absorbed there. The 3% absorbed in the stratosphere is due mostly to ozone and molecular oxygen, while carbon dioxide and water vapor contribute about 0.5%. The 17% solar absorption in the troposphere is due primarily to water vapor (13%) and clouds (3%), while carbon dioxide, ozone, and oxygen together contribute the remaining 1%.

A striking feature of Fig. 2.4 is that the internal exchanges between the surface and the atmosphere by longwave radiative fluxes have the largest magnitudes of all, larger even than the insolation at the top of the atmosphere. This indicates the significance of the greenhouse effect in the atmosphere of Earth. The main contributors to the trapping of longwave radiation in the troposphere are water vapor, clouds, carbon dioxide, ozone, nitrous oxide, methane, and several other minor constituents. Water vapor and clouds provide about 80% of the current greenhouse effect. A good indicator of the efficiency of the atmospheric greenhouse effect is the left-hand column in the longwave

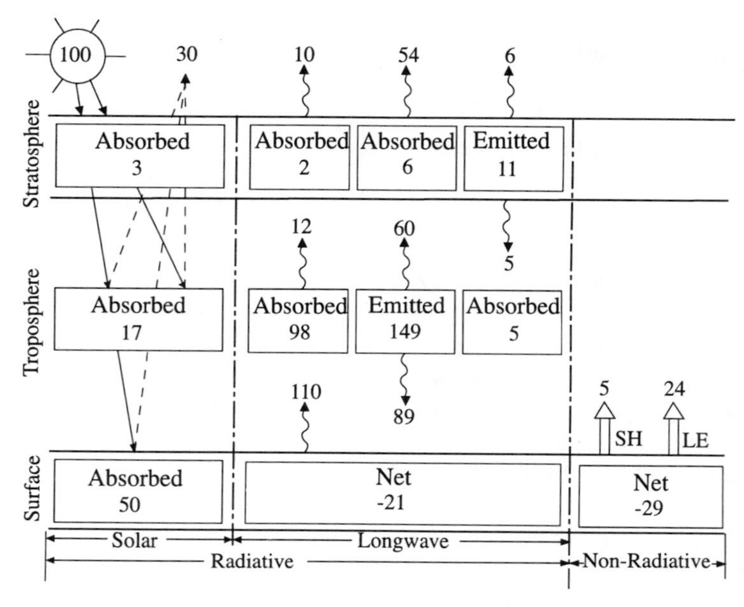

Fig. 2.4 Radiative and nonradiative energy flow diagram for Earth and its atmosphere. Units are percentages of the global-mean insolation (100 units = 342 W m^{-2}).

section of the diagram, which shows that only 10 of the 110 units emitted by the ground pass directly into space without being absorbed and reemitted by the atmosphere. The troposphere receives 149 units from radiative and nonradiative sources: 103 units of longwave energy absorption, 24 of latent heat release, 17 of solar absorption, and 5 of sensible heat transfer from the ground. The troposphere re-emits 89 units back to the ground and emits 60 units upward to the stratosphere and space.

The very strong downward emission of terrestrial radiation from the atmosphere is essential for maintaining the relatively small diurnal variations in surface temperature. If the downward longwave were not larger than the solar heating of the surface, then the land surface temperature would cool very rapidly at night and would warm exceedingly rapidly during the day. The greenhouse effect not only maintains relatively warm surface temperatures, but also limits the amplitude of the diurnal variation in surface temperature over land.

2.7 Distribution of Insolation

Seasonal and latitudinal variations in temperature are driven primarily by variations of insolation and average solar zenith angle. The amount of solar radiation incident on the top of the atmosphere depends on the latitude, season, and time of day. The amount of solar energy that is reflected to space without absorption depends on the solar zenith angle and the properties of the local surface and atmosphere. The climate depends on the insolation and zenith angle averaged over a 24-hour period, over a season, and over a year. In this section the geometric factors that determine insolation and solar zenith angle will be described.

The 1367 W m^{-2} mean solar flux per unit area at the mean position of Earth is measured for a surface that is perpendicular to the solar beam. Because Earth is approximately spherical, most of the planet's surface is inclined at an oblique angle to the solar beam. In such cases the solar flux density is spread over a surface area that is larger than the perpendicular area, so that the flux per unit surface area is smaller than the solar flux density. We define the *solar zenith angle,* θ_s, as the angle between the local normal to Earth's surface and a line between a point on Earth's surface and the sun. Figure 2.5 indicates that the ratio of the shadow area to the surface area is equal to the cosine of the solar zenith angle. We may write the solar flux per unit surface area as

$$Q = S_0 \left(\frac{\overline{d}}{d} \right)^2 \cos \theta_s \tag{2.14}$$

where $\overline{d}$ is the mean distance for which the flux density S_0 is measured, and d is the actual distance from the sun.

The solar zenith angle depends on the latitude, season, and time of day. The season can be expressed in terms of the *declination angle* of the sun, which is the latitude of the point on the surface of Earth directly under the sun at noon. The declination angle *(*δ*)* currently varies between +23.45° at northern summer solstice (June 21) to –23.45° at northern winter solstice (December 21). The *hour angle, h,* is defined as the longitude of the subsolar point relative to its position at noon. If these definitions are made, then the cosine of the solar zenith angle can be derived for any

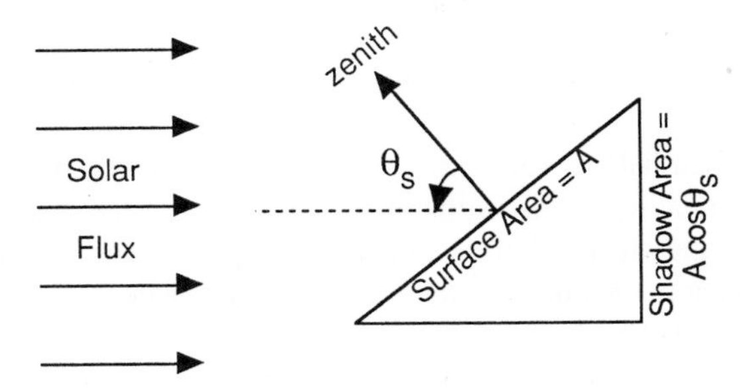

Fig. 2.5 Diagram showing the relationship of solar zenith angle to insolation on a plane parallel to the surface of a planet.

latitude (ϕ), season, and time of day from spherical trigonometry formulas (see Appendix A).

$$\cos\theta_s = \sin\phi\,\sin\delta + \cos\phi\,\cos\delta\,\cos h \tag{2.15}$$

If the cosine of the solar zenith angle is negative, then the sun is below the horizon and the surface is in darkness. Sunrise and sunset occur when the solar zenith angle is 90°, in which case (2.15) gives

$$\cos h_0 = -\tan\phi\,\tan\delta \tag{2.16}$$

where h_0 is the hour angle at sunrise and sunset.

Near the poles special conditions prevail. When the latitude and the declination angle are of the same sign (summer), latitudes poleward of $90 - \delta$ are constantly illuminated. At the pole the sun moves around the compass at a constant angle of δ above the horizon. In the winter hemisphere, where ϕ and δ are of the opposite sign, latitudes poleward of $90 - |\delta|$ are in polar darkness. At the poles, six months of darkness alternate with six months of sunlight. At the equator day and night are both twelve hours long throughout the year.

The average daily insolation on a level surface at the top of the atmosphere may be obtained by substituting (2.15) into (2.14), integrating the result between sunrise and sunset, and then dividing by 24 hours. The result is

$$\overline{Q}^{\text{day}} = \frac{S_0}{\pi}\left(\frac{\overline{d}}{d}\right)^2 \left[h_0 \sin\phi\sin\delta + \cos\phi\cos\delta\sin h_0\right] \tag{2.17}$$

where the hour angle at sunrise and sunset, h_0, must be given in radians. The daily average insolation is plotted in Fig. 2.6 as a function of latitude and season. Earth's orbit is not exactly circular, and currently Earth is somewhat closer to the sun during Southern Hemisphere summer than during Northern Hemisphere summer. As a result the maximum insolation in the Southern Hemisphere is about 6.9% higher than that in the Northern Hemisphere. Note that at the summer solstice the insolation in high latitudes is actually greater than that near the equator. This results from the very long days during summer and in spite of the relatively large solar zenith angles at high latitudes.

If the daily insolation is averaged over the entire year, the distribution given in Fig. 2.7 is obtained. The annual-average insolation at the top of the atmosphere at the poles is less than half its value at the equator, where it reaches a maximum. By comparing the annual mean insolation at the equator with the insolation at the solstices one can see that the insolation at the equator goes through a semiannual variation with maxima at the equinoxes and minima at the solstices.

Because the local albedo of Earth depends on the solar zenith angle, the zenith angle enters in determining both the available energy per unit of surface area and the albedo. It is therefore of interest to consider the average solar zenith angle during the daylight hours as a function of latitude and season. In calculating a daily average zenith angle it is appropriate to weight the average with respect to the insolation,

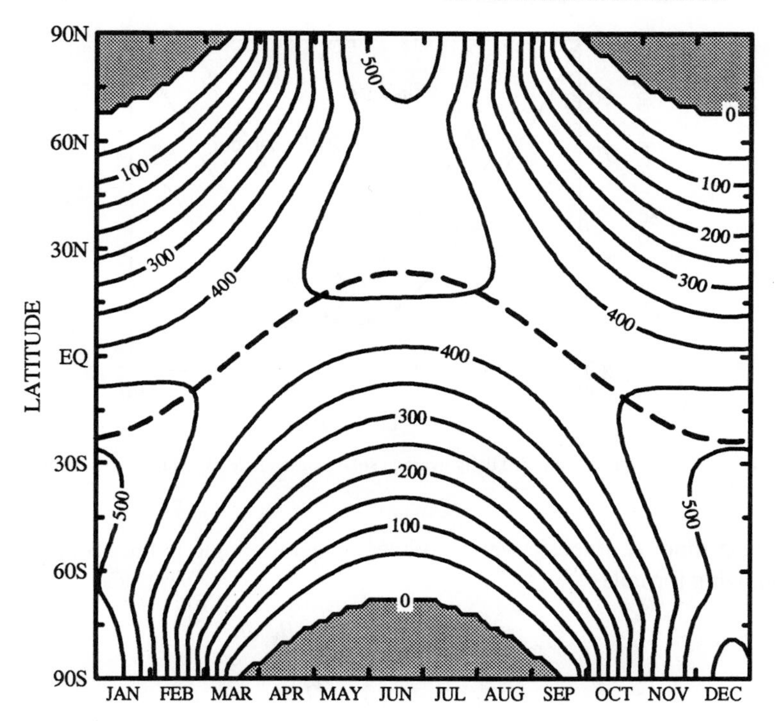

Fig. 2.6 Contour graph of the daily average insolation at the top of the atmosphere as a function of season and latitude. The contour interval is 50 W m^{-2}. The heavy dashed line indicates the latitude of the subsolar point at noon.

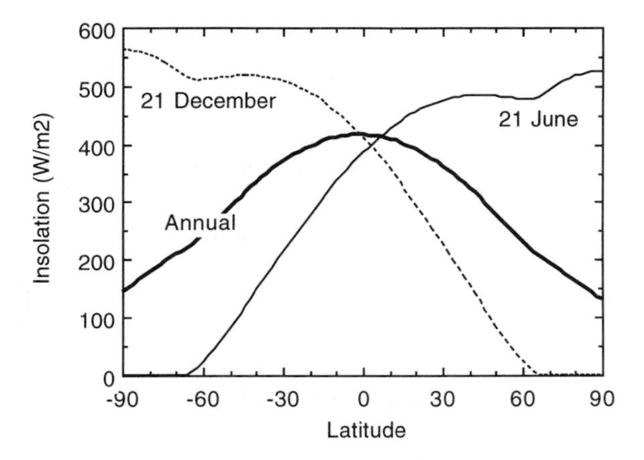

Fig. 2.7 Annual-mean and solstice insolation as functions of latitude.

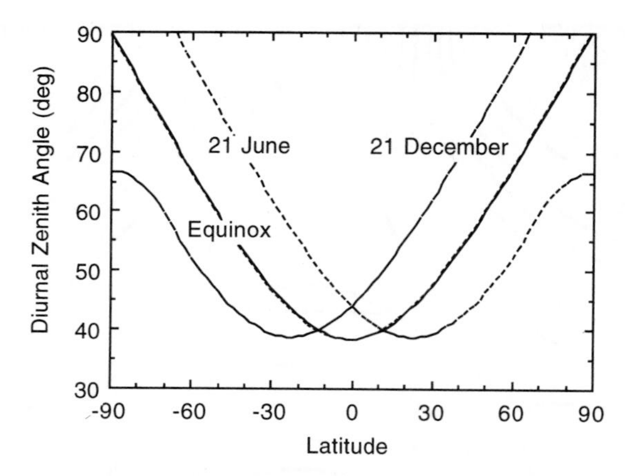

Fig. 2.8 Insolation-weighted daily average solar zenith angle as a function of latitude.

rather than time. A properly weighted average zenith angle is thus computed with the following formula,

$$\overline{\cos\,\theta_s}^{\text{day}} = \frac{\int_{-h_0}^{h_0} Q \cos\theta_s \, dh}{\int_{-h_0}^{h_0} Q \, dh} \tag{2.18}$$

where Q is the instantaneous insolation given by (2.14).

The daily average solar zenith angle calculated according to (2.18) varies from a minimum of 38.3° at the subsolar latitude and increases to 90° at the edge of the polar darkness (Fig. 2.8). At the pole the minimum value of the daily average solar zenith angle is achieved at the summer solstice, when it equals $\phi - \delta = 66.55°$. Because of the much higher average solar zenith angles in high latitudes, more solar radiation is reflected than would be from a similar scene in tropical latitudes.

2.8 The Energy Balance at the Top of the Atmosphere

The amount of energy absorbed and emitted by Earth varies geographically and seasonally, depending on atmospheric and surface conditions as well as the distribution of insolation. The energy balance at the top of the atmosphere is purely radiative and can be measured accurately from Earth-orbiting satellites. The albedo is estimated by measuring the solar radiation reflected from a region of Earth and comparing that with the insolation. The albedo shows interesting geographic structure (Fig. 2.9).[2] It

[2]We use a Hammer equal-area projection at many points in this book. In displaying flux density, it is especially important that the fraction of the area of the projection occupied by a particular feature be proportional to the fraction of the area of Earth's surface that the feature occupies.

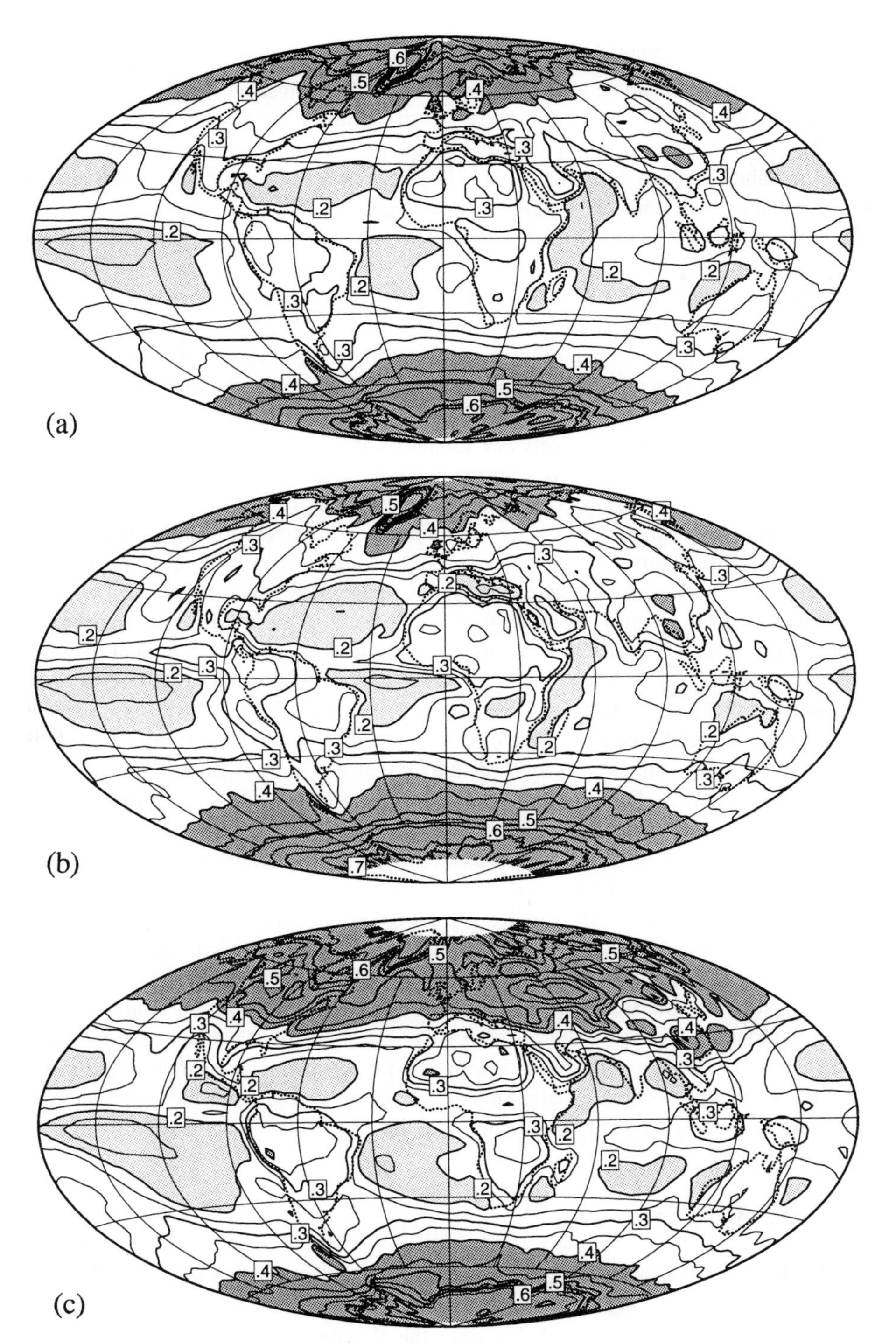

Fig. 2.9 Global maps of planetary albedo in a Hammer equal-area projection averaged for (a) annual mean, (b) JJA season, and (c) DJF season. Values are given as fractions. Contour interval is 0.05. Values greater than 0.4 are heavily shaded; values less than 0.2 are lightly shaded.

is highest in the polar regions where cloud and snowcover are plentiful and where average solar zenith angles are large. Secondary maxima of albedo occur in tropical and subtropical regions where thick clouds are prevalent or over bright surfaces such as the Sahara Desert. The smallest albedos occur over those tropical ocean regions where clouds are sparsely distributed. The ocean surface has an intrinsically low albedo, so that when clouds and sea ice are absent the planetary albedo over ocean areas is only 8–10%.

Outgoing longwave radiation (OLR) is greatest over warm deserts and over tropical ocean areas where clouds occur infrequently (Fig. 2.10). It is lowest in polar regions and in regions of persistent high cloudiness in the tropics. The OLR is controlled by the temperature of the emitting substance, so that the cold poles and the cold cloud tops produce the lowest values. The highest values occur when a warm surface is overlaid by a relatively dry, cloudless atmosphere.

The net radiation is negative near the poles and positive in the tropics (Fig. 2.11). The highest positive values of about 120 W m^{-2} occur over the subtropical oceans in the summer hemisphere, where large insolation and relatively low albedos both contribute to large absorbed solar radiation. The greatest losses of energy occur in the polar darkness of the winter hemisphere, where the OLR is uncompensated by solar absorption. Dry desert areas such as the Sahara in northern Africa are interesting because, despite their subtropical latitude, they lose energy in the annual average. Their energy loss is related to the relatively high surface albedo of dry deserts, combined with the large OLR loss associated with a dry atmosphere above a warm surface.

When averaged around latitude circles, the components of the energy balance at the top of the atmosphere clearly show the influence of the latitudinal gradient of insolation (Fig. 2.12). The poleward decrease in absorbed solar energy is even greater than that of insolation because the albedo increases with latitude. Thus a smaller fraction of the available insolation is absorbed at higher latitudes than at the equator. The albedo increases with latitude because solar zenith angle, cloud coverage, and snowcover all increase with latitude. The energy emitted to space by the atmosphere does not decrease with latitude as rapidly as the absorbed solar radiation, because the atmosphere and ocean transport heat poleward and support a net energy loss to space in polar regions. The absorbed solar radiation exceeds the OLR in the tropics, so that the net radiation is positive there. Poleward of about 40° the absorbed solar radiation falls below the OLR and the net radiation balance is negative, so that the climate system loses energy to space. The latitudinal gradient in annual-mean net radiation must be balanced by a poleward flux of energy in the climate system of Earth.

2.9 Poleward Energy Flux

In the lowest curve of Fig. 2.12 we see that the annual mean net radiation is positive equatorward of about 40° of latitude and negative poleward of that latitude. As illustrated in Fig. 2.13, the energy balance for the climate system involves only the ex-

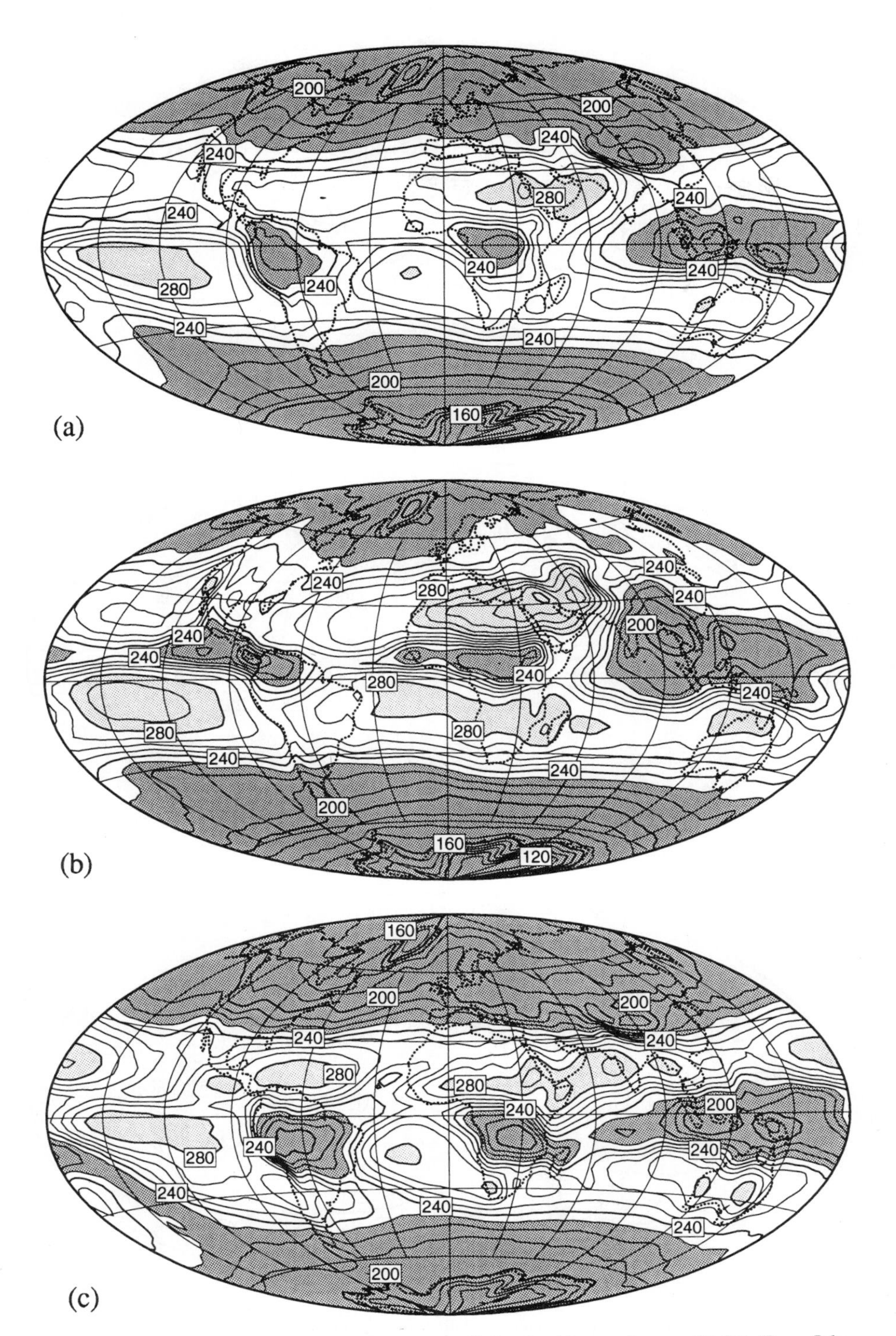

Fig. 2.10 Global maps of outgoing longwave radiation for (a) annual mean, (b) JJA (June–July–August) season, and (c) DJF (December–January–February) season. Contour interval is 10 W m^{-2}. Values greater than 280 W m^{-2} are lightly shaded, and values less than 240 W m^{-2} are heavily shaded.

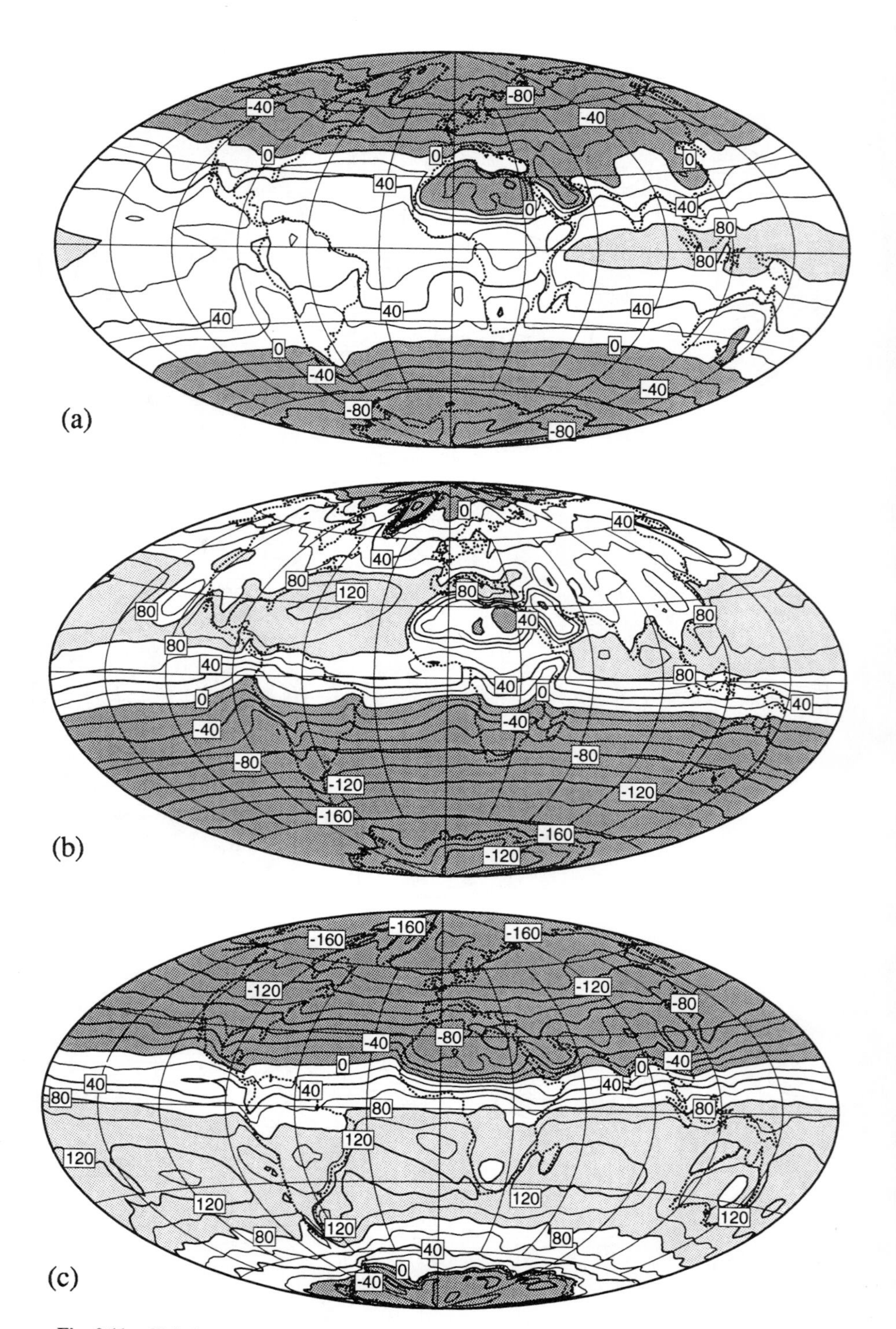

Fig. 2.11 Global maps of net incoming radiation at the top of the atmosphere for (a) annual mean, (b) JJA season, and (c) DJF season. Contour interval is 20 W m^{-2}. Values greater than 80 W m^{-2} are lightly shaded, and values less than zero are heavily shaded.

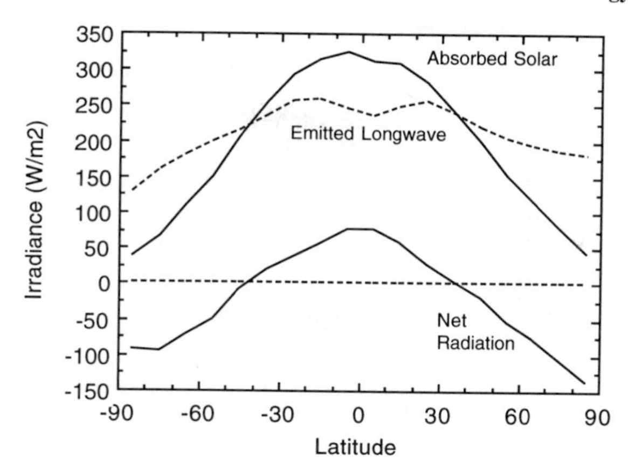

Fig. 2.12 Graphs of annual-mean absorbed solar radiation, OLR, and net radiation averaged around latitude circles.

change at the top of the atmosphere, the transport through the lateral boundaries of the region in question by the atmosphere and ocean, and the time rate of change of energy within the region. Energy exchange with the solid earth can be neglected. We can write the energy balance for the climate system as

$$\frac{\partial E_{\mathrm{ao}}}{\partial t} = R_{\mathrm{TOA}} - \Delta F_{\mathrm{ao}} \tag{2.19}$$

where $\partial E_{\mathrm{ao}}/\partial t$ is the time rate of change of the energy content of the climate system, R_{TOA} is the net incoming radiation at the top of the atmosphere, and ΔF_{ao} is the divergence of the horizontal flux in the atmosphere and ocean.

If we average over a year, then the storage term becomes small, and we have an approximate balance between net flux at the top of the atmosphere and horizontal transport.

$$R_{\mathrm{TOA}} = \Delta F_{\mathrm{ao}} \tag{2.20}$$

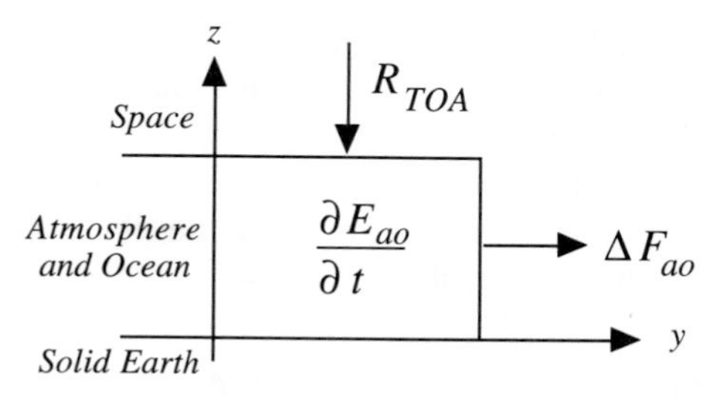

Fig. 2.13 Diagram of the energy balance for the climate system.

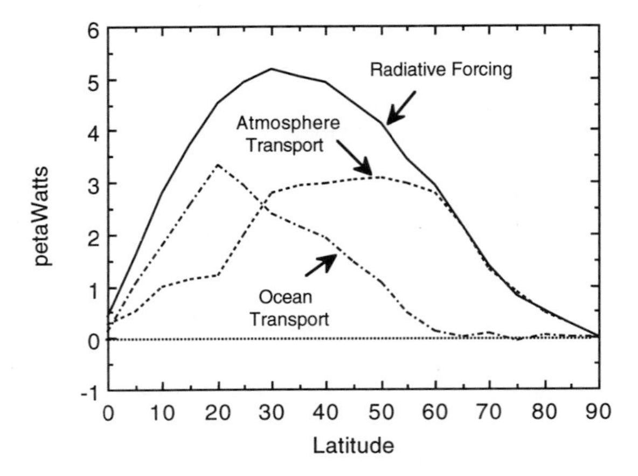

Fig. 2.14 Meridional transport of energy for annual-mean conditions. Net radiation and atmospheric transport are estimated from observations; ocean transport is calculated as a residual in the energy balance. [Adapted from Vonder Haar and Oort (1973). Used with permission from the American Meteorological Society.]

We can then use the observation of net radiation in Fig. 2.12 to derive the required annual-mean energy transport in the north–south direction. If we integrate the net radiation over a polar cap area, we can calculate the total energy flux across each latitude belt in the following way:[3]

$$\int_{-\frac{\pi}{2}}^{\phi} \int_{0}^{2\pi} R_{\mathrm{TOA}} a^2 \cos\phi \; d\lambda \; d\phi = F_{\phi} \tag{2.21}$$

The total northward energy flux inferred from the radiation imbalance peaks in midlatitudes at about 5×10^{15} W, or 5 petawatts (PW) (Fig. 2.14). This flux includes the contributions from both the atmosphere and the ocean. The flux of energy in the atmosphere can be estimated from balloon and satellite observations of wind, temperature, and humidity. If this flux is subtracted from the total flux, it gives an estimate of the oceanic energy flux, which is much more difficult to estimate from direct measurements. At 30° of latitude, the contributions from the atmospheric and oceanic poleward fluxes are each about 2.5 PW. The ocean flux peaks at about 20°N in the subtropics, whereas the atmospheric flux has a broad maximum between 30°N and 60°N. In Chapters 6 and 7 we investigate in more detail how the atmosphere and the oceans accomplish this poleward flux of energy. If the fluid envelope of Earth did not transport heat poleward, the tropics would be warmer and the polar regions would be much colder. Transport of heat by the atmosphere and the oceans makes the climate of Earth much more equable than it would otherwise be.

[3] Vonder Haar and Oort (1973); Oort and Vonder Haar (1976).

If we integrate over the globe, the horizontal transport is zero, since energy cannot be removed from spherical Earth through horizontal transport. In the annual global average then, the net radiation is very close to zero. Any global mean net radiative imbalance will lead to a heating or cooling of Earth.

Exercises

1. Use the data in Table 2.3 to calculate the emission temperatures for all of the planets. The actual emission temperature of Jupiter is about 124 K. How must you explain the difference between the number you obtain for Jupiter and 124 K?
2. Calculate the emission temperature of Earth if the solar luminosity is 30% less, as it is hypothesized to have been early in Solar System history. Use today's albedo and Earth–sun distance.
3. Calculate the emission temperature of Earth, if the planetary albedo is changed to that of ocean areas without clouds, about 10%.
4. Using the model illustrated in Fig. 2.3, calculate the surface temperature if the insolation is absorbed in the atmosphere, rather than at the surface.
5. Using the data in Fig. 2.4, estimate the blackbody temperature of Earth's surface and the blackbody temperature of the atmosphere as viewed from Earth's surface. How well do these temperatures agree with the temperatures derived from the model shown in Fig. 2.3?
6. Calculate the solar zenith angle at 9AM local sun time at Seattle (47°N, 122°W) at summer solstice and at winter solstice. Do the same for your hometown.
7. A mountain range is oriented east–west and has a slope on its north and south faces of 15 degrees. It is located at 45°N. Calculate the insolation per unit surface area on its south and north faces at noon at the summer and winter solstices. Compare the difference between the insolation on the north and south faces in each season with the seasonal variation on each face. Ignore eccentricity and atmospheric absorption.
8. In his science fiction novel, *Ringworld,* Larry Niven imagines a manufactured world consisting of a giant circular ribbon of superstrong material that circles a star. The width of the ribbon normal to a radius vector originating at the star's center is much greater than the thickness of the ribbon in the radial direction. The width of the ribbon is much less than the distance of the ribbon from the star, however. If the diameter of the ribbon is the same as the diameter of Earth's orbit about the sun, the luminosity is the same as Earth's sun, and the albedo of the ribbon is 0.3, what is the emission temperature of the sunlit side of the ribbon? Consider both the case in which heat is conducted very efficiently from the sunlit to the dark side of the ribbon, and the case in which no heat is conducted from the sunlit to the dark side of the ribbon. Explain why your emission temperature is different from Earth's.

Chapter 3 Atmospheric Radiative Transfer and Climate

3.1 Photons and Minority Constituents

The energy that sustains warmth and life at the surface of Earth travels from the sun in the form of radiation. The interaction of incoming solar radiation with the atmosphere and the surface determines the total amount of solar energy absorbed and the distribution of solar heating among atmospheric layers and the surface. Because the atmosphere is relatively transparent to solar radiation, about half of the incoming solar radiation is absorbed at the ocean or land surface (Fig. 2.4). To achieve energy balance, heat provided by the absorption of solar radiation must be returned to space by emission from Earth. In this process it is the transmission of thermal infrared radiation through the atmosphere and the upward transport of heat by atmospheric motions that are of importance. The transmission properties of the atmosphere are determined by its gaseous composition and the nature of aerosols and water clouds present in it. The composition of the atmosphere is such that it efficiently absorbs and emits thermal infrared radiation. The efficient absorption and emission of thermal infrared radiation by the atmosphere, combined with its relative transparency to solar radiation, causes the surface to be much warmer than it would be in the absence of an atmosphere.

The absorption of thermal radiation in air is accomplished by molecules that comprise a small fraction of the atmosphere's mass. The dependence of the climate on the abundance of these minority constituents makes the climate sensitive to natural and human-induced changes in atmospheric composition. Relatively small changes in composition can affect the flow of energy through the climate system and thereby produce surprisingly large climate changes.

To understand how climate depends on atmospheric composition, it is necessary to understand the nature of electromagnetic radiation and the physical processes through which it interacts with gases and particles. The equation of radiative transfer forms the mathematical basis for keeping track of these physical processes in an aggregate sense so that radiative energy fluxes can be computed. One-dimensional models, in which the vertical fluxes of energy by radiation are calculated from the radiative transfer equation, can be used to estimate the effect of trace-gas concentrations, clouds, and aerosols on the global mean surface temperature.

3.2 The Nature of Electromagnetic Radiation

Electromagnetic radiation can be thought of either as a wave or as a particle that represents the movement of energy through space. For scattering of light by particles and surfaces, electromagnetic wave theory is most helpful. When considering absorption and emission of radiation, it is useful to think of radiant energy as discrete parcels of energy that we call photons.

The speed of electromagnetic radiation in a vacuum is a constant $c^* = 3 \times 10^8$ m s^{-1}. This means that frequency ν and wavelength λ are inversely related in a one-to-one correspondence.

$$\nu = \frac{c^*}{\lambda}; \qquad \lambda = \frac{c^*}{\nu} \tag{3.1}$$

High frequencies correspond to short wavelengths and low frequencies, to long wavelengths. Most of the time we will describe radiation in terms of its wavelength, which we will give in millimeters (mm = 10^{-3} m), micrometers (μm = 10^{-6} m), or nanometers (nm = 10^{-9} m).

In explaining the photoelectric effect, Einstein postulated that radiant energy exists and propagates in quantum bits called photons. If we think of light as photons, then a photon has an energy, E_ν, that is proportional to its frequency

$$E_\nu = \hbar \nu \tag{3.2}$$

where $\hbar = 6.625 \times 10^{-34}$ J s is Planck's constant. Therefore, high-frequency, short-wavelength radiation has more energy per photon than low-frequency radiation.

Most of the sun's radiant energy output is contained between wavelengths of 100 nm and 4 μm, and consists of ultraviolet, visible, and near infrared radiation. Ninety-nine percent of the sun's emission comes from the visible (0.4–0.75 μm) and near infrared (0.75–5 μm) portions of the spectrum. Ultraviolet radiation makes up less than 1% of the total, but is nonetheless important because of its influence in the upper atmosphere, and because it is harmful to life if it reaches the surface. Earth's energy emission is almost all contained between about 4 and 200 μm, and is therefore entirely thermal infrared.

3.3 Description of Radiative Energy

The energy of radiation is measured by its intensity or radiance. The monochromatic intensity describes the amount of radiant energy (dF_ν) within a frequency interval (ν to $\nu + d\nu$) that will flow through a given increment of area (dA) within a solid angle (dω) of a particular direction in a time interval (dt).

$$dF_\nu = I_\nu \cos\theta \, d\omega \, dA \, d\nu \, dt \tag{3.3}$$

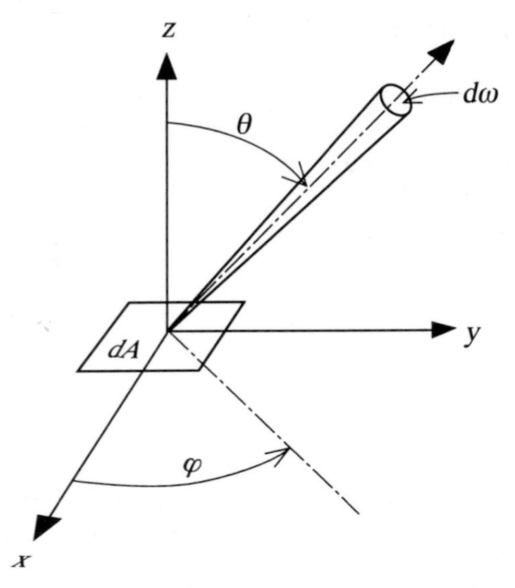

Fig. 3.1 Diagram showing the angles that define the radiance flowing through a unit area dA in the x–y plane, in the direction defined by the zenith angle θ, and the azimuth angle φ, and within the increment of solid angle $d\omega$.

The direction is defined by the zenith angle, θ, and the azimuth angle, φ, as shown in Fig. 3.1. The magnitude of the radiant intensity, I_ν, is given in energy per unit time, per unit area, per unit of frequency interval, per unit of solid angle, or watts per meter squared per hertz per steradian. In this book, we will be mostly concerned with the total energy per unit frequency passing across a unit area of a plane surface from one side to the other. To obtain this quantity we integrate the radiant intensity over all solid angles in a hemisphere. To do this we need to make use of the definition of an increment of solid angle.

$$d\omega = \sin\theta \; d\theta \; d\varphi \tag{3.4}$$

Inserting (3.4) into (3.3) and integrating over the upper hemisphere we obtain

$$F_\nu = \int_0^{2\pi} \int_0^{\pi/2} I_\nu(\theta,\phi) \cos\theta \sin\theta \; d\theta \; d\varphi \tag{3.5}$$

This quantity is called the *spectral flux density.* If F_ν is integrated over all frequencies we obtain the flux density, which has units of watts per meter squared.

$$F = \int_0^\infty F_\nu \; d\nu \tag{3.6}$$

When a beam of radiation encounters an object such as a molecule, an aerosol particle, or a solid surface, several possible interactions between the radiation and the object can take place. The radiation can pass the object unchanged, which is called *perfect transmission*. The radiation can change direction without a change in energy, which is *pure scattering*. The radiation can be absorbed, with its energy transferred to the object. The probability that a photon will be scattered, absorbed, or transmitted depends on the frequency of the radiation and the physical properties of the object in question. Pure water droplets in clouds scatter visible radiation very effectively with relatively little absorption taking place. Water vapor and carbon dioxide are very effective absorbers of thermal infrared radiation at certain frequencies. Matter can also add to the intensity of a beam of radiation by emitting radiation in the direction of the beam. Emission of radiation by matter depends on the substance's physical properties and temperature.

3.4 Planck's Law of Blackbody Emission

The intensity of radiation in a cavity in thermodynamic equilibrium is given uniquely as a function of frequency and temperature by Planck's law. An object that absorbs all radiation incident on it is called a *blackbody*. A blackbody with temperature T emits radiation at frequency ν with an intensity given by Planck's law:

$$B_\nu(T) = \frac{2\hbar\nu^3}{c*^2}\frac{1}{\left(e^{\hbar\nu/kT} - 1\right)} \tag{3.7}$$

where $\hbar$ = 6.625×10^{-34} J s (Planck's constant), k = 1.37×10^{-23} J K^{-1} (Boltzmann's constant), $c* = 3 \times 10^8$ m s^{-1} (speed of light), ν is the frequency of radiation in s^{-1}, T is the temperature in kelvins.

The Stefan–Boltzmann law is an integral of Planck's law over all frequencies and over all angles in a hemisphere, and expresses the strong dependence of energy emission on temperature.

$$\pi\int_0^\infty B_\nu(T)d\nu = \sigma T^4 \tag{3.8}$$

The factor of π arises from the integration over one hemisphere, assuming that the emission is independent of angle (isotropic). The Stefan–Boltzmann constant defined in (2.5) can be expressed in terms of the more fundamental constants of Planck's law:

$$\sigma = \frac{2\pi^5 k^4}{15c*^2\,\hbar^3} \tag{3.9}$$

Planck's law of blackbody radiation contains within it Wien's law of displacement, which states that the wavelength of maximum emission is inversely proportional to temperature. That is, the hotter the object, the higher the frequency and the shorter the wavelength of emitted radiation. Note that for terrestrial temperatures (~255 K) the emission peaks around 10 μm, while for the temperature of the sun's

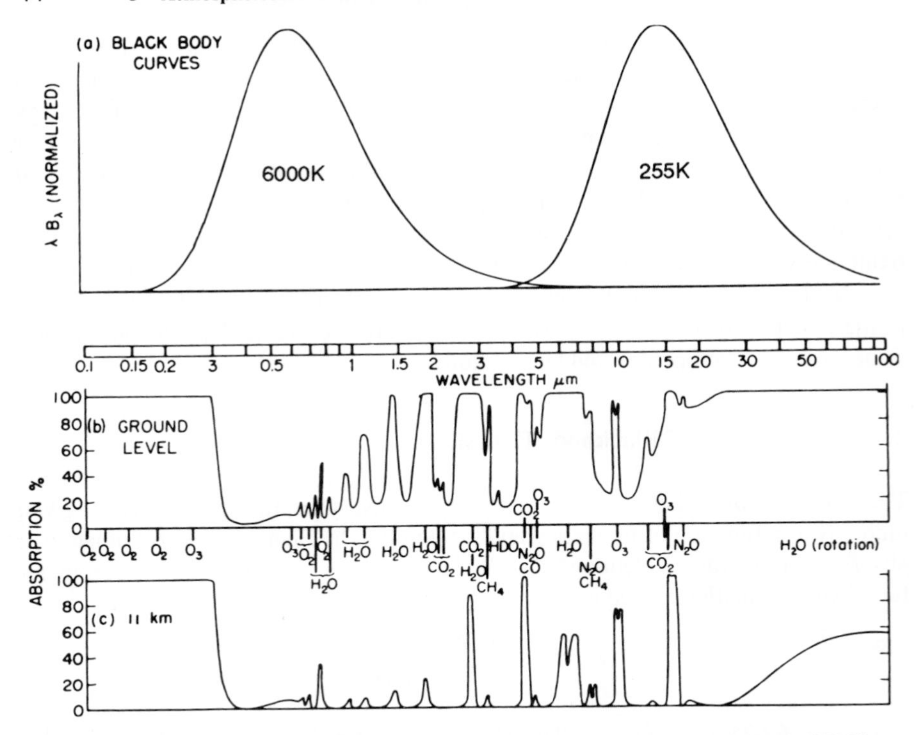

Fig. 3.2 The normalized blackbody emission spectra for the sun (6000 K) and Earth (255 K) as a function of wavelength (top). The fraction of radiation absorbed while passing from the surface to the top of the atmosphere as a function of wavelength (middle). The fraction of radiation absorbed from the tropopause to the top of the atmosphere as a function of wavelength (bottom). The atmospheric molecules contributing the important absorption features at each frequency are indicated. [Taken from Goody and Yung (1989). Reprinted with permission from Oxford University Press.]

photosphere (~6000 K) the emission peaks around 0.6 μm (Fig. 3.2). The energy emission from both the sun and Earth become negligible near 4 μm, so that the frequencies at which they emit are almost completely distinct for energetic purposes. We can thus speak of solar and terrestrial radiation as separate entities. In climatology, solar radiation is often called *shortwave* radiation and terrestrial radiation is called *longwave*.

3.5 Selective Absorption and Emission by Atmospheric Gases

In considering the global energy balance of Earth we found that the effective emission temperature is 255 K, which is much less than the observed global mean surface temperature of 288 K. The explanation for this disparity is found in the different transmission properties of the atmosphere for terrestrial and solar radiation. The atmosphere is relatively transparent to solar radiation, whereas it is nearly opaque to terrestrial radiation. To understand the fundamental reasons behind this "greenhouse effect," we must understand a little about the interaction of radiation with matter.

In deriving his law of blackbody radiation, Planck found it necessary to postulate that the energy levels of an atomic or molecular oscillator are limited to a discrete set of values that satisfy

$$E_\nu = n\hbar\nu \qquad (n = 0, 1, 2, \ldots) \tag{3.10}$$

Equation (3.10) describes a set of discrete energy levels that each differ by an amount $\hbar\nu$. The oscillator may represent the periodic motions of an atom or molecule. A transition from one energy level to another, called a quantum jump, corresponds to the release or capture of an amount of energy equal to $\hbar\nu$. One way to accomplish this energy transition is for the molecule to emit or absorb a photon of energy $\hbar\nu$. This quantization effect only becomes apparent for very small oscillators like molecules or atoms. For macroscopic oscillators like a pendulum, or a spring and weight apparatus, the amount of energy represented by $\hbar\nu$ is too small to be noticed in comparison to the total energy of the system.

A photon is emitted from a substance in some finite amount of time ($\leq 10^{-8}$ s) and then travels through space until it is absorbed. If it approaches a mass such as an air molecule or a solid particle, the photon can change phase or direction, a process called *scattering*, or it can be absorbed. When a photon is absorbed it ceases to exist and its energy is transferred to the substance that absorbed it. This energy can appear as increased internal energy of a molecule or atom, or as heat. The energy of a molecule can be stored in vibrational, rotational, electronic, or translational forms.

$$E_{\text{total}} = E_{\text{translational}} + E_{\text{rotational}} + E_{\text{vibrational}} + E_{\text{electronic}} \tag{3.11}$$

A molecule in the atmosphere can absorb a photon only if the energy of the photon corresponds to the difference between the energy of two allowable states of the molecule. Each mode of energy storage in a molecule corresponds to a range of energies, with electronic transitions corresponding to the largest energy differences and rotational transitions corresponding to the smallest. Allowable transitions between energy levels of the molecules making up the atmosphere determine the frequencies of radiation that will be efficiently absorbed and emitted by the atmosphere. If no transitions correspond to the energy of a photon, then it will have a good chance to pass through the gaseous atmosphere without absorption.

3.5.1 Translational or Kinetic Energy (Temperature)

Translational energy corresponds to the gross movement of molecules or atoms through space and is not quantized. At terrestrial temperatures the kinetic energy of a molecule is generally small compared to the energy required for a vibrational transition. Collisions between molecules can carry away or supply energy to interactions between photons and matter, and this plays an important role in broadening the range of frequencies of radiation that can be absorbed by a particular transition between molecular energy levels. The Doppler effect associated with the movement of molecules can also broaden the range of frequencies absorbed or emitted by a particular energy transition of the molecule.

3.5.2 Rotational Energy

A macroscopic object like a child's top can store an amount of energy that, in effect, varies continuously with its rotation rate. For tiny objects like molecules in the atmosphere, the energy of rotation is quantized and can take on only discrete values. *Rotational energy* transitions involve energy changes that correspond to the energy of photons with wavelengths shorter than about 1 cm.

3.5.3 Vibrational Energy

Atoms are bonded together into stable molecules when the forces of attraction and repulsion are in balance at the appropriate interatomic distance. Molecular energy can be stored in the vibrations about this stable point. The states of the molecule and the allowable energy levels are again quantized. Vibrational transitions require a photon with a wavelength of less than 20 μm. There are three independent modes of vibration for a triatomic molecule like CO_2: two stretching modes and one bending mode (Fig. 3.3).

A symmetric, linear molecule like CO_2 has no permanent dipole moment, since it looks the same from both ends. For this reason it has no pure rotational transitions. During vibrational transitions the CO_2 molecule develops temporary dipole moments so that rotational transitions can accompany a vibrational transition. The combination of possible vibrational and rotational transitions allows the molecule to absorb and emit photons at a large number of closely spaced frequencies, composing an absorption band.

Water vapor is a good absorber of terrestrial radiation because it is a bent triatomic molecule. Because it is bent, it has a permanent dipole moment and therefore has pure rotation bands in addition to vibration–rotation bands.

Vibration–rotation bands and pure rotation bands of polyatomic molecules account for the longwave absorption of the clear atmosphere that is shown in Fig. 3.2. The bending mode of CO_2 produces a very strong vibration–rotation absorption band near 15 μm, which is very important for climate and critical for the stratosphere where it accounts for much of the longwave absorption. This absorption feature is particularly important because it occurs near the peak of the terrestrial emission spectrum. Water vapor has an important vibration–rotation band near 6.3 μm, and a densely spaced band of pure rotational lines of water vapor strongly absorbs terrestrial emission at wavelengths in excess of about 12 μm. Between these two water

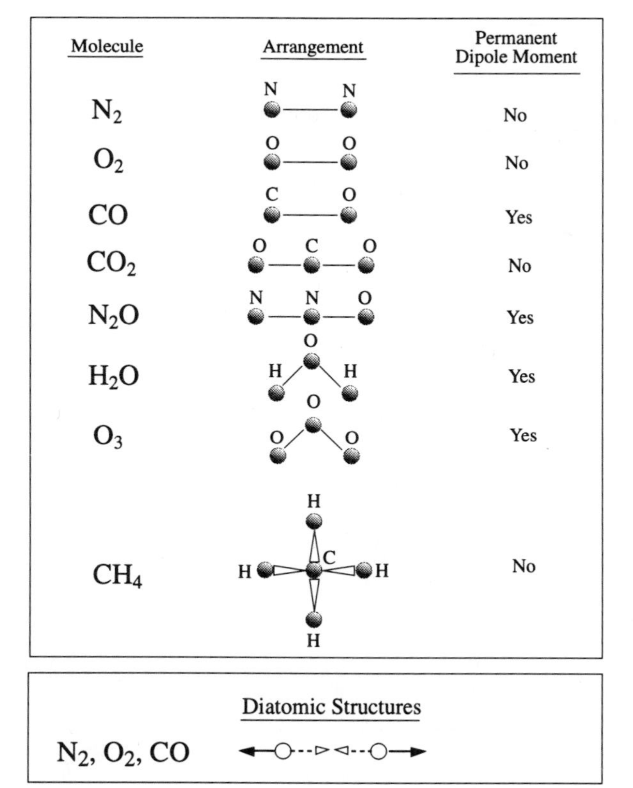

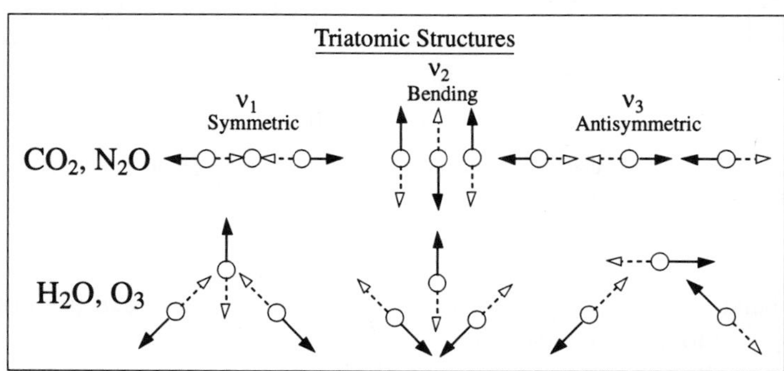

Fig. 3.3 Schematic diagrams showing the vibrational modes of diatomic and triatomic molecules. [From McCartney (1983). Reprinted with permission from Wiley and Sons, Inc.]

vapor features there is relatively weak absorption by water vapor, and so this wavelength region is called the *water vapor window* since only in this frequency range can longwave radiation pass relatively freely through the atmosphere (see middle panel of Fig. 3.2 and Fig. 3.4). In the middle of this atmospheric window sits the

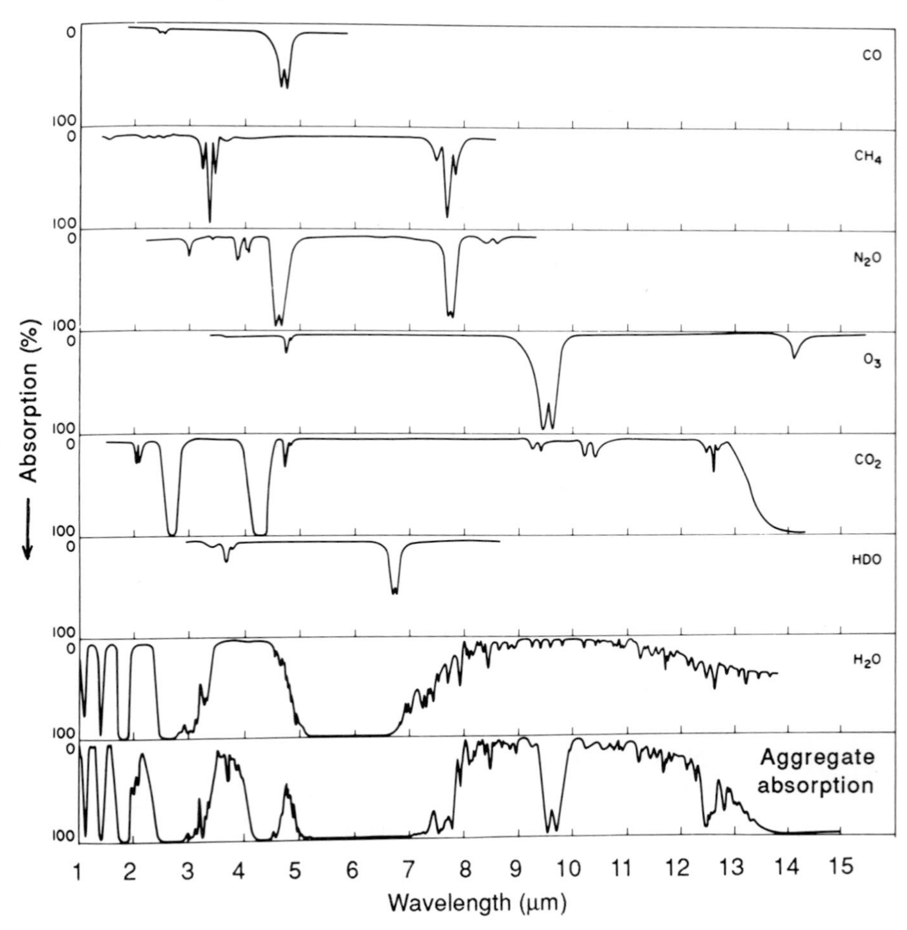

Fig. 3.4 Infrared absorption spectra for various atmospheric gases. [From Valley (1965). Used with permission from McGraw-Hill, Inc.]

9.6-μm band of ozone. Together, these absorption features make the troposphere nearly opaque to longwave radiation.

Some significant absorption by vibration-rotation bands of atmospheric gases also occurs at near-infrared wavelengths between about 1 and 4 μm, mostly by water vapor and carbon dioxide (Figs. 3.2 and 3.4). These absorption features account for most of the absorption of solar radiation by air molecules in the troposphere. Visible wavelengths (~0.3–0.8 μm) are virtually free of gaseous absorption features, giving the atmosphere its transparency in these wavelengths. Because a large fraction of the sun's radiant energy is contained in visible wavelengths (Fig. 3.2), solar radiation penetrates relatively freely to the surface, where it provides heat and light.

3.5.4 Photodissociation

If a photon is sufficiently energetic, it can break the bond that holds together the atoms in a molecule. For this photodissociation to occur for the molecules in the atmosphere, wavelengths shorter than ~1 μm are required. Oxygen is dissociated in the upper atmosphere by radiation with wavelengths shorter than ~200 nm. When a photon participates in the dissociation of a molecule it ceases to exist and its energy is absorbed by the atmosphere.

Ozone is a bent molecule made up of three oxygen atoms and is more loosely bonded than molecular oxygen. It can be dissociated by radiation between 200 and 300 nm, and absorbs most of the radiation in this band before it reaches the surface (Fig. 3.2), where it would otherwise do considerable damage to life.

3.5.5 Electronic Excitation

Photons with energies in excess of that corresponding to a wavelength of 1 μm can also excite the electrons in the outer shell of an atom. Sometimes when oxygen or ozone is dissociated by solar radiation, one or both of the resulting oxygen atoms are in an electronically excited state.

3.5.6 Photoionization

If the wavelength of radiation is shorter than about 100 nm, then it can actually remove electrons from the outer shell surrounding the nucleus and produce ionized atoms. This process is responsible for the ionosphere. If the photon is even more energetic (shorter than ~10 nm), it can also photoionize the inner shell of electrons.

3.5.7 Absorption Lines and Line Broadening

The atmosphere is a very effective absorber at those discrete frequencies corresponding to an energy transition of an atmospheric gas. We may call each of these discrete absorption features an *absorption line*. The collection of such absorption lines in a particular frequency interval can be called an *absorption band*. Vibrational and rotational transitions are of primary interest for absorption and emission of terrestrial radiation in the atmosphere, since the energy levels associated with these transitions correspond to the energies of photons of thermal infrared radiation. Polyatomic molecules, like H_2O, CO_2, O_3, CH_4, N_2O, and many others, have vibration–rotation bands of importance in the thermal infrared portion of the electromagnetic spectrum (Fig. 3.4, Table 3.1). Since lower energies are required for rotational transitions, molecules with pure rotational transitions can give a densely packed band of rotation lines. This is the case for water vapor, which has many rotational absorption lines at closely spaced frequencies, which form a rotation band that absorbs much of Earth's emission at wavelengths between 12 and 200 μm.

Table 3.1

Wavelengths of Vibrational Modes of Some Important Atmospheric Molecules

Species	Vibrational modes ν_1	ν_2	ν_3
CO	4.67		
CO_2		15.0	4.26
N_2O	7.78	17.0	4.49
H_2O	2.73	6.27	2.65
O_3	9.01	14.2	9.59
NO	5.25		
NO_2	7.66	13.25	6.17
CH_4	3.43	6.52	3.31
CH_4	5.25		

Units are in micrometers (μm). [From Herzberg and Herzberg (1957), © McGraw Hill, Inc. and from Shimanouchi (1967a, 1967b, 1968)].

A portion of a line absorption spectrum might look like Fig. 3.5(a). Lines are not always evenly spaced or equally strong, since some transitions are more probable than others. The line width depends on broadening processes, which include natural, pressure, and Doppler broadening. After broadening, the absorption spectrum may have substantial absorption between the line centers [Fig. 3.5(b)].

Natural broadening is associated with the finite time of photon emission or absorption and with the uncertainty principle. If we know the energy exactly, then we can only know the frequency to a finite precision. This mechanism is usually less important than pressure or Doppler broadening.

Pressure broadening (also called *collision broadening*) is brought about by collisions between molecules or atoms, which can supply or remove small amounts of energy during radiative transitions, thereby allowing photons with a broader range of frequencies to produce a particular transition of a molecule. This is the primary broadening mechanism in the troposphere.

Doppler broadening results from the movement of molecules relative to a photon, which can cause the frequency of radiation to be Doppler-shifted. This again allows a broader range of frequencies of radiation to effect a particular transition. Doppler broadening becomes the dominant mechanism at high altitudes, where collisions are less frequent.

Pressure broadening is dominant in the troposphere and gives the lines a characteristic shape and width (Fig. 3.6). The line shape is important. Absorption is most probable at the frequency corresponding to the change in the energy of the molecule between two states, and this frequency locates the line center. The probability of

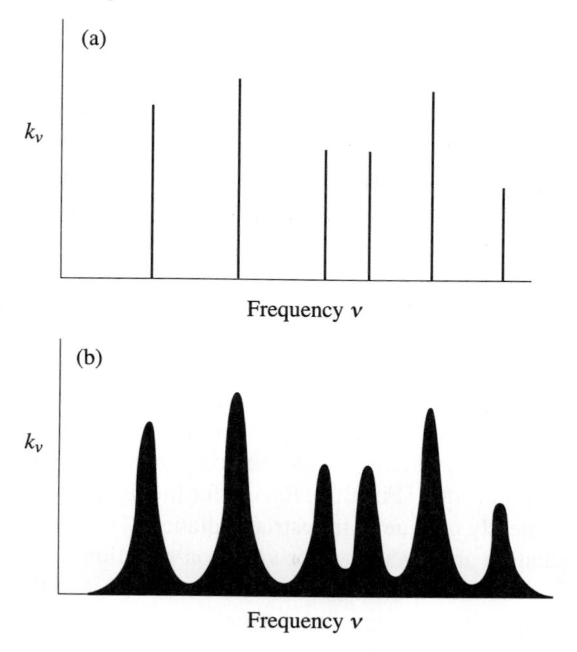

Fig. 3.5 Hypothetical line spectrum (a) before broadening, (b) after broadening.

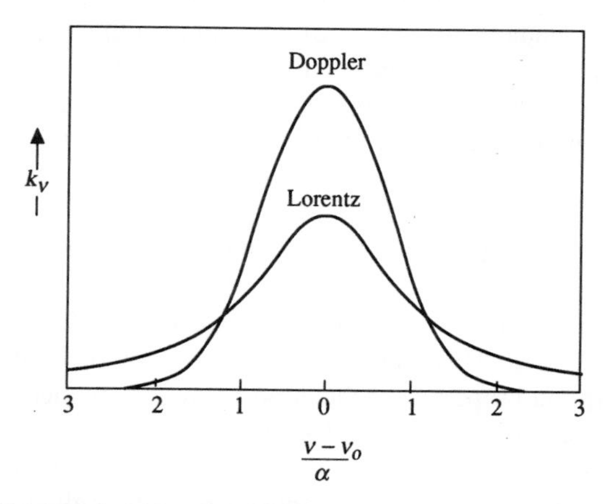

Fig. 3.6 Line shapes produced by pressure (Lorentz line shape) and Doppler broadening for the same line width (α) and intensity at some central frequency ν_o. [From Goody and Yung (1989). Reprinted with permission from Oxford University Press.]

absorption decreases away from the line center but remains significant at some frequency interval away from it. These weak absorption features away from the line center are called the *wings* of the absorption line. For strong absorption lines, the radiation at frequencies near the center of the line is completely absorbed after traveling a relatively small distance through the atmosphere. Under these conditions, most of the transmission of energy is carried by frequencies in the wings of the absorption lines, where absorption is weaker. Pressure broadening, which produces the Lorentz line shape and has broad wings, is particularly effective at producing absorption and emission far from the line centers.

Most of the atmosphere is made up of molecular nitrogen and oxygen. These are diatomic molecules, which have no dipole moment even when vibrating. Therefore they have no vibration–rotation transitions at the small energies corresponding to terrestrial radiation. Thus it is the minor trace concentrations of polyatomic molecules that determine the infrared transmissivity of the atmosphere. The most important gases are water vapor, carbon dioxide, and ozone (in that order), but many other gases contribute significantly (Fig. 3.2). Except for the region between 8 and 12 μm, the atmosphere is nearly opaque to terrestrial radiation. The key absorption features for terrestrial radiation are a water vapor vibration–rotation band near 6.3 μm, the 9.6-μm band of ozone, the 15-μm band of carbon dioxide, and the dense rotational bands of water vapor that become increasingly important at wavelengths longer than 12 μm.

Visible radiation is too energetic to be absorbed by most of the gases in the atmosphere and not energetic enough to photodissociate them, so that the atmosphere is almost transparent to it. Solar radiation with wavelengths between about 0.75 and 5 μm, which we will call near-infrared radiation, is absorbed weakly by water, carbon dioxide, ozone, and oxygen (Fig. 3.2). Most of the ultraviolet radiation from the sun with wavelengths shorter than 0.2 μm is absorbed in the upper atmosphere through the photodissociation and ionization of nitrogen and oxygen. Radiation at frequencies between 0.2 and 0.3 μm is absorbed by ozone in the stratosphere.

3.6 The Lambert–Bouguet–Beer Law: Formulation of Flux Absorption

In the preceding section the physical processes whereby molecules absorb and emit radiation were discussed. This section and the following one illustrate how knowledge of the absorption properties of the atmosphere can be incorporated into a formula to calculate the flux of radiation. To simplify the initial discussion, we will not consider emission by the atmosphere. Suppose, for example, that we are interested in the absorption of solar radiation in the atmosphere. Solar radiation can be absorbed by atmospheric gases, but because the atmosphere is much colder than the sun, the energy reemitted from the atmospheric gases is at longer wavelengths. For the moment we will ignore scattering and consider that the atmosphere can only transmit or absorb solar radiation.

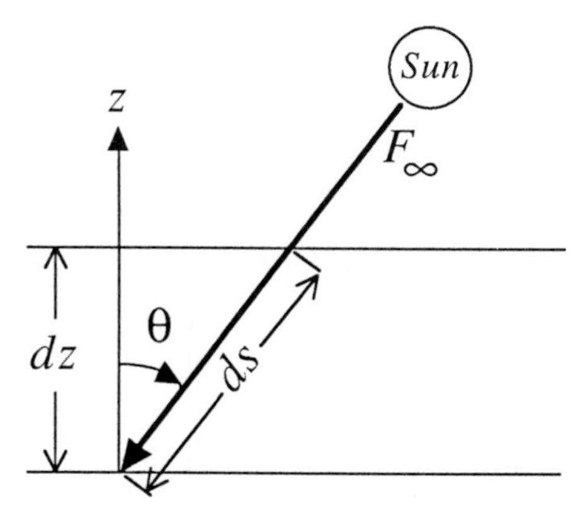

Fig. 3.7 Diagram showing the extinction path of a solar beam through a plane-parallel atmosphere.

For many applications the plane-parallel approximation is accurate and greatly simplifies radiative transfer calculations. Under this approximation the sphericity of Earth is ignored and atmospheric properties are assumed to be functions only of the vertical coordinate. This situation is illustrated in Fig. 3.7. Since unscattered solar radiation can be considered to be a parallel beam of radiation, we need consider only one direction of radiation, which is characterized fully in this case by the zenith angle θ. The *Lambert–Bouguet–Beer law of extinction* states that absorption is linear in the intensity of radiation and the absorber amount. The absorption by a layer of depth dz is proportional to the radiation flux (F) times the mass of absorber along the path the radiation follows. The proportionality constant derives from the quantum-mechanical considerations outlined in the previous section, which determine the probability that a photon with a particular energy will be absorbed by a particular molecule. We call this constant the absorption coefficient and give it the symbol k_{abs}. In general, it depends on the pressure and temperature, since these affect the strength and shape of the absorption lines of the absorber in question.

The change in the radiation flux (dF) along a path of length ds, where the density of the absorber is ρ_a and the absorption coefficient is k_{abs}, may be written

$$dF = -k_{\text{abs}}\, \rho_a\, F\, ds \tag{3.12}$$

In (3.12) F and ds are both measured positive downward. F and k_{abs} depend on frequency, but we have dropped the frequency subscript for economy. The units of k_{abs} in (3.12) must be $\text{m}^2\ \text{kg}^{-1}$. Because its units are area per unit mass, k_{abs} is sometimes also called the absorption cross section of the gas in question. From Fig. 3.7, the pathlength is related to altitude according to

$$dz = -\cos\theta\ ds \tag{3.13}$$

Therefore (3.12) becomes

$$\cos\theta \frac{dF}{dz} = k_{\text{abs}} \rho_a F \tag{3.14}$$

We can define the optical depth (τ) along a vertical path.

$$\tau = \int_z^\infty k_{\text{abs}} \rho_a \, dz \tag{3.15}$$

Note that (3.15) implies that $d\tau = -k_{\text{abs}} \rho_a dz$, so that we can write (3.14) as

$$\cos\theta \frac{dF}{d\tau} = -F \tag{3.16}$$

This equation has a very simple solution,

$$F = F_\infty \, e^{-\tau/\cos\theta} \tag{3.17}$$

where F_∞ is, in this case, the downward flux density at the top of the atmosphere. The incident flux thus decays exponentially along the slant path *ds* where the optical depth is given by $\tau/\cos\theta$.

3.6.1 Absorption Rate

In an isothermal atmosphere in hydrostatic balance, the density (ρ_a) of an absorber with constant mass mixing ratio is given by (Section 1.4):

$$\rho_a = \rho_{\text{as}} \, e^{(-z/H)} \tag{3.18}$$

Here $H = RT/g$ is the scale height, $R = 287 \text{ J K}^{-1} \text{ kg}^{-1}$, $g = 9.81 \text{ m s}^{-2}$, and ρ_{as} is the density of the absorber at the surface.

If we introduce this into the equation for optical depth (3.15), assume that k_{abs} is a constant, and perform the integration, we obtain

$$\tau = \frac{p_s}{g} M_a \, k_{\text{abs}} \left[e^{-z/H} \right] \tag{3.19}$$

where $M_a = \rho_a/\rho$ is the mass mixing ratio of the absorber. From (3.19) we can see that the total optical depth is $(p_s/g) M_a k_{\text{abs}}$, which is the total mass of the absorber times the absorption cross section. We can use (3.19) to show that

$$\frac{d\tau}{dz} = -\frac{\tau}{H} \tag{3.20}$$

To calculate the energy absorption rate per unit volume, we multiply the flux times the density times the absorption coefficient. Using (3.14), (3.17), and (3.20), and defining $\mu = \cos\theta$, we obtain that

$$\text{Absorption rate} = \frac{dF}{dz} = \frac{k_{\text{abs}} \rho_a F}{\mu} = -\frac{d\tau}{dz} \frac{F}{\mu} = \frac{F_\infty}{\mu} \, e^{-\tau/\mu} \, \frac{\tau}{H} \tag{3.21}$$

The absorption rate per unit volume peaks where the product of flux and absorbing mass cross section reaches a maximum. We can find this point by differentiating the absorption rate in (3.21) with respect to optical depth and setting the derivative to zero to find the maximum.

$$\frac{d}{d\tau}\left[\frac{F_\infty}{\mu}\, e^{-\tau/\mu}\,\frac{\tau}{H}\right] = \frac{F_\infty}{\mu H}\, e^{-\tau/\mu}\left[1-\frac{\tau}{\mu}\right] = 0 \tag{3.22}$$

Thus we derive that the absorption rate peaks at $\tau/\mu = 1$. One may show that the pressure level where the absorption rate per unit volume is maximum is given by

$$\frac{p_{\text{max abs}}}{p_s} = \frac{\cos\theta}{H\, k_{\text{abs}}\, \rho_{\text{as}}} \tag{3.23}$$

where p_s is the surface pressure. The pressure of maximum absorption is proportional to the cosine of the solar zenith angle, so that as the sun moves toward the horizon the absorption occurs higher in the atmosphere. The pressure at the level of maximum absorption is inversely proportional to the mass of absorber per unit surface area, $H\rho_{\text{as}}$, and to the absorption coefficient k_{abs}.

The heating rate associated with absorption of a downward flux of radiant energy is given by

$$\left.\frac{\partial T}{\partial t}\right|_{\text{rad}} = \frac{1}{c_p \rho}\frac{\partial F}{\partial z} \tag{3.24}$$

where c_p is the specific heat at constant pressure and ρ is the air density. Utilizing (3.14), (3.24) can be written as

$$\left.\frac{\partial T}{\partial t}\right|_{\text{rad}} = \frac{k_{\text{abs}}\, M_a}{c_p \mu} F \tag{3.25}$$

If the mass mixing ratio of the absorber, M_a, is independent of altitude, then the heating rate will be proportional to the flux itself, which is maximum at the outer extremity of the atmosphere. This is the case for absorption of ultraviolet radiation by molecular oxygen and nitrogen in the upper atmosphere, which produces maximum heating rates at very high altitudes and accounts for the rapid increase of temperature with altitude in the thermosphere shown in Fig. 1.2. For an absorber like ozone, whose mixing ratio peaks sharply in the stratosphere, the heating rate will also peak in the stratosphere. This local maximum in the ozone heating rate produces the local maximum in the climatological temperature at the stratopause near 50 km.

3.7 Infrared Radiative Transfer Equation: Absorption and Emission

We will next develop the radiative transfer equation for a plane, parallel atmosphere in which both emission and absorption by gases occur. The goal of this section is to provide an intuitive understanding of the transfer of longwave radiation through the atmosphere and its importance for climate. Toward this end some simplifications will be made that cannot be made in accurate calculations of radiative transfer. Readers not interested in the details of infrared radiative transfer in the atmosphere may skip ahead to Sections 3.8 and 3.9, where simple heuristic models are described. In Section 3.10 the results of more accurate calculations will be presented.

We again make the plane, parallel atmosphere approximation under which Earth is considered to be a flat plane, and the properties of the atmosphere depend only on altitude. These are good approximations if the properties of the atmosphere vary slowly in the horizontal compared to the vertical. This is generally true for temperature and humidity because the atmosphere is so thin compared to the radius of Earth. It may not be a good approximation when clouds are present whose horizontal dimension is comparable to their vertical dimension.

3.7.1 Schwarzchild's Equation

Consider the situation depicted in Fig. 3.8 in which a beam of intensity I_ν passes upward through a layer of depth dz, making an angle of θ with the vertical direction. The change of intensity along the path through this layer will be equal to the emission from the gas along the path minus the absorption

$$dI_\nu = E_\nu - A_\nu \tag{3.26}$$

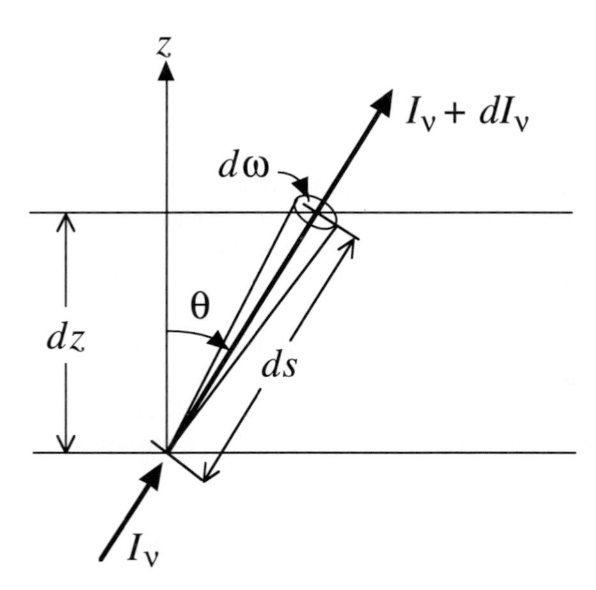

Fig. 3.8 Diagram showing the path of upward-directed terrestrial radiance through a plane-parallel atmosphere.

The absorption can be assumed to follow the Lambert–Bouguet–Beer law.

$$dI_\nu = E_\nu - \rho_a \, ds \, k_\nu I_\nu \tag{3.27}$$

The emissivity ε_ν is defined as the ratio of the emission of a substance to the intensity of radiation in a cavity at equilibrium. The latter is given by Planck's function (3.7), so that the emission may be written as follows:

$$E_\nu = \varepsilon_\nu B_\nu(T) \tag{3.28}$$

When collisions with other molecules are much more frequent than radiative transitions of a molecule, the assumption of local thermodynamic equilibrium may be made. This assumption is warranted for most absorption lines in the relatively high pressure environment of the troposphere and lower stratosphere. In thermodynamic equilibrium, according to Kirchhoff's law, the emissivity of a substance must be equal to its absorptivity.

With the assumption of local thermodynamic equilibrium, we have $\varepsilon_\nu = \rho_a ds \, k_\nu$ so that (3.27) becomes

$$dI_\nu = \rho_a \, ds \, k_\nu \left(B_\nu(T) - I_\nu\right) \tag{3.29}$$

Utilizing (3.13) and taking the limit of small dz, we obtain a radiative transfer equation

$$\cos\theta \, \frac{dI_\nu}{dz} = \rho_a \, k_\nu \left(B_\nu(T) - I_\nu\right) \tag{3.30}$$

In this equation I_ν is a function of both altitude z and zenith angle θ, and all other dependent variables are functions of only z. For infrared radiation we may reasonably assume that the intensity is independent of azimuth angle, φ (Fig. 3.1), and that the blackbody emission, $B_\nu(T(z))$, is isotropic, being independent of both θ and φ, but still dependent on z.

To proceed toward a solution of (3.30) we again introduce the optical depth, defined this time from the surface upward.

$$\tau_\nu(z) = \int_0^z \rho_a \, k_\nu \, dz \tag{3.31}$$

Using the definition (3.31) in (3.30) we obtain

$$\cos\theta \, \frac{dI_\nu\left(\tau_\nu(z), \theta\right)}{d\tau_\nu} = B_\nu\left(T\left(\tau_\nu(z)\right)\right) - I_\nu\left(\tau_\nu(z), \theta\right) \tag{3.32}$$

Parentheses in (3.32) show the dependence of the radiance on zenith angle and altitude, the latter being represented parametrically by vertical optical depth.

Multiply (3.32) by the integrating factor $e^{\tau_\nu/\mu}$, where $\mu = \cos\,\theta$, to convert it to a form that can be integrated directly.

$$\mu \frac{d}{d\tau_\nu} \left\{ I_\nu(\tau_\nu(z),\theta)\, e^{\{\tau_\nu(z)/\mu\}} \right\} = B_\nu\left(T\left(\tau_\nu(z)\right)\right) e^{\{\tau_\nu(z)/\mu\}} \tag{3.33}$$

We integrate (3.33) from the surface, where the optical depth is zero, to some arbitrary altitude where we wish to calculate the upward intensity and where the optical depth is $\tau_\nu(z)$. The integral is performed using τ'_ν as the dummy variable of integration.

$$I_\nu(\tau_\nu(z),\mu) = I_\nu(0,\mu)\; e^{\{-\tau_\nu(z)/\mu\}} + \int_0^{\tau_\nu(z)} \mu^{-1}\, B_\nu\left(T\left(\tau'_\nu\right)\right)\; e^{\{\left(\tau'_\nu - \tau_\nu(z)\right)/\mu\}}\, d\tau'_\nu \tag{3.34}$$

The first term on the right in (3.34) represents the emission from the surface, reduced by the extinction along the path from the surface to the altitude z. The second term represents the summation of the emission from all of the layers of the atmosphere below the level z, which reaches the level z without absorption.

3.7.2 Simple Flux Forms of the Radiative Transfer Equation Solution

In Appendix D simplified flux forms of the solution (3.34) are derived that can be used to gain a physical understanding of how fluxes of thermal infrared radiation in the atmosphere depend on the vertical temperature profile and the infrared opacity of the atmosphere. A number of approximations are necessary to derive the following simplified equations for the upward $F^{\uparrow}(z)$ and downward $F^{\downarrow}(z)$ flux of terrestrial radiation at some height z in the atmosphere.

$$F^{\uparrow}(z) = \sigma T_s^4\, \mathcal{T}\{z_s, z\} + \int_{\mathcal{T}\{z_s,z\}}^{1} \sigma T(z')^4\; d\mathcal{T}\{z', z\} \tag{3.35}$$

$$F^{\downarrow}(z) = \int_{\mathcal{T}\{z,\infty\}}^{1} \sigma T(z')^4\; d\mathcal{T}\{z', z\}\,. \tag{3.36}$$

These equations represent the fluxes of terrestrial energy at all frequencies and integrated over all angles, and would be given in W m^{-2} (watts per square meter). The fluxes depend on the temperature and on the flux transmission, $\mathcal{T}\{z',z\}$, which is the fraction of the total terrestrial energy flux that can pass between altitudes z and z' without absorption. The transmission is always between zero, which indicates no transmission, and one, which indicates complete transmission. The transmission approaches one as the two altitudes get arbitrarily close together. As the two altitudes move apart the transmission decreases at a rate that depends on the absorber amount between them.

The net flux of terrestrial radiation is given by the difference between the upward and downward flux:

$$F(z) = F^{\uparrow}(z) - F^{\downarrow}(z) \tag{3.37}$$

The heating rate associated with the divergence of the terrestrial flux density is then

$$\frac{\partial T}{\partial t} = -\frac{1}{\rho c_p} \frac{\partial F}{\partial z} \tag{3.38}$$

The first term on the right in (3.35) is a boundary term and assumes that the ground emits like a blackbody at the temperature of the surface T_s. At the level z the flux density associated with the surface emission is reduced to the fraction $\mathcal{T}\{z_s, z\}$ of its surface value. The integral in (3.35) represents the contribution to the upward flux from the atmosphere below the level z, and in (3.36) the integral represents the contribution to the downward flux at z from the atmosphere above that level. The contributions to these integrals are greatest where the transmission function is changing most rapidly. This occurs when the difference in optical depth between the two locations is close to one.

Of particular interest for climate is the upward flux of terrestrial radiation at the top of the atmosphere and the downward flux at the surface. From (3.35) and (3.36) these are

$$F^{\uparrow}(\infty) = \sigma T_s^4 \, \mathcal{T}\{z_s, \infty\} + \int_{\mathcal{T}\{z_s,\infty\}}^{1} \sigma T(z')^4 \, d\mathcal{T}\{z', \infty\} \tag{3.39}$$

$$F^{\downarrow}(z_s) = \int_{\mathcal{T}\{z_s,\infty\}}^{1} \sigma T(z')^4 \, d\mathcal{T}\{z', z_s\}. \tag{3.40}$$

Figure 3.9 illustrates what the two transmission functions in (3.39) and (3.40) might look like, together with a representation of the temperature profile in the atmosphere. The primary contributions to the integrals come from the levels where the transmission functions are changing most rapidly. In the case of outgoing longwave radiation (OLR), $F^{\uparrow}(\infty)$, only a small amount of the emission from the surface is able to escape to space under average terrestrial conditions. Under typical conditions most of the OLR originates in the troposphere at levels where the temperature is significantly less than the surface value. From (3.39) it can be seen that the emission temperature defined in Chapter 2 is a weighted mean of the temperature profile from the surface to the top of the atmosphere, where the weight is the change in the transmission function.

When an absorbing atmosphere is present, the average emission temperature is less than the surface value, and the loss of energy by emission to space is much less than the infrared emission from the surface. This is one component of the greenhouse

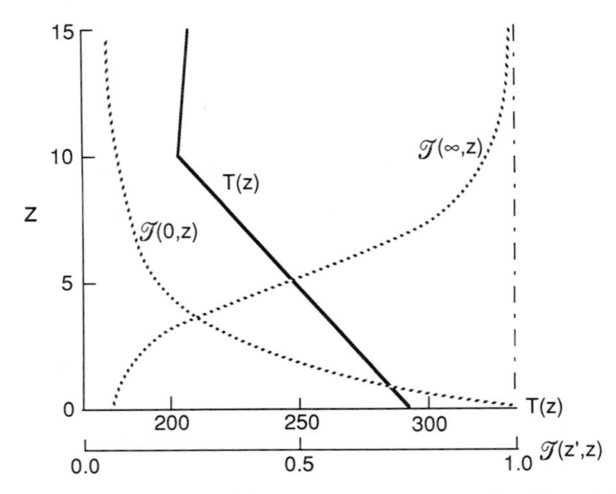

Fig. 3.9 Transmission functions and air temperature (K) as functions of altitude.

effect. Another aspect is the supply of energy to the surface by downward emission of terrestrial radiation from the atmosphere. The downward flux of terrestrial radiation at the surface originates in the lower troposphere, where most of the water vapor resides. The temperatures there are relatively warm. In the global mean, this downward flux is larger by nearly a factor of two than the input of energy to the surface from the sun (Fig. 2.4).

It is interesting to consider the case of an isothermal atmosphere at temperature T_A, overlying a surface of temperature T_s. In this case the temperature can be taken outside the integrals in (3.39) and (3.40).

$$F^{\uparrow}(\infty) = \sigma T_s^4 \mathcal{T}\{z_s, \infty\} + \sigma T_A^4 \left[1 - \mathcal{T}\{z_s, \infty\}\right] \tag{3.41}$$

$$F^{\downarrow}(z_s) = \sigma T_A^4 \left[1 - \mathcal{T}\{z_s, \infty\}\right] \tag{3.42}$$

From these expressions we obtain the limiting cases for an opaque and a transparent atmosphere:

$$\mathcal{T}\{z_s, \infty\} = 0 \quad \Rightarrow F^{\downarrow}(z_s) = \sigma T_A^4, \qquad F^{\uparrow}(\infty) = \sigma T_A^4 \tag{3.43}$$

$$\mathcal{T}\{z_s, \infty\} = 1 \quad \Rightarrow F^{\downarrow}(z_s) = 0, \qquad F^{\uparrow}(\infty) = \sigma T_s^4 \tag{3.44}$$

In the first limit (3.43) the atmosphere is perfectly opaque to longwave radiation and emits both upward and downward like a blackbody. In the next section we consider a model atmosphere made up of layers that are opaque to longwave radiation. In the case of unit transmissivity (3.44) the atmosphere has no effect on terrestrial radiation, so that the downward longwave is zero and the surface emission escapes directly to space.

Clouds strongly affect the transmission of terrestrial radiation through the atmosphere. If a cloud is reasonably thick and has a sharp top and bottom, then it can be treated accurately as a perfect absorber of longwave radiation. If such a cloud is present with a bottom at z_{cb} and a top at z_{ct}, then (3.39) and (3.40) are changed.

$$F^{\uparrow}(\infty) = \sigma T_{z_{ct}}^4 \, \mathcal{T}\{z_{ct},\infty\} + \int_{\mathcal{T}\{z_{ct},\infty\}}^{1} \sigma T(z')^4 \, d\mathcal{T}\{z',\infty\} \tag{3.45}$$

$$F^{\downarrow}(z_s) = \sigma T_{z_{cb}}^4 \, \mathcal{T}\{z_{cb},z_s\} + \int_{\mathcal{T}\{z_{cb},z_s\}}^{1} \sigma T(z')^4 \, d\mathcal{T}\{z',z_s\} \tag{3.46}$$

In this case the downward flux at the ground (3.46) also has a boundary term coming from the bottom of the cloud. It can be shown using the mean value theorem of calculus that if the temperature decreases with altitude, then the outgoing flux at the top of the atmosphere will be decreased by a cloud, and the downward flux at the ground will be increased. The amount of the change will depend on the lapse rate and the position of the cloud top and base relative to the transmission function for the clear atmosphere.

3.8 Heuristic Model of Radiative Equilibrium

A layer of atmosphere that is almost opaque for longwave radiation can be crudely approximated as a blackbody that absorbs all terrestrial radiation incident on it and emits like a blackbody at its temperature. For an atmosphere with a large infrared optical depth, the radiative transfer process can be represented with a series of blackbodies arranged in vertical layers. Two layers centered at 0.5- and 2.0-km altitudes provide a simple approximation for Earth's atmosphere.[1] If we assume that the atmospheric layers are transparent to solar radiation, we have the schematic energy flow diagram shown in Fig. 3.10.

We can solve for all of the unknown temperatures by using the energy balance at each of the layers. If no net energy gain or loss occurs at any of the levels, then the temperatures obtained are the radiative equilibrium values. At the top of the atmosphere we must have energy balance, so that

$$\frac{S_0}{4}\left(1-\alpha_p\right) = \sigma T_e^4 = \sigma T_1^4 \tag{3.47}$$

We thus know immediately that the top layer temperature must equal the emission temperature of the planet, since, in this approximation, the only longwave emission that escapes to space comes from the upper layer. The energy balance at layer 1 is

$$\sigma T_2^4 = 2\sigma T_1^4 \tag{3.48}$$

The balance at layer 2 yields

$$\sigma T_1^4 + \sigma T_s^4 = 2\sigma T_2^4 \tag{3.49}$$

[1]Goody and Walker (1972).

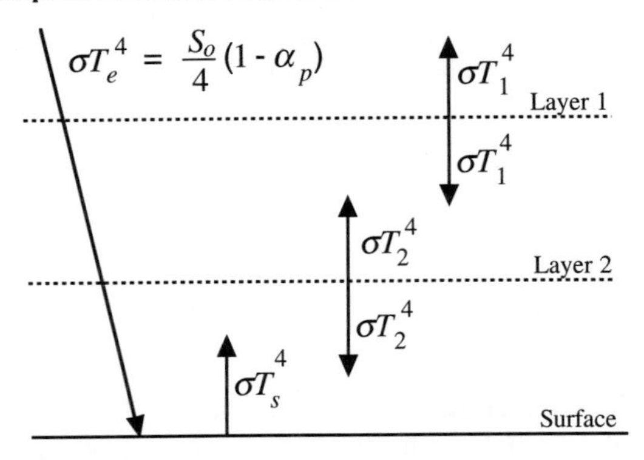

Fig. 3.10 Diagram of simple two-layer radiative equilibrium model for the atmosphere–Earth system, showing the fluxes of radiant energy.

And the balance at the surface is

$$\frac{S_0}{4}\left(1-\alpha_p\right)+\sigma T_2^4=\sigma T_s^4 \tag{3.50}$$

The critical effect of an atmosphere that absorbs and emits longwave radiation appears in (3.50). The energy supplied to the surface by the sun is augmented by a downward flux of longwave radiation from the atmosphere. This allows the surface temperature to rise significantly above the value it would have in the absence of an atmosphere.

We can use (3.47) through (3.50) to solve for the surface temperature.

$$T_s^4=3\frac{\left(S_0/4\right)\left(1-\alpha_p\right)}{\sigma}=3T_e^4 \tag{3.51}$$

By extension, if such a model atmosphere has an arbitrary number of layers, n, the surface temperature in equilibrium will be

$$T_s=\sqrt[4]{n+1}\;T_e \tag{3.52}$$

The radiative equilibrium surface temperature for a two-layer atmosphere is 335 K, which is much hotter than Earth's surface temperature. Radiative equilibrium is not a good approximation for the surface temperature, since we know that latent and sensible heat fluxes remove substantial amounts of energy from the surface.

If the atmosphere absorbs no solar radiation, the energy balance for a thin layer of emissivity ε at the top of the atmosphere is between absorption of the flux of terrestrial radiation from below and the emission from the layer itself.

$$\varepsilon\sigma T_e^4=2\varepsilon\sigma T_{\text{strat}}^4 \tag{3.53}$$

where T_{strat} is the temperature at the outer edge of the atmosphere, which we may take to be the stratosphere.

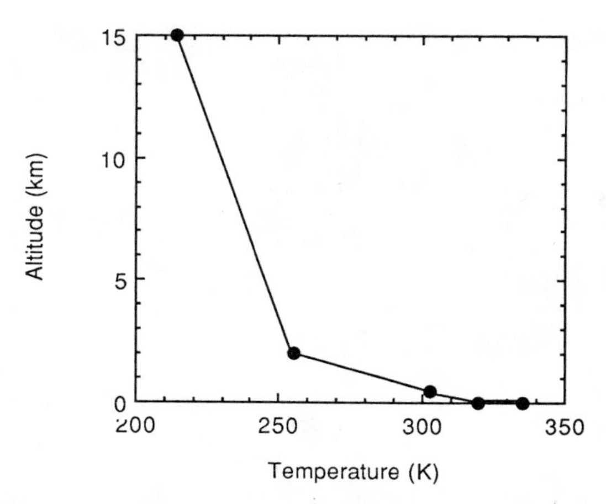

Fig. 3.11 Plot of temperature profile obtained from the simple two-level atmosphere radiative equilibrium model.

A thin layer of atmosphere near the surface absorbs a fraction ε of the emission from above and below and emits in both directions. The temperature of the air adjacent to the surface, T_{SA}, may be derived from the energy balance there.

$$\varepsilon\sigma T_s^4 + \varepsilon\sigma T_2^4 = 2\varepsilon\sigma T_{SA}^4 \tag{3.54}$$

We can solve for all of the temperatures and obtain the following values.

$T_1 = 255$ K $\quad T_2 = 303$ K $\quad T_s = 335$ K

$T_{strat} = 214$ K $\quad T_{SA} = 320$ K

These temperatures are plotted in Fig. 3.11. In pure radiative equilibrium the temperatures of the surface and the air in contact with the surface are different. This discontinuity is caused by the absorption of solar radiation at the surface. Such discontinuities are usually greatly suppressed in reality because of efficient heat transport by conduction and convection.

3.9 Clouds and Radiation

Clouds consist of liquid water droplets or ice particles suspended in the atmosphere. They are formed by the condensation of atmospheric water vapor when the temperature falls below the saturation temperature. Water droplets and ice particles have substantial interactions with both solar and terrestrial radiation (Fig. 3.12). The nature of these interactions depends on the total mass of water, the size and shape of the droplets or particles, and their distribution in space. The problem is often simplified by assuming that clouds are uniform and infinite in the horizontal, which is called

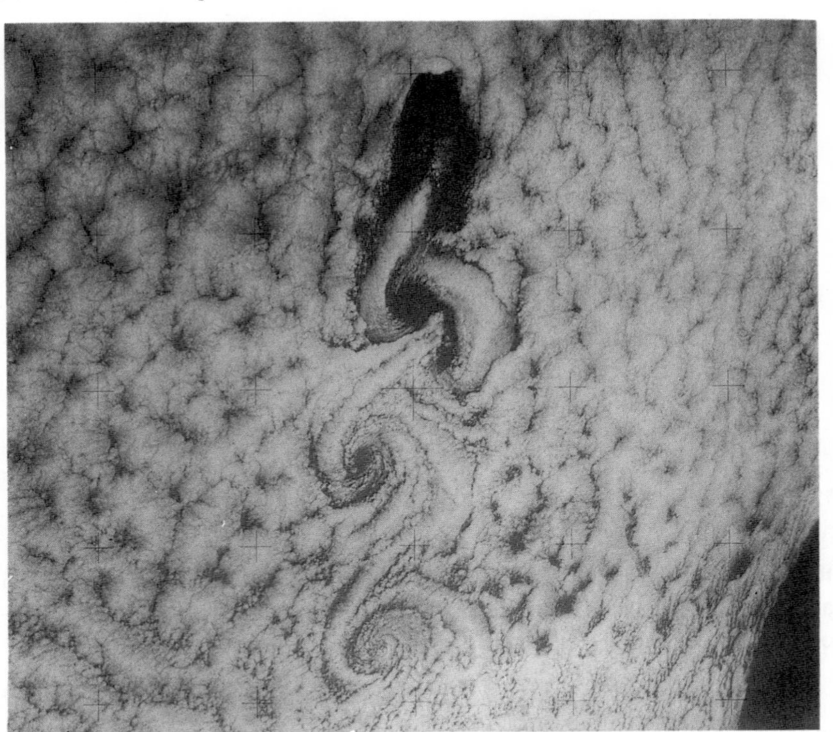

Fig. 3.12 Stratocumulus clouds moving past Guadalupe Island off the west coast of Mexico. Note the vortex rings shed downstream of the island. Stratocumulus clouds have a large negative effect on the radiative energy budget of Earth because they are good reflectors of solar radiation, but are confined near the surface and so do not provide strong trapping of outgoing longwave emission. The vortex centers are about 60 km apart. (Skylab 3, NASA, August 1973.)

the plane-parallel cloud assumption. If the droplet size distribution and the vertical distribution of humidity are assumed, then the cloud albedo and absorption depend on the total liquid water content of the cloud and the solar zenith angle. Cloud liquid water content is defined as the total mass of cloud water in a vertical column of atmosphere per unit of surface area.

Model calculations of the cloud albedo and absorption for plane-parallel water clouds are shown in Fig. 3.13. The albedo increases with the total water content or depth of the cloud and also with the solar zenith angle. The albedo increase with liquid water content is most rapid for smaller amounts of liquid water. As the cloud becomes very thick the cloud albedo slowly approaches a limiting upper value and becomes insensitive to further increases in cloud mass. In very thick clouds most of the solar radiation is scattered before it can reach the particles deep in the cloud, and radiation that is scattered from particles deep in the cloud is unlikely to find its way

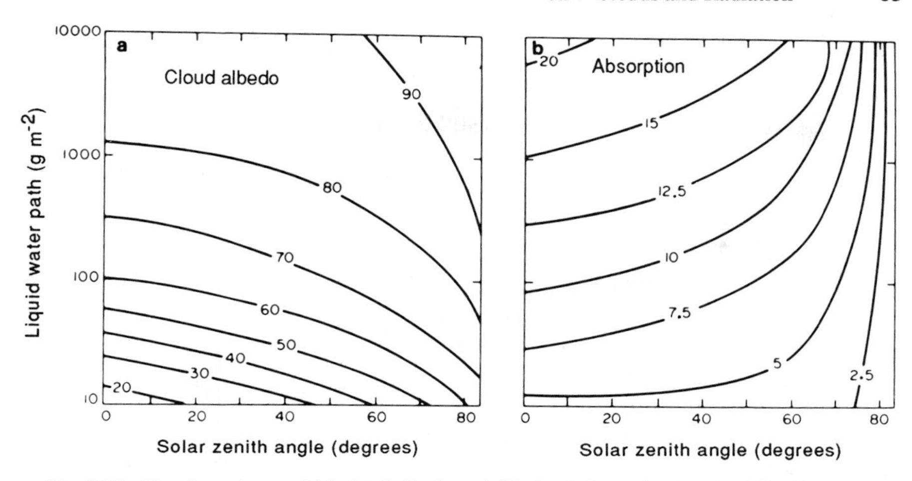

Fig. 3.13 The dependence of (a) cloud albedo and (b) cloud absorption on cloud liquid water path and solar zenith angle. Values are given in percent. [From Stephens (1978). Reprinted with permission from the American Meteorological Society.]

back out to space. For these reasons, increases in the optical thickness of a cloud eventually cease to make much difference in the reflectivity of the cloud. The variation of albedo with zenith angle is most rapid when the sun is near the horizon and least when the sun is overhead. Absorption of solar radiation by plane-parallel clouds decreases with increasing zenith angle because radiation that is reflected to space at the higher zenith angles penetrates less deeply into the cloud and is therefore less likely to be absorbed. The absorption increases approximately linearly with liquid water content for an overhead sun.

The variations of the albedo of typical clouds in the atmosphere are dominated by variations in the column amount of liquid water and ice in the cloud. Nonetheless the albedo of clouds is sensitive to the droplet size. Figure 3.14 shows how the calculated albedo of clouds changes as the radius of the droplets is varied, while keeping the liquid water content fixed. The albedo is greatest for smaller droplets, principally because these present a larger surface area for the same mass.

Clouds absorb terrestrial radiation very effectively. Figure 3.15 shows the emissivity of water and ice clouds as a function of liquid water content. Clouds become opaque to longwave radiation when the liquid water path exceeds about 20 g m^{-2}. If this liquid water path is achieved in an altitude range where the temperature is essentially uniform, then cloud surfaces can be assumed to absorb and emit terrestrial radiation essentially like blackbodies. This assumption produces accurate results except for very thin clouds such as cirrus, which may be partially transparent to longwave radiation. Comparison of Figs. 3.13 and 3.15 shows that the albedo of clouds continues to increase with additional liquid water content long after the cloud has become opaque to longwave radiation.

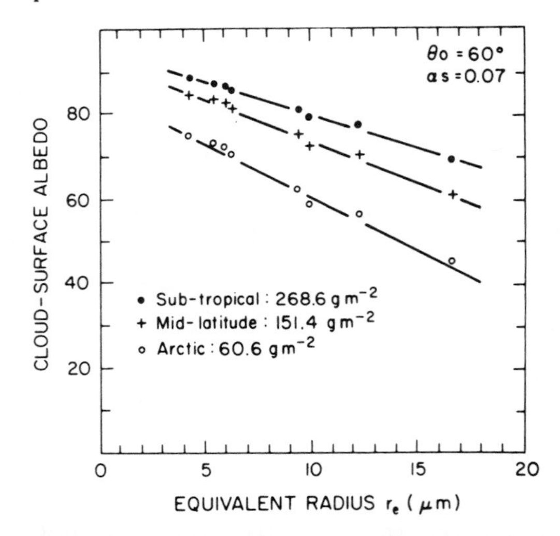

Fig. 3.14 The dependence of planetary albedo on the size of cloud droplets. [From Slingo and Schrecker (1982). Reprinted with permission from the Royal Meteorological Society.]

3.10 Radiative–Convective Equilibrium Temperature Profiles

As a reasonably straightforward method of attempting to understand the effects of radiative transfer on climate, one can solve the radiative transfer equation for global-mean terrestrial conditions. This involves construction of appropriate models for the

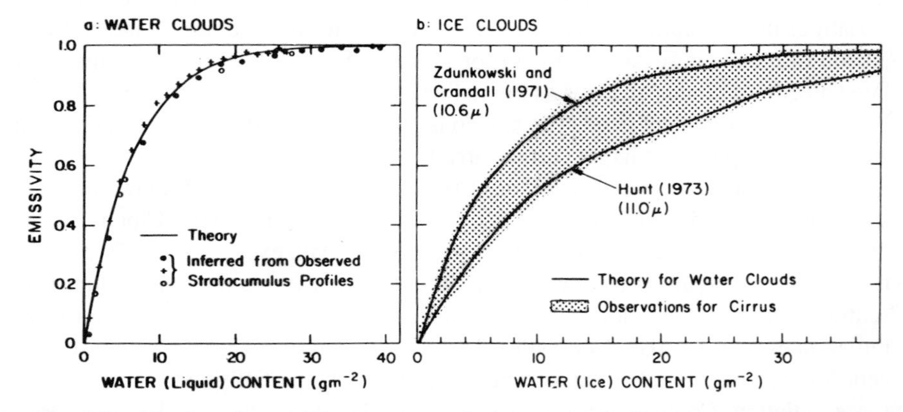

Fig. 3.15 The dependence of the longwave emissivity on (a) liquid water content [from Slingo *et al.* (1982); reprinted with permission from the Royal Meteorological Society] and (b) ice content [from Griffith *et al.* (1980); reprinted with permission from the American Meteorological Society].

transmission of the various band systems of importance in the atmosphere, insertion of these into a computational analog of the radiative transfer equation, and iteration to obtain a steady balance solution. Such models are a much more sophisticated version of the simple radiative equilibrium model discussed in Section 3.8.

The variables that determine the fluxes of radiant energy in the atmosphere include the atmospheric gaseous composition, the aerosol and cloud characteristics, the surface albedo, and the insolation. Since horizontal transport of energy by atmospheric and oceanic motion affects the local climate, it is of most interest to calculate the radiative equilibrium for conditions averaged over the globe. In a global-mean model the temperature and all other variables depend only on altitude, and the globally averaged insolation and solar zenith angle are appropriate. To understand the basic radiative energy balance of Earth, we need to specify the following:

1. H_2O: Water vapor is the most important gas for the transfer of radiation in the atmosphere. Its distribution is highly variable. The sources and sinks (evaporation and condensation) are determined by the climate itself, and they are fast compared to the rate at which the atmosphere's motion mixes moist and dry air together. Water vapor has a vibration–rotation band near 6.3 μm and a rotation continuum at wavelengths longer than about 12 μm. It is also the principal absorber of solar radiation in the troposphere.

2. CO_2: The mixing ratio of carbon dioxide is increasing about 0.4% per year primarily because of coal and oil combustion. In 1990 the value was about 350 ppmv. Because sources and sinks of CO_2 are slow compared to the time it takes the atmosphere to mix thoroughly, its mixing ratio can be assumed constant with latitude and altitude up to about 100 km. The strong vibration-rotation band of CO_2 at 15 μm is important for longwave radiative transfer. A significant amount of solar radiation is absorbed by carbon dioxide.

3. O_3: Ozone has fast sources and sinks in the stratosphere where most of the atmospheric ozone resides. Near the surface ozone is produced in association with photochemical smog. Its concentration in the middle and upper stratosphere is dependent on temperature, insolation, and a host of photochemically active trace species. Ozone has a vibration–rotation band near 9.6 μm that is important for longwave energy transfer, and also has a dissociation continuum that absorbs solar radiation between 200 and 300 nm. Absorption of solar radiation by ozone heats the middle atmosphere and causes the temperature increase with height that defines the stratosphere and tropopause.

4. *Aerosols:* Atmospheric aerosols of various types affect the transmission of both solar and terrestrial radiation. A layer of sulfuric acid aerosols exists near 25 km in the midstratosphere. Sulfate aerosols in the troposphere are also radiatively important and seem to be increasing as a result of human activity, primarily fossil fuel combustion.

5. *Surface albedc:* The surface albedo is highly variable from location to location in land areas, depending on the type and condition of the surface material and

vegetation. Over open ocean it is mostly a function of solar zenith angle, but it also depends on sea state. When the surface is snow covered, its albedo is generally much higher than when surface ice is not present.

6. *Clouds:* Clouds vary considerably in amount and type over the globe. They have very important effects on longwave and solar energy transfer in the atmosphere. The distribution in time and space and the optical properties of clouds are important for climate. For a global-mean radiative equilibrium calculation, cloud radiative properties must be specified. The simplest approach is to assume plane-parallel clouds and specify their distribution in the vertical. The optical properties of the clouds must also be specified. For solar radiation one must specify the fractions of a beam of radiation that are absorbed and reflected by clouds, which can be called the *absorptivity* and *reflectivity,* respectively. Normally, water clouds have relatively weak solar absorption, but they effectively scatter solar radiation back toward space and so have high reflectivities. Thick clouds can be assumed to be black bodies for longwave radiation, so that they absorb all incident longwave radiation and emit like a blackbody with the temperature of the atmosphere at the same level as the cloud. A simple approach is to specify the properties of three types of clouds, as in Table 3.2. The albedo values are based on old estimates and are not necessarily the most representative set, but they are the values used in the calculations shown here.

Figure 3.16 shows a calculated temperature profile that is in radiative equilibrium. Atmospheric temperatures in radiative equilibrium decrease rapidly with altitude near the surface. In the troposphere, radiative equilibrium temperature profiles are hydrostatically unstable in the sense that parcels of air that are elevated slightly will become buoyant and continue to rise. In the real atmosphere, atmospheric motions move heat away from the surface and mix it through the troposphere. Figure 2.4 indicates that 60% of the energy removal from the surface is done by the transport of heat and water vapor by atmospheric motions and only 40% by net longwave radiation emission. The global mean temperature profile of Earth's atmosphere is not in radiative equilibrium, but rather in radiative–convective equilibrium. To obtain a realistic global-mean vertical energy balance, the vertical flux of energy by atmospheric motions must be included.

Table 3.2

Values of Cloud Shortwave Reflectivity and Absorptivity and Fractional Area Coverage Assumed in Manabe and Strickler (1964)

Type	SW reflectivity	SW absorptivity	% of area
High (cirrus)	0.21	0.005	0.228
Medium (cumulus)	0.48	0.020	0.090
Low (stratus)	0.69	0.035	0.313

(Reprinted with permission from the American Meteorological Society.)

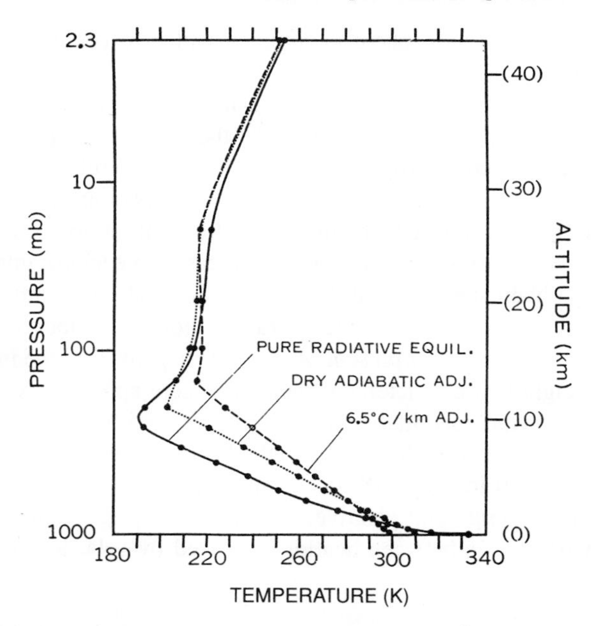

Fig. 3.16 Calculated temperature profiles for radiative equilibrium, and thermal equilibrium with lapse rates of 9.8°C km^{-1} and 6.5°C km^{-1}. [From Manabe and Strickler (1964). Reprinted with permission from the American Meteorological Society.]

The simplest artifice by which the effect of vertical energy transports by motions can be included in a global-mean radiative transfer model is a procedure called *convective adjustment*. Under this constraint the lapse rate is not allowed to exceed a critical value, say, 6.5 K km^{-1}. Where radiative processes would make the lapse rate greater than the specified maximum value, a nonradiative upward heat transfer is assumed to occur that maintains the specified lapse rate while conserving energy. This artificial vertical redistribution of energy is intended to represent the effect of atmospheric motions on the vertical temperature profile without explicitly calculating nonradiative energy fluxes or atmospheric motions. In a global mean model, this "adjusted" layer extends from the surface to the tropopause.

A temperature profile that is in energy balance when radiative transfer and convective adjustment are taken into account may be called a *radiative–convective equilibrium* or *thermal equilibrium* profile. Thermal equilibrium profiles for assumed maximum lapse rates of 6.5°C km^{-1} and the dry adiabatic lapse rate of 9.8°C km^{-1} are shown in Fig. 3.16. The thermal equilibrium profile obtained with a lapse rate of 6.5°C km^{-1} is close to the observed global mean temperature profile. No a priori reason exists for choosing a 6.5°C km^{-1} adjustment lapse rate other than that it corresponds to the observed global-mean value. The maintenance of the lapse rate of the atmosphere is complex and involves many processes and scales of motion.

One use of a climate model is to understand what factors are most important and how changes in these factors will affect the climate. In particular, the one-dimensional radiative–convective equilibrium model is useful for understanding the role of trace gases and clouds in determining the temperature profile. Figure 3.17 shows three equilibrium profiles obtained with different gaseous compositions, but without clouds. With only water vapor present a reasonable approximation to the observed profile is obtained except that the stratosphere is absent. Carbon dioxide with a mixing ratio of 300 ppm raises the temperature about 10 K above the equilibrium obtained with only water vapor present. A sharp tropopause and the increase of temperature with height that characterizes the stratosphere appear only when solar absorption by ozone is included in the model.

The contributions of individual gases to the heating rate in radiative–convective equilibrium are shown in Fig. 3.18. In the stratosphere the lapse rate for radiative equilibrium is never large and positive, so convective adjustment is not required. The first-order balance is between heating produced by solar absorption by ozone

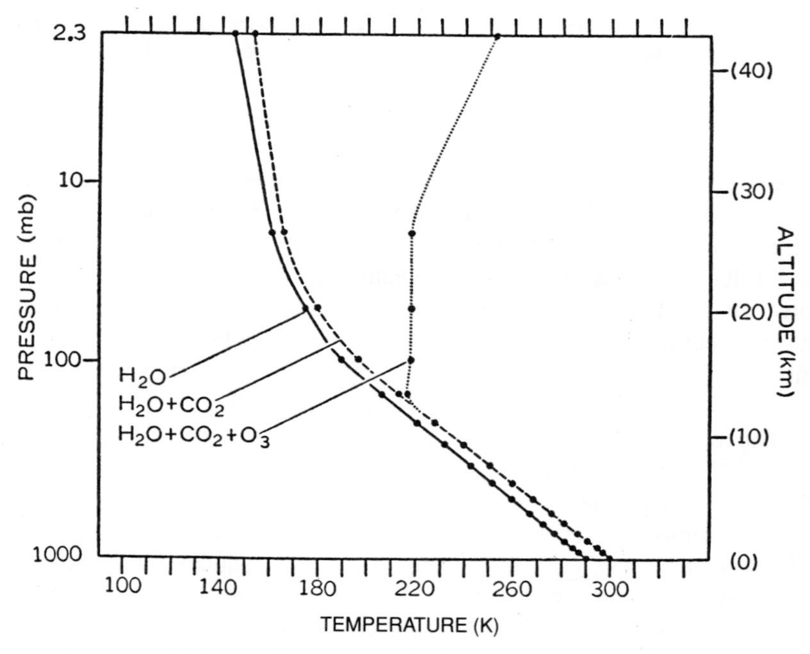

Fig. 3.17 Thermal equilibrium profiles for three cloudless atmospheres obtained with a critical lapse rate of 6.5 K km^{-1}. One atmosphere has water vapor only; one includes water vapor and carbon dioxide; and the third contains water vapor, carbon dioxide, and ozone. [From Manabe and Strickler (1964). Reprinted with permission from the American Meteorological Society.]

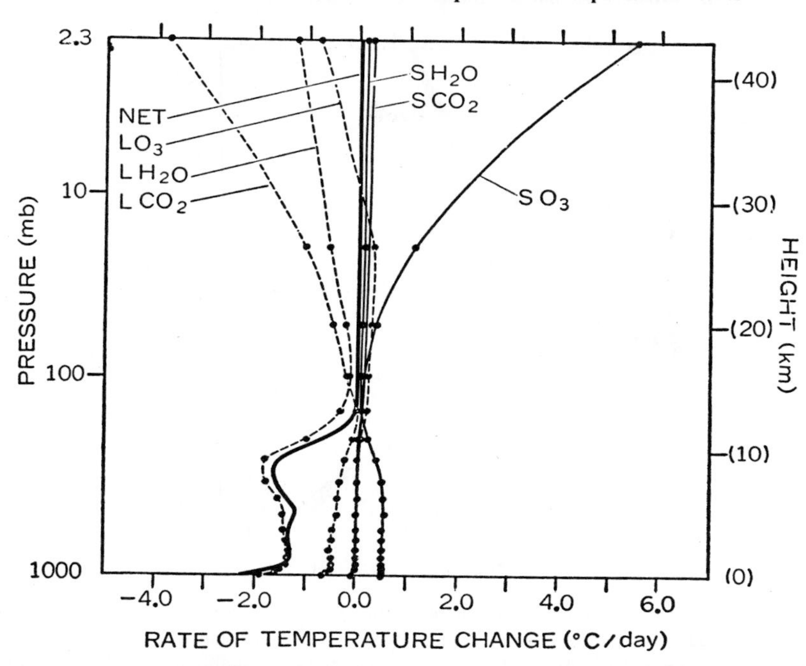

Fig. 3.18 Radiative heating rate profiles for a clear atmosphere. LH_2O, LCO_2, and LO_3 show the heating rates associated with longwave cooling by water vapor, carbon dioxide, and ozone, respectively. The S prefix indicates the heating rate associated with solar absorption by each of these gases. NET is the sum of the solar and longwave radiative heating rates contributed by all gases. [From Manabe and Strickler (1964). Reprinted with permission from the American Meteorological Society.]

and cooling produced by longwave emission from carbon dioxide. The troposphere is not in radiative equilibrium, and a net radiative cooling rate of about 1.5 K day^{-1} is balanced by convective heat transfer from the surface, where a positive net radiative imbalance exists. In a clear atmosphere this net longwave cooling is closely approximated by the cooling from water vapor emission. In the troposphere, longwave cooling from carbon dioxide is approximately balanced by solar absorption by water vapor. From the results of radiative–convective equilibrium calculations presented in Figs. 3.17 and 3.18, we conclude that water vapor is by far the preeminent greenhouse gas in the natural atmosphere.

Radiative–convective equilibrium models can also be used to examine the effect of simple clouds on the temperature profile. Figure 3.19 shows the effect of inserting clouds with various characteristics. Low clouds greatly reduce the temperature at the surface and in the troposphere, whereas the addition of high clouds can cause the surface temperature to exceed the value obtained for cloud-free conditions. The albedos assumed for the lower clouds are higher, so that the greater reflection of solar radiation from these clouds explains a good part of their stronger cooling effect

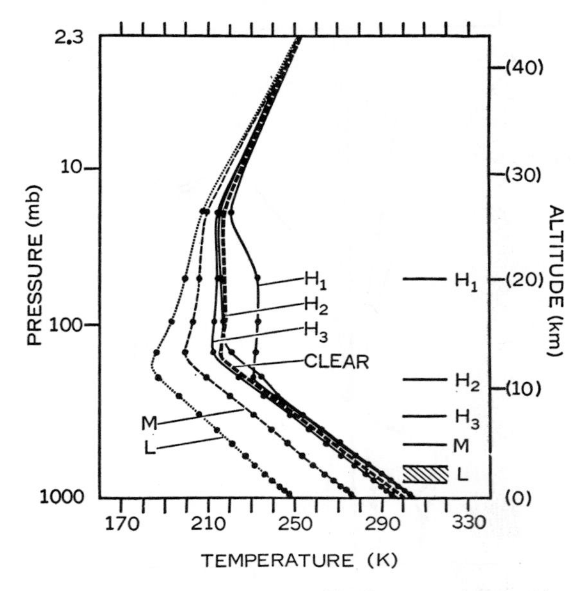

Fig. 3.19 Thermal equilibrium temperature profiles for atmospheres with various cloud distributions. The cloud heights corresponding to each type of cloud are shown on the right (L = low, M = medium, and H = high cloud). The heavy dashed line shows the equilibrium profile for clear skies. [From Manabe and Strickler (1964). Reprinted with permission from the American Meteorological Society.]

(Table 3.2). Lower clouds have a weaker effect on the escaping longwave radiation, however, since their top temperatures are warmer, and this also explains part of the greater cooling effect of low clouds in these calculations.

3.11 A Simple Model for the Net Radiative Effect of Cloudiness

Clouds are potentially very important for the sensitivity of climate, since they can affect both solar and longwave radiative transfer in the atmosphere. Clouds of sufficient thickness are typically almost perfect absorbers of terrestrial radiation and are at the same time excellent reflectors of solar radiation. These two properties of clouds produce opposite effects on the radiation balance. The reflection of solar radiation tends to cool Earth. Because of the decrease in temperature with altitude in the atmosphere, clouds reduce the outgoing terrestrial radiative flux at the top of the atmosphere, which tends to warm the climate.

We can illustrate the relative roles of the reflection of solar radiation and trapping of longwave radiation by clouds with a very simple model of their effect on the global energy balance at the top of the atmosphere. The energy balance at the top of the atmosphere is the difference between the absorbed solar radiation and the outgoing longwave radiation (OLR).

$$R_{\mathrm{TOA}} = \frac{S_0}{4}\left(1-\alpha_p\right) - F^{\uparrow}(\infty) \tag{3.55}$$

where S_0 is the solar constant and α_p is the albedo, so that $(S_0/4)(1 - \alpha_p) = Q_{\mathrm{abs}}$ is the absorbed solar radiation.

We wish to calculate the difference in the net radiation that results from adding a cloud layer with specified properties to a clear atmosphere.

$$\Delta R_{\mathrm{TOA}} = R_{\mathrm{cloudy}} - R_{\mathrm{clear}} = \Delta Q_{\mathrm{abs}} - \Delta F^{\uparrow}(\infty) \tag{3.56}$$

Suppose that we can specify the albedo for both clear and cloudy conditions, so that the difference in absorbed solar radiation is

$$\begin{aligned}\Delta Q_{\mathrm{abs}} &= \frac{S_0}{4}\left(1-\alpha_{\mathrm{cloudy}}\right) - \frac{S_0}{4}\left(1-\alpha_{\mathrm{clear}}\right) \\ &= -\frac{S_0}{4}\left(\alpha_{\mathrm{cloudy}} - \alpha_{\mathrm{clear}}\right) = -\frac{S_0}{4}\Delta\alpha_p\end{aligned} \tag{3.57}$$

To calculate the change in OLR we subtract (3.39) from (3.45):

$$\Delta F^{\uparrow}(\infty) = F^{\uparrow}_{\mathrm{cloudy}}(\infty) - F^{\uparrow}_{\mathrm{clear}}(\infty) \tag{3.58}$$

$$\Delta F^{\uparrow}(\infty) = \sigma T_{z_{\mathrm{ct}}}^4\, \mathcal{T}\left\{z_{\mathrm{ct}},\infty\right\} - \sigma T_s^4\, \mathcal{T}\left\{z_s,\infty\right\} \; - \int_{\mathcal{T}\{z_s,\infty\}}^{\mathcal{T}\{z_{\mathrm{ct}},\infty\}} \sigma T(z')^4 \; d\mathcal{T}\left\{z',\infty\right\} \tag{3.59}$$

If the top of the cloud is above most of the gaseous absorber of longwave radiation, which is water vapor, then we may make the approximation

$$\mathcal{T}\left\{z_{\mathrm{ct}},\infty\right\} \approx 1.0 \tag{3.60}$$

in which case (3.59) becomes

$$\Delta F^{\uparrow}(\infty) = \sigma T_{z_{\mathrm{ct}}}^4 - \sigma T_s^4\, \mathcal{T}\left\{z_s,\infty\right\} \; - \int_{\mathcal{T}\{z_s,\infty\}}^{1} \sigma T(z')^4 \; d\mathcal{T}\left\{z',\infty\right\} \tag{3.61}$$

or

$$\Delta F^{\uparrow}(\infty) = \sigma T_{z_{\mathrm{ct}}}^4 - F^{\uparrow}_{\mathrm{clear}}(\infty) \tag{3.62}$$

Inserting (3.57) and (3.62) into (3.56) gives an approximate formula for the change in net radiation at the top of the atmosphere that is produced by the addition of clouds to a clear atmosphere.

$$\Delta R_{\mathrm{TOA}} = -\frac{S_0}{4}\Delta\alpha_p + F^{\uparrow}_{\mathrm{clear}}(\infty) - \sigma T_{z_{\mathrm{ct}}}^4 \tag{3.63}$$

If the cloud top is above most of the longwave absorber, then (3.63) indicates that the change in net radiation produced by the cloud depends on the albedo contrast between clear and cloudy conditions and on the temperature at the cloud top. Since most of the water vapor is in the first few kilometers of the atmosphere, the approximation (3.60) is qualitatively correct for cases with cloud tops above 4 or 5 km.

From (3.63) it is possible that the albedo contrast and the cloud top temperature can be such that the cloud produces no change in the net radiation. The condition for this is obtained by setting $\Delta R_{\mathrm{TOA}} = 0$ in (3.63) and solving for the cloud top temperature.

$$T_{z_{\mathrm{ct}}} = \left\{ \frac{-(S_0/4)\Delta\alpha_p + F_{\mathrm{clear}}^{\uparrow}(\infty)}{\sigma} \right\}^{1/4} \tag{3.64}$$

If we assume that the temperature decreases with a lapse rate of Γ from a surface value of T_s, then the temperature of the cloud top can be related to its altitude.

$$T_{z_{\mathrm{ct}}} = T_s - \Gamma z_{\mathrm{ct}} \tag{3.65}$$

We can use (3.64) and (3.65) to solve for the cloud top altitude for which the reduction in OLR will just cancel the reduction in absorbed solar radiation associated with the presence of a cloud. For numeric values we can use a solar flux of 1367 W m^{-2}, a clear-sky OLR of 265 W m^{-2}, a surface temperature of 288 K, and a lapse rate of 6.5 K km^{-1}. These are all reasonable global-mean values. The resulting curve of cloud top altitude versus albedo contrast is shown as the heavy curve in Fig. 3.20. Clouds with albedo contrasts and altitudes that fall along the heavy line have no net

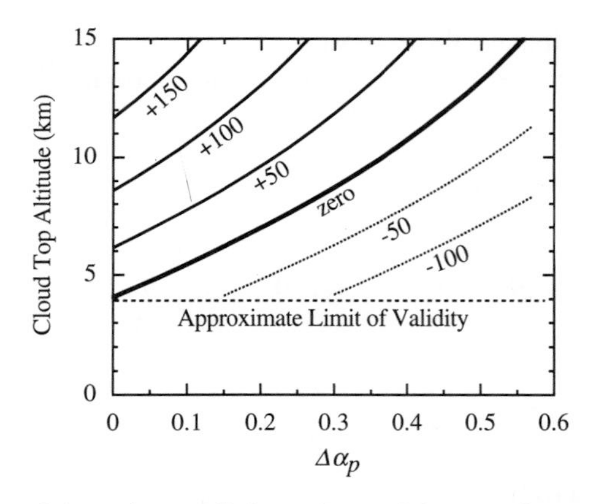

Fig. 3.20 Contours of change in net radiation at the top of the atmosphere caused by the insertion of a cloud into a clear atmosphere, plotted against cloud top altitude and the planetary albedo contrast between cloudy and clear conditions. The net radiation changes are calculated with the approximate model described in Section 3.8 that is invalid for clouds with tops lower than about 4 km. Contours from −100 to +150 W m^{-2} are shown at an interval of 50 W m^{-2}. The zero contour where the cloud has no net effect on the radiation budget at the top of the atmosphere is bold and negative contours are dashed.

effect on the energy balance at the top of the atmosphere. Those that fall below the line will produce a reduction in net radiation, or a cooling, and those above will produce warming. One can obtain the approximate cloud albedo for global average conditions by adding the clear-sky albedo, which is about 15%. As the cloud top rises, the albedo contrast between cloudy and clear conditions that will just balance the OLR change also increases. Clouds with high cold tops and low albedos can cause a significant positive change in net radiation, while low bright clouds can cause a large negative change in net radiation at the top of the atmosphere.

3.12 Observed Role of Clouds in the Energy Balance of Earth

It is possible to measure the radiative fluxes of energy entering and leaving Earth from orbiting satellites. If the spatial resolution of the measurements of the energy fluxes provided by the instrument on the satellite is great enough, then cloud-free scenes may be identified. These cloud-free scenes can be averaged together to estimate the clear-sky radiation budget. If these cloud-free scenes are taken to represent the atmosphere in the absence of clouds, then the difference between the cloud-free radiation budget and the average of all scenes represents the effect of clouds on the radiation budget. We can call the effect of clouds on the radiation budget the cloud radiative forcing of the energy balance.

Table 3.3 shows estimates of the globally and annually averaged radiation budget components for average conditions, cloud-free conditions, and the difference between them, which is called *cloud forcing*. The uncertainty of these estimates is about 5 W m^{-2}, as indicated by the average net radiation of 5 W m^{-2}, which should logically be zero. In round numbers the observations indicate that clouds increase the albedo from 15 to 30%, which results in a reduction of absorbed solar radiation of 50 W m^{-2}. This cooling is offset somewhat by the greenhouse effect of clouds, which reduces the OLR by about 30 W m^{-2}. The net cloud forcing of the radiation budget is thus a loss of about 20 W m^{-2}. The meaning of this number is that, if clouds could suddenly be removed without changing any other climate variable, then

Table 3.3

Cloud Radiative Forcing as Estimated from Satellite Measurements

	Average	Cloud-free	Cloud forcing
OLR	234	266	+31
Absorbed solar radiation	239	288	−48
Net radiation	+5	+22	−17
Albedo	30%	15%	+15%

Radiative flux densities are given in W m^{-2} and albedo in percent. [From Harrison *et al.* (1990), © American Geophysical Union.]

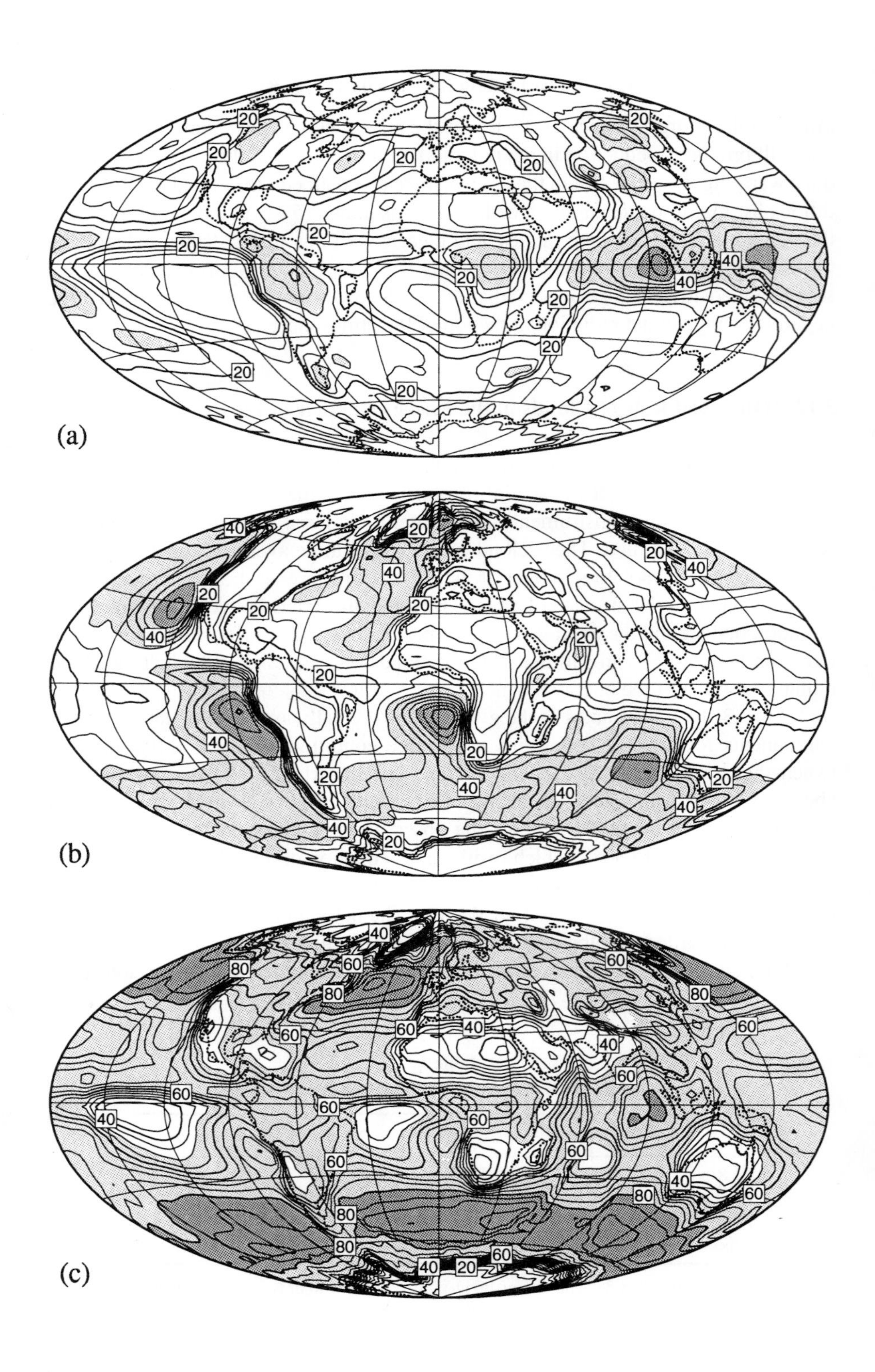

20
20
20
20
20
20
20
20
20
40
40
20
20
20
(a)
40
20
20
40
20
20
40
20
40
20
40
40
20
40
40
20
40
20
(b)
40
60
60
80
80
60
60
40
40
60
60
60
40
60
60
60
60
40
80
60
80
80
60
40
20
(c)

Earth would begin to gain 20 W m^{-2} in net radiation and consequently begin to warm up. The size of the temperature increase that might result from such a change in the radiation balance is the subject of Chapter 9.

The observed distribution of clouds has been estimated in two ways. Surface observers have recorded the type and fractional distribution of clouds and a long record of such observations has been compiled into a cloud climatology.[2] In more recent years attempts have been made to systematically characterize cloud distributions from the observations of visible and infrared radiation taken from meteorological satellites.[3] Each of these data sets has its strengths and weaknesses related to the viewing geometry (up versus down) and the instrumentation (the human eye versus a radiometer). Surface observations have a much better view of cloud base, whereas satellite measurements see the tops of the highest clouds very well and provide a more direct means of estimating the visible optical depth of the clouds.

Figure 3.21 shows global maps of the fractional area coverage of clouds with tops at pressures lower than 440 mb (high clouds), clouds with tops at pressures greater than 680 mb (low clouds) and clouds with tops at any pressure (total cloud amount). High clouds are concentrated in the convection zones of the tropics over equatorial South America and Africa, and a major concentration exists over Indonesia and the adjacent regions of the eastern Indian and western Pacific Oceans. Low clouds are most prevalent in the subtropical eastern ocean margins and in middle latitudes. The low cloud concentrations in the eastern subtropical oceans are associated with lower than average sea surface temperature (SST) (Fig. 7.11) and consist of stratocumulus clouds trapped below an inversion. Low clouds are heavily concentrated over the oceanic regions and are less commonly observed over land. The total cloud cover also shows a preference for oceanic regions, particularly in midlatitudes where the total cloud cover is greatest. Minima in total cloud cover occur in the subtropics in desert regions, but regions with low total cloud amounts also occur over the Caribbean Sea and over the southern subtropical zones of the Pacific, Atlantic, and Indian oceans.

Estimates of the effect of clouds on the radiative energy budget at the top of the atmosphere can be derived from satellite measurements of the broadband energy flux.[4] The longwave cloud forcing is the reduction of the OLR by the clouds, and so

[2]Warren *et al.* (1986, 1988).
[3]Rossow and Schiffer (1991).
[4]Harrison *et al.* (1990).

Fig. 3.21 Annual average cloud fractional area coverage in percent estimated from satellite data under the International Satellite Cloud Climatology Project [ISCCP, Rossow and Schiffer (1991)]. (a) Clouds with tops higher than 440 mb, (b) clouds with tops lower than 680 mb, and (c) all clouds. In (a) and (b) the contour interval is 5%, with values greater than 30% lightly shaded, and greater than 50% heavily shaded. In (c) the contour interval is also 5%, but light shading is applied for values greater than 50% and heavy shading for values greater than 80%.

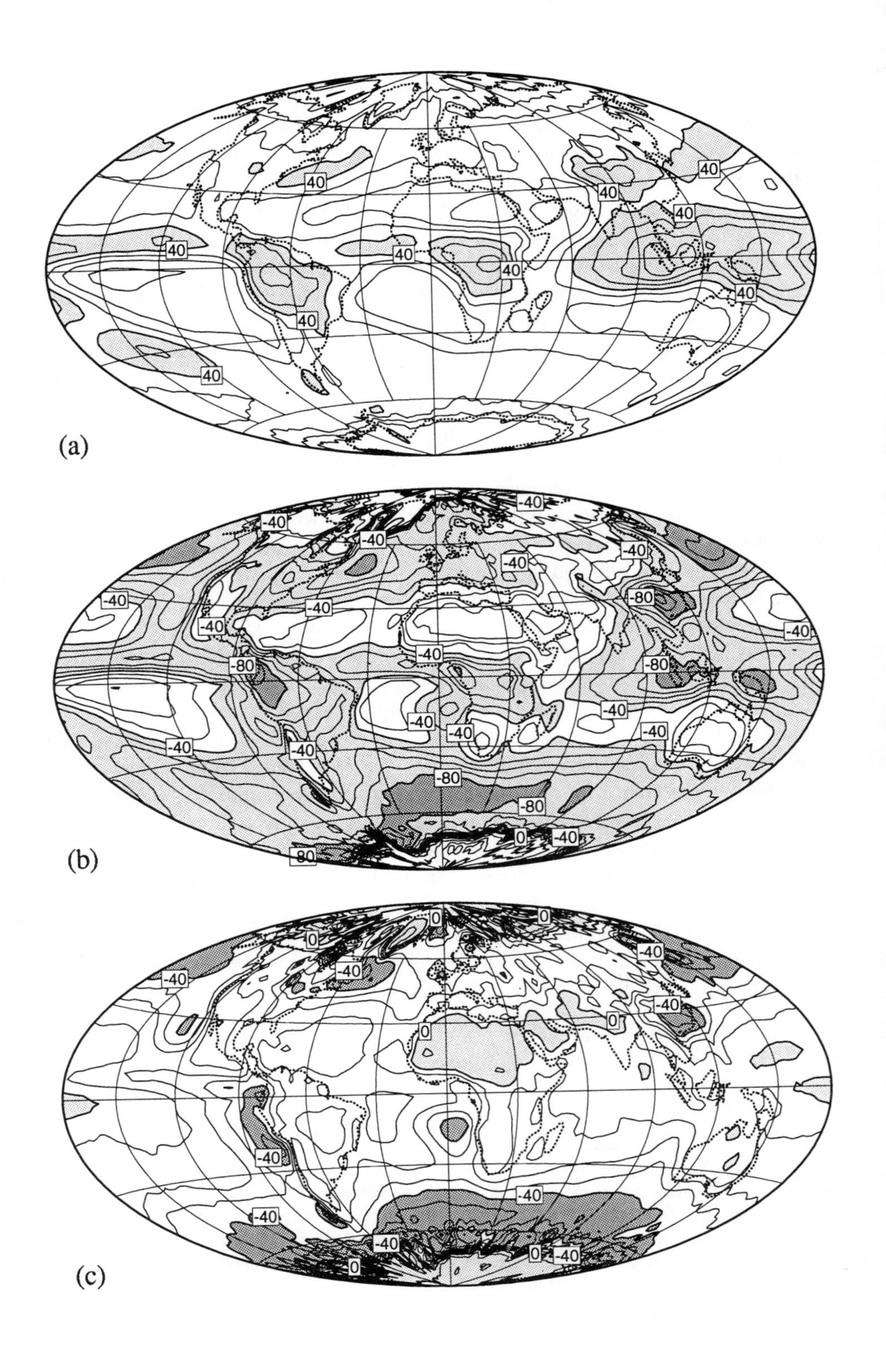
40
40
40
40
40
40
40
40
40
40
40
(a)
-40
-40
-40
-40
-40
-80
-40
-40
-40
-40
-80
-40
-80
-40
-40
-40
-40
-40
-40
-80
-80
0
-40
-80
(b)
0
0
0
-40
-40
-40
-40
-40
0
0
-40
-40
-40
-40
0
-40
0
(c)

is a positive contribution to the radiation budget or warming influence on the surface climate. The largest contributions are made in the convective regions and the Intertropical Convergence Zone (ITCZ) in the tropics where high clouds with cold tops are abundant [Fig. 3.22(a)]. The reduction of absorbed solar radiation is also relatively large in these regions, since the deep convective clouds also have high albedos, but low clouds in high latitudes are also very effective in reducing the absorbed solar radiation [Fig. 3.22(b)]. The net effect of clouds on the energy budget at the top of the atmosphere is generally smaller than its longwave and shortwave components because in most cases they are of opposite sign. The largest net contributions are reductions in net radiation by low clouds in high latitudes and in the stratus cloud regions in the eastern subtropical oceans [Fig. 3.22(c)].

Exercises

1. Suppose a gas that absorbs solar radiation has a uniform mixing ratio of 1 g kg^{-1} and an absorption cross section of 5 m^2 kg^{-1}. At what altitude will the maximum rate of energy absorption per unit volume occur? Assume an isothermal atmosphere with $T = 260$ K, and a surface pressure of 1.025×10^5 Pa, and that the sun is directly overhead.
2. In problem 1, if the frequency range at which the absorption is taking place contains 5% of the total solar energy flux, what is the heating rate in degrees per day at the level of maximum energy absorption? What is the heating rate one scale height above and below the level of maximum energy absorption? Use the globally averaged insolation.
3. Do problem 1 with a solar zenith angle of 45°. Discuss the difference the angle makes.
4. Use the model of Fig. 3.10, but distribute the solar heating such that $0.3\sigma T_e^4$ is absorbed in each of the two atmospheric layers and the remaining $0.4\sigma T_e^4$ is absorbed at the surface. Calculate the new radiative equilibrium temperature profile. How does it differ from the case where all of the solar heating is applied at the surface?
5. Place the two layers in the model of Fig. 3.10 at 2.5 and 5.0 km. Assume a fixed lapse rate of 6.5 K km^{-1}. Derive energy balance equations that include an unknown convective energy flux from the surface to the lower layer and from the lower layer to the upper layer. Solve for the temperature profile in

Fig. 3.22 Annual average cloud forcing of the radiation budget at the top of the atmosphere estimated from satellite data under the Earth Radiation Budget Experiment [ERBE; Harrison *et al.* (1990)]. (a) Reduction of OLR caused by clouds, (b) increase in absorbed solar radiation caused by clouds (note values are negative), and (c) increase in net radiation caused by clouds. Contour interval is 10 W m^{-2} in each plot. Shading is applied for values greater than +40 in (a), less than −40 (light) and −80 (heavy) in (b), and less than −40 (heavy) and greater than zero (light) in (c).

thermal equilibrium and find the required convective energy fluxes from the surface and the lower layer. (*Hint:* Start from the top and work down.) How do the radiative and convective fluxes compare with the proportions given in Fig. 2.4?

6. For the conditions of problem 5, calculate the pure radiative equilibrium temperature profile with no convective adjustment. Plot the thermal equilibrium and pure radiative equilibrium temperature profiles from this model with the levels at 2.5 and 5.0 km and compare them with the profiles shown in Fig. 3.16. What happens to this comparison if the top level is moved up or down by 2 km? Is the dependence on the height of the layer reasonable? How do the required convective fluxes change as you move the top level?
7. Suppose that tropical convective clouds give an average planetary albedo of about 0.6 compared to the cloud-free albedo of about 0.1. The insolation is about 400 W m^{-2}, and the cloud-free OLR is about 280 W m^{-2}. Use the simplified model of Section 3.11 to find the cloud top temperature for which the net radiative effect of these clouds will be zero. Are such temperatures observed in the tropical troposphere, and, if so, where? (Refer to Fig. 1.3.) If the surface temperature is 300 K and the average lapse rate is 5 K km^{-1}, at what altitude would the cloud top be to make the longwave and shortwave effects of the cloud just equal and opposite?
8. For the conditions of problem 7, what is the rate of net radiative energy loss if the cloud albedos are 0.7 rather than 0.6? By how much would you need to lower the cloud tops to produce an equal reduction in net radiation?

Chapter 4 The Energy Balance of the Surface

4.1 Contact Point

The surface of Earth is the boundary between the atmosphere and the land or ocean. Defining the location of this boundary can be difficult over a highly disturbed sea or over land surfaces with a variable plant canopy. We will assume that the location of the surface can be appropriately defined, and we will treat it as a simple interface between two media, but in considering the important energy exchange processes we must include the atmosphere and oceanic boundary layers and the first few meters of soil. The energy fluxes across the surface are as important to the climate as the fluxes at the top of the atmosphere, especially since the climate at the surface is of most practical significance. The surface energy balance determines the amount of energy flux available to evaporate surface water and to raise or lower the temperature of the surface. Surface processes also play an important role in determining the overall energy balance of the planet.

The energy budget at the surface is more complex than the budget at the top of the atmosphere, because it requires consideration of fluxes of energy by conduction and by convection of heat and moisture through fluid motion, as well as by radiation. The local surface energy budget depends on the insolation, the surface characteristics such as wetness, vegetative cover, and albedo, and on the characteristics of the overlying atmosphere. The energy budget of the surface is intimately related to the hydrologic cycle, since evaporation from the surface is a key component in the budgets of both energy and water. Understanding the energy budget of the surface is a necessary part of understanding climate and its dependence on external constraints.

4.2 The Surface Energy Budget

The energy budget can be written in terms of energy flux per unit area passing vertically through the air-surface interface and is measured in watts per square meter. The processes that determine energy transfer between the surface and atmosphere include solar and infrared radiative transfer, and fluxes of energy associated with fluid motions of the atmosphere and oceans. The storage and transport of energy below the surface are also important. For the purposes of energy budget computations, surface storage takes place in that volume between the boundary with the atmosphere and a depth below the surface where energy fluxes and the storage rate of energy are considered negligible. This depth can be as little as a few meters in dry land areas or as much as several kilometers in oceanic areas where deep water is formed. For water surfaces the horizontal energy fluxes accomplished by the fluid motions under the surface can be very important. The surface energy balance can be written symbolically as (4.1).

$$\frac{\partial E_s}{\partial t} = G = R_s - \mathrm{LE} - \mathrm{SH} - \Delta F_{\mathrm{eo}} \tag{4.1}$$

where $\partial E_s / \partial t = G$ is the storage of energy in the surface soil and water, R_s is the net radiative flux of energy into the surface, LE is the latent heat flux from the surface to the atmosphere, SH is the sensible heat flux from the surface to the atmosphere, and ΔF_{eo} is the horizontal flux out of the column of land–ocean below the surface.

Under steady-state conditions in which the storage of energy is small, such as one might assume to hold for annual averages or for daily averages over land, the energy balance is between radiative heating and the processes that remove energy from the surface.

$$R_s = \mathrm{LE} + \mathrm{SH} + \Delta F_{\mathrm{eo}} \tag{4.2}$$

Under most conditions radiation heats the surface and latent and sensible heat fluxes cool it, so that the radiative, latent, and sensible heat flux terms in (4.1) and (4.2) are most often positive (Fig. 4.1).

The physical meaning of (4.1) is that the storage of energy below the surface is equal to the net radiative input minus the heat lost from the surface by evaporation, sensible heat flux, and horizontal heat transport to other latitudes or longitudes. In constructing (4.1) we have left out a multitude of other terms that can be important locally or for brief periods. These terms include the following:

- The latent heat of fusion required for melting ice and snow in spring may require 10% of the radiative imbalance for limited periods.

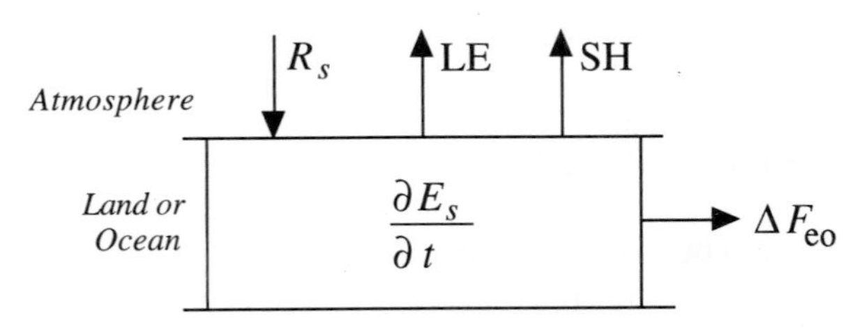

Fig. 4.1 Diagram showing the relationship of the various terms in the surface energy balance (R_s = net radiation, LE = evaporative cooling, SH = sensible cooling, $\partial E_s / \partial t$ = heat storage below the surface, ΔF_{eo} = divergence of horizontal energy flux below the surface).

- Conversion of the kinetic energy of winds and waves to thermal energy is generally small.
- Heat transfer by precipitation can occur if the precipitation is at a different temperature than the surface. This mechanism is particularly important during summer showers, because the precipitation can be much cooler than the surface, and the thermal capacity of water is large.
- Some solar energy is not realized as heat, but is stored in the chemical bonds formed during photosynthesis. This is less than 1% globally, but can reach ~5% locally for limited periods of time.
- Heat release by oxidation of biological substances, as in biological decay or forest fires, is the reverse of photosynthesis. Energy bound up in biological matter during photosynthesis is returned to the physical climate system through oxidation. This process takes place year round, but proceeds most rapidly when the surface is warm and moist.
- Geothermal energy release in hot springs, earthquakes, and volcanoes is small in a global sense.
- Heat released by fossil fuel burning or nuclear power generation can be important locally, but is not significant for the global energy balance.

4.3 Storage of Heat in the Surface

Energy storage in the surface is very important for the seasonal cycle of temperature over the oceans and the diurnal cycle over land and ocean. Using the simplest description, the amount of energy in the surface may be written as the product of an effective heat capacity for the Earth–ocean system and a corresponding mean temperature.

$$E_s = \overline{C}_{eo} T_{eo} \tag{4.3}$$

where $\overline{C}_{eo}$ is the effective heat capacity of the land or ocean system (J m^{-2} K^{-1}) and T_{eo} is the effective temperature of the land or ocean energy-storing material (K).

The heat capacity depends on the physical properties of the surface materials and the depth of the surface layer that communicates with the atmosphere on the time scale of interest. It is generally only the first few meters of soil that respond to seasonal forcing of the surface energy balance, but the temperature of the top 50–100 m of ocean changes with the seasons. The depth of ocean that exchanges energy with the surface varies seasonally, so that possible time dependence of the effective heat capacity must be considered.

The heat capacity of the atmosphere is estimated by including only the energy associated with the motion of the molecules, which is related to the temperature. The heat capacity is the amount of energy that is required to raise the temperature by one degree. We may approximate the specific heat of air by the specific heat at constant pressure for dry air. To obtain the heat capacity for the entire atmosphere, we integrate over the mass of the atmosphere to get

$$\overline{C}_a = c_p \frac{p_s}{g} = \frac{1004 \text{ J K}^{-1} \text{kg}^{-1} \bullet 10^5 \text{ Pa}}{9.81 \text{ m s}^{-2}} = 1.02 \times 10^7 \text{ J K}^{-1} \text{m}^{-2} \tag{4.4}$$

One can estimate the thermal capacity of the ocean by using the thermal capacity of pure liquid water at 0°C. The thermal capacity for an arbitrary depth of water, d_w, can be obtained from the density ρ_w, and the specific heat c_w.

$$\begin{aligned} \overline{C}_o &= \rho_w c_w d_w = 10^3 \text{kg m}^{-3} \bullet 4218 \text{ J K}^{-1} \text{kg}^{-1} \bullet d_w \\ &= d_w \bullet 4.2 \times 10^6 \text{ J K}^{-1} \text{m}^{-2} \text{m}^{-1} \end{aligned} \tag{4.5}$$

Comparing (4.4) and (4.5) we see that the thermal capacity of the atmosphere is equal to that of a little over 2 m of water. As discussed in Chapter 7, about the top 70 m of ocean interact with the atmosphere on the time scale of a year, so that on the seasonal time scale the thermal capacity of the ocean is about 30 times that of the atmosphere.

4.3.1 Heat Storage in Soil

The land has a much smaller effective heat capacity than the ocean. Because the surface is solid, the efficient heat transport by fluid motions that occurs in the atmosphere and the ocean does not take place. Heat is transferred through the soil mostly by the less efficient process of conduction. Only the top meter or two of the soil is affected by seasonal variations. The heat capacity of a land surface is typically slightly smaller than that of the atmosphere.

The vertical flux of energy by conduction in the soil is proportional to the vertical temperature gradient in the soil.

$$F_s = -K_T \frac{\partial T}{\partial z} \tag{4.6}$$

where K_T is the thermal conductivity. The heat balance in the soil is between storage in the soil and convergence of the diffusive heat flux.

$$C_s \frac{\partial T}{\partial t} = -\frac{\partial}{\partial z}(F_s) = \frac{\partial}{\partial z}\left(K_T \frac{\partial T}{\partial z}\right) \tag{4.7}$$

The volumetric heat capacity of the surface material C_s is the product of the specific heat of the soil c_s, and the soil density ρ_s. The heat capacity of the soil depends on the volume fractions of soil f_s, organic matter f_c, water f_w, and air f_a, and the density and specific heat of each component of the surface material.

$$C_s = \rho_s c_s f_s + \rho_c c_c f_c + \rho_w c_w f_w + \rho c_p f_a \tag{4.8}$$

Table 4.1

Properties of Soil Components at 293 K

	Specific heat (c_p) (J kg^{-1} K^{-1})	Density (ρ) (kg m^{-3})	$\rho\, c_p$ (J m^{-3} K^{-1})
Soil inorganic material	733	2600	1.9×10^6
Soil organic material	1921	1300	2.5×10^6
Water	4182	1000	4.2×10^6
Air	1004	1.2	1.2×10^3

[After Brutsaert (1982). Reprinted with permission from Kluwer Academic Publishers.]

From Table 4.1 it can be seen that the heat capacity of the air in the soil is very small, so that when water replaces air in the open spaces in the soil the heat capacity greatly increases. Porosity is the volumetric fraction of the soil that can be occupied by air or water.

Thermal conductivity of soils depends on the material, the porosity, and the soil water content. The thermal conductivity increases with water content for soils with relatively high porosity. Values vary from 0.1 W m^{-1} K^{-1} for dry peat to 2.5 W m^{-1} K^{-1} for wet sand. Under the condition that the thermal conductivity, K_T, is independent of depth, (4.7) simplifies to the heat equation

$$\frac{\partial T}{\partial t} = D_T \frac{\partial^2 T}{\partial z^2} \tag{4.9}$$

where $D_T = K_T / C_s$ is the thermal diffusivity of the surface material. A simple scale analysis of (4.9) can be used to determine the depth through which a temperature anomaly applied at the surface will penetrate in a given time. One can show that the penetration depth, h_T, of temperature anomalies associated with a periodic forcing of temperature at the surface is given by

$$h_T = \sqrt{D_T\, \tau} \tag{4.10}$$

where τ is the time scale of the periodic forcing at the surface. Taking a typical value of soil diffusivity of $D_T = 5 \times 10^{-7}$ m^2 s^{-1}, we obtain a penetration depth of about 10 cm for diurnal forcing and about 1.5 m for annual forcing. A surface temperature variation with a time scale of 10,000 years would penetrate the surface material to a depth of about 150 m. Because of the slow rates at which heat can be transported through soil and rock by conduction, horizontal heat transport under the land surface is entirely negligible.

Figure 4.2 shows temperatures at various depths in the soil as a function of time during a clear day in summer. The near surface soil experiences a large diurnal variation in temperature with minimum temperatures just before sunrise and maximum

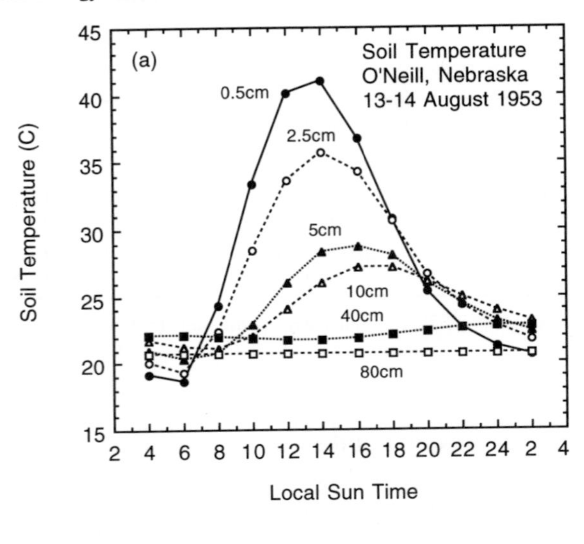

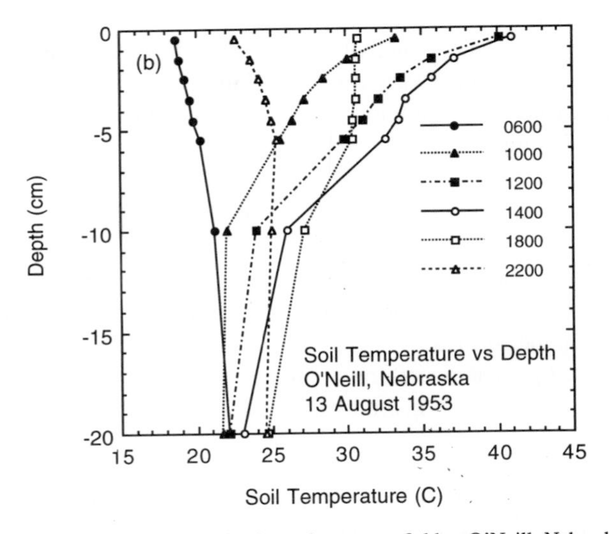

Fig. 4.2 Soil temperature at various depths under a grass field at O'Neill, Nebraska on August 13, 1953: (a) temperature at various depths as a function of local time; (b) temperature as a function of depth at various times. Measured thermal diffusivities on the day illustrated range from 2.5×10^{-7} m^2 s^{-1} at 1 cm to 6×10^{-7} m^2 s^{-1}, at 5-cm depth in the soil. [Data from Lettau and Davidson (1957).]

temperatures shortly after noon. Deeper in the soil the temperature variations are smaller and occur later in the day, because of the time it takes the temperature pulse to diffuse into the soil. The amplitude of the temperature perturbation decreases to about e^{-1} of its surface value at about 10 cm below the surface and is very small by

40 cm below the surface. This observed decrease in the amplitude of the diurnal temperature perturbation is consistent with the simple dimensional arguments given above.

The temperature data can be used to infer the thermal conductivity as a function of depth in the soil through the use of (4.7). Alternatively, if the vertical profiles of C_s and K_T are known and assumed constant with time, then measured temperature profiles in a deep layer of Earth can be used to estimate past variations in surface temperature on time scales of hundreds to thousands of years.[1]

4.4 Radiative Heating of the Surface

The net input of radiative energy to the surface is the sum of the net solar and longwave flux densities at the surface.

$$R_s = S^{\downarrow}(0) - S^{\uparrow}(0) + F^{\downarrow}(0) - F^{\uparrow}(0) \tag{4.11}$$

where $S^{\downarrow}(0)$ and $S^{\uparrow}(0)$ are the downward and upward flux density of solar radiation at the surface, respectively, and $F^{\downarrow}(0)$ and $F^{\uparrow}(0)$ represent similarly defined longwave flux densities. Solar heating is the basic driver of the climate system.

4.4.1 Absorption of Solar Radiation at the Surface

The net downward solar energy flux can be written as the product of the downward solar flux at the surface multiplied by the absorptivity of the surface.

$$S^{\downarrow}(0) - S^{\uparrow}(0) = S^{\downarrow}(0)(1 - \alpha_s) \tag{4.12}$$

The surface albedo, α_s, is defined as the fraction of the downward solar flux density that is reflected by the surface.

The surface albedo varies widely depending on the surface type and condition, ranging from values as low as 5% for oceans under light winds to as much as 90% for fresh, dry snow. The numbers in Table 4.2 are characteristic, but each surface type can exhibit a range of albedos (Fig. 4.3). The most common surface is that of water, and its albedo depends on solar zenith angle, cloudiness, wind speed, and impurities in the water. The dependence of water surface albedo on solar zenith angle and cloudiness is shown in Fig. 4.4. The fraction of an incident beam of solar radiation that is reflected from a water surface depends on the angle of incidence of the beam with the water surface. The surface albedo of ocean under clear skies increases dramatically as the sun approaches the horizon. Clouds scatter radiation very effectively, so that the solar radiation under a cloud is no longer a parallel beam but is scattered in all directions. Under a cloud the photons that reach the surface come from all possible directions with about equal probability, so that beneath a sufficiently

[1]Lachenbruch and Marshall (1986).

Table 4.2

Albedos for Various Surfaces in Percent

Surface type	Range	Typical value
Water		
Deep water: low wind, low altitude	5–10	7
Deep water: high wind, high altitude	10–20	12
Bare surfaces		
Moist dark soil, high humus	5–15	10
Moist gray soil	10–20	15
Dry soil, desert	20–35	30
Wet sand	20–30	25
Dry light sand	30–40	35
Asphalt pavement	5–10	7
Concrete pavement	15–35	20
Vegetation		
Short green vegetation	10–20	17
Dry vegetation	20–30	25
Coniferous forest	10–15	12
Deciduous forest	15–25	17
Snow and ice		
Forest with surface snowcover	20–35	25
Sea ice, no snowcover	25–40	30
Old, melting snow	35–65	50
Dry, cold snow	60–75	70
Fresh, dry snow	70–90	80

thick cloud it is impossible to tell where in the sky the sun is located. Therefore, the surface albedo under overcast skies is insensitive to solar zenith angle. The amount of solar energy that reaches the surface under overcast skies is sensitive to solar zenith angle, however, since clouds are very effective reflectors of solar radiation and their albedo is sensitive to solar zenith angle.

The reflectivities of various surfaces depend on the frequency of radiation (Fig. 4.5). Clouds and snow are most reflective for visible radiation, and become less reflective at near-infrared wavelengths, where substantial absorption by water occurs. Green plants have a very low albedo for photosynthetically active radiation, where chlorophyll absorbs radiation efficiently. Radiation in the wavelength band from about 0.4 to 0.7 μm is effective for photosynthesis and growing plants absorb more than 90% of it. At about 0.7 μm the albedo of green plants increases dramatically, so their albedo for near-infrared radiation can be as high as 50%. Since nearly half of the solar energy that reaches the surface is at wavelengths longer than 0.7 μm, this increase in albedo is significant for the energy budget of the surface. Plants need wavelengths shorter than 0.7 μm for photosynthesis, but the near-infrared energy absorption at

Fig. 4.3 Aerial photograph of the Antarctic ice sheet where it meets the Indian Ocean at Dumont d'Urville, Terre Adélie, Antarctica. Note the high albedo of the ice compared to the ocean water. (Photo by R. Guillard. Reprinted with permission from I.F.R.T.P.)

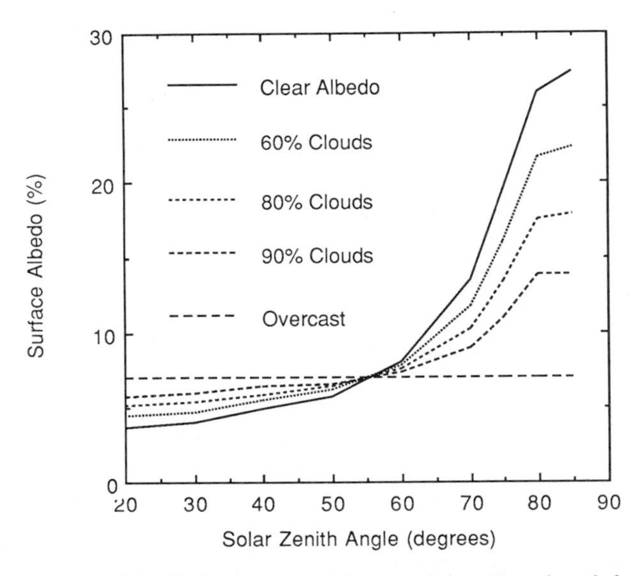

Fig. 4.4 Dependence of the albedo of a water surface on solar zenith angle and cloud cover. [Data from Mirinova (1973).]

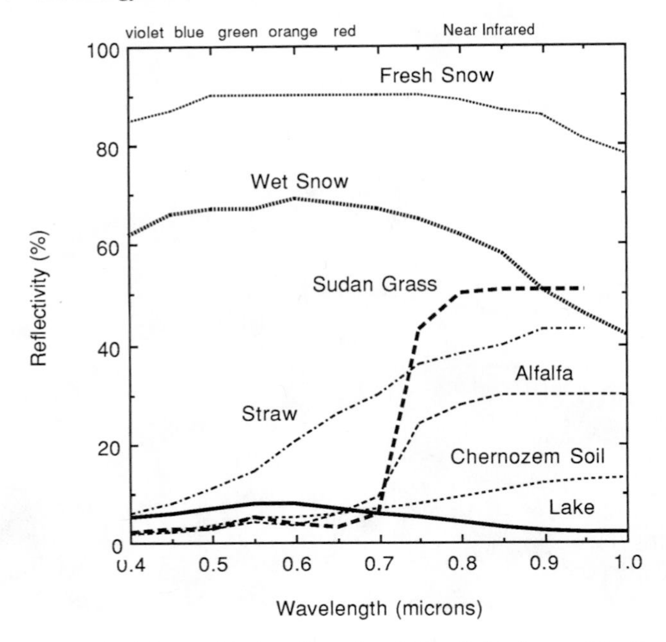

Fig. 4.5 Surface reflectivity as a function of wavelength of radiation for a variety of natural surfaces. Human eyesight is sensitive to wavelengths from 0.4 μm (violet) to 0.7 μm (red). Alfalfa and sudan grass appear green because their albedo is higher for green light (~0.55 μm) than for other visible wavelengths. [Data from Mirinova (1973).]

wavelengths longer than 0.7 μm heats the leaves without any conversion of energy to plant tissue. The higher albedos at wavelengths longer than 0.7 μm thus help the leaves to stay cool. When green plants die and dry out, their chlorophyll content decreases and their albedo at visible wavelengths increases, as shown by the example of a field of straw.

The albedo of vegetated surfaces depends on the texture and physiological condition of the plant canopy. Leaf canopies with complex geometries and many cavities can have albedos that are lower than the albedo of an individual leaf. The ratio of near-infrared to visible radiation decreases with depth below the top of the plant canopy, because the visible wavelengths are more effectively absorbed by leaves. The higher albedo of leaves for near-infrared radiation allows it to be scattered down through the plant canopy and heat the soil.

Pure water is most reflective for blue light. Natural water bodies contain many impurities and may be most reflective for green light, but their reflectance is generally higher at visible than at near-infrared wavelengths. Soils have higher reflectivities at near-infrared than at visible wavelengths. Soils have significantly higher albedos when they are dry than when they are wet, and smooth soil surfaces have higher albedos than rough surfaces (Table 4.3).

Table 4.3

Albedos for Dry and Moist Soil Surfaces

	Even surface		Tilled surface	
	Dry	Moist	Dry	Moist
Chernozem of dark gray color	13	8	8	4
Light chestnut soil of gray color	18	10	14	6
Chestnut soil of grayish red color	20	12	15	7
Gray sandy soil	25	18	20	11
White sand	40	20	—	—
Dark blue clay	23	16	—	—

[From Mironova (1973).]

Because surface albedo is highly variable and has a strong effect on absorbed solar radiation, it can have a large effect on surface temperature. Surface albedo can also have a strong effect on the sensitivity of climate, if it changes systematically with climatic conditions. Feedback processes involving surface albedo are discussed in Chapter 9.

4.4.2 Net Longwave Heating of the Surface

To calculate the net downward longwave radiation at the surface, one must know the downward longwave radiation coming from the atmosphere, the temperature of the surface, and the longwave emissivity of the surface, ε. If the frequencies of downward longwave radiation and radiation emitted from the surface are essentially the same, then the effective absorptivity of the surface is equal to its emissivity. Since this is approximately true, a fraction ε of the downward longwave radiation at the ground is absorbed, so that the upward longwave at the surface can be written

$$F^{\uparrow}(0) = (1-\varepsilon)F^{\downarrow}(0) + \varepsilon\sigma T_s^4 \tag{4.13}$$

We may thus write

$$F^{\downarrow}(0) - F^{\uparrow}(0) = \varepsilon\left(F^{\downarrow}(0) - \sigma T_s^4\right) \tag{4.14}$$

Because of the strong greenhouse effect at work in Earth's atmosphere, the downward longwave from the atmosphere and the emission from the surface are both relatively large and tend to offset each other. The longwave emissivities of most natural surfaces are between 90 and 98% (Table 4.4) and do not play a key role in the determination of surface climate. Inaccuracy in the estimation of surface emissivity can cause errors in the calculation of net longwave flux at the surface of about 5%. The errors in estimates of the surface temperature in equilibrium are much smaller, however, because in deriving the temperature from the energy flux balance, a one-fourth

Table 4.4

Infrared Emissivities (percent) of Some Surfaces

Surface	Emissivity	Surface	Emissivity
Water and soil surfaces		Vegetation	
Water	92–96	Alfalfa, dark green	95
Snow, fresh fallen	82–99.5	Oak leaves	91–95
Snow, ice granules	89	Leaves and plants	
Ice	96	0.8 μm	5–53
Soil, frozen	93–94	1.0 μm	5–60
Sand, dry playa	84	2.4 μm	70–97
Sand, dry light	89–90	10.0 μm	97–98
Sand, wet	95	Miscellaneous	
Gravel, coarse	91–92	Paper, white	89–95
Limestone, light gray	91–92	Glass pane	87–94
Concrete, dry	71–88	Bricks, red	92
Ground, moist, bare	95–98	Plaster, white	91
Ground, dry plowed	90	Wood, planed oak	90
Natural surfaces		Paint, white	91–95
Desert	90–91	Paint, black	88–95
Grass, high dry	90	Paint, aluminum	43–55
Field and shrubs	90	Aluminum foil	1–5
Oak woodland	90	Iron, galvanized	13–28
Pine forest	90	Silver, highly polished	2
		Skin, human	95

[Data from Sellers (1965). Reprinted with permission from the University of Chicago Press.]

root is taken. This reduces the variation in surface temperature associated with emissivity variations to about 1%.

4.5 The Atmospheric Boundary Layer

The *atmospheric boundary layer* is the lowest part of the troposphere, where the wind, temperatures, and humidity are strongly influenced by the surface. The wind speed decreases from its value in the free atmosphere to near zero at the surface. Fluxes of momentum, heat, and moisture by small-scale turbulent motions in the boundary layer communicate the presence of the lower boundary to the atmosphere and are critical to the climate. Aerosols and gaseous chemical constituents of the atmosphere are also exchanged with the surface through the atmospheric boundary layer. A characteristic of the atmospheric boundary layer is its quick response to changes in surface conditions. The response of the surface to the daily variation of insolation is felt strongly throughout the boundary layer, but in the free troposphere diurnal changes are usually small, unless thermal convection associated with daytime heating of the land surface penetrates deep into the atmosphere.

The depth of the atmospheric boundary layer can vary between about 20 m and several kilometers, depending on the conditions, but a typical boundary-layer depth is about a kilometer. The boundary layer is generally deeper when the surface is being heated, when the winds are strong, when the surface is rough, and when the mean vertical motion in the free troposphere is upward.

Transports of mass, momentum, and energy through the boundary layer are accomplished by turbulent motions. If it were not for the chaotic swirls of turbulent motion in the boundary layer, the exchange between the surface and the atmosphere would be extremely slow. The turbulent motions that carry the vertical fluxes in the boundary layer range in scale from the depth of the boundary layer to the smallest scales where molecular diffusion becomes an important transport mechanism. Turbulence can be generated thermally or mechanically. *Mechanical turbulence* is generated by the conversion of the mean winds to turbulent motions, and is strongest when the mean wind in the lower atmosphere is large. *Convective turbulence* is generated when warm air parcels near the surface are accelerated upward by their buoyancy. Convective turbulence is most easily observed over land surfaces during daylight hours, when strong solar heating of the surface provides a source of buoyant energy, but is also very common over the oceans. When the boundary layer is relatively unstable, so that buoyancy forces or shearing instabilities are generating turbulence, the boundary layer may contain a well-mixed layer, where momentum, heat and moisture are almost independent of height.

The structure of the planetary boundary layer varies widely, depending on the meteorological conditions and whether the surface is being heated or cooled. When a surface heat source is present, such as over land during the daytime, the boundary layer is often unstable and has a structure that is generally like that shown in Fig. 4.6. The lowest part of the boundary layer is called the *surface layer,* where the vertical

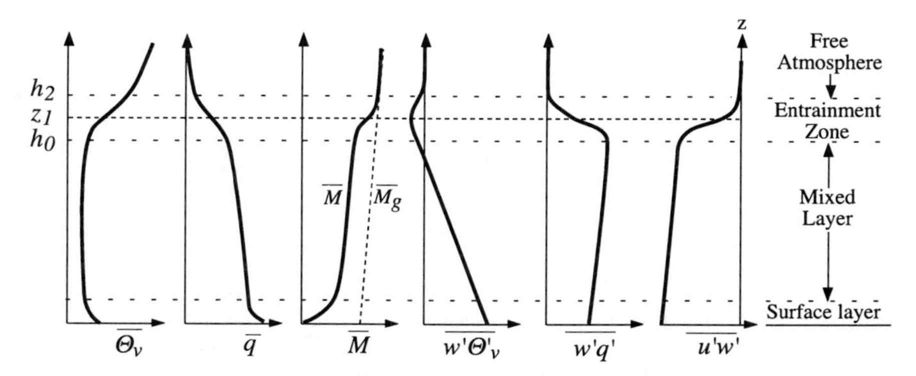

Fig. 4.6 Structure of a convective boundary layer showing the distributions of mean virtual potential temperature $\overline{\Theta}_v$, water vapor mixing ratio $\overline{q}$, momentum $\overline{M}$, geostrophic momentum $\overline{M}_g$, and the vertical eddy fluxes of potential temperature, humidity, and momentum. [From Stull (1988) after Dreidonks and Tennekes (1984). Reprinted with permission from Kluwer Academic Publishers.]

fluxes of momentum, heat, and moisture are almost constant with height. In the mixed layer buoyancy drives turbulent motions that maintain the potential temperature, $\overline{\Theta}_v$, the humidity, $\overline{q}$, and the momentum, $\overline{M}$, at values that are almost independent of height. Heat and moisture are transported upward in the mixed layer and momentum is transported downward toward the surface. The top of the boundary layer is a transition zone between the boundary layer and the free atmosphere, which is often called the *entrainment zone*. Across this transition the air properties change rapidly from those of the mixed layer to those of the free atmosphere above, generally marked by a decrease in humidity, an increase in potential temperature, and a decrease in the magnitude of the vertical fluxes of heat, moisture, and momentum by turbulent motions. *Entrainment* is the process whereby air from the free atmosphere is incorporated into the boundary layer. Entrainment is measured by a small downward eddy potential temperature flux in the entrainment zone. Entrainment adds mass to the boundary layer and is necessary when the boundary layer is deepening, or when the large-scale flow is downward into the boundary layer.

At night longwave emission cools the land surface more rapidly than the air above it and the boundary layer can become very stable, with cold, dense air trapped near the surface (Fig. 4.7). Under these conditions turbulence and the vertical fluxes it produces can be greatly suppressed, and the surface becomes mechanically uncoupled from the free atmosphere, although radiative transports can still occur. The potential temperature increases rapidly with height near the surface, and the transport of potential temperature is downward, so that less dense air is being forced

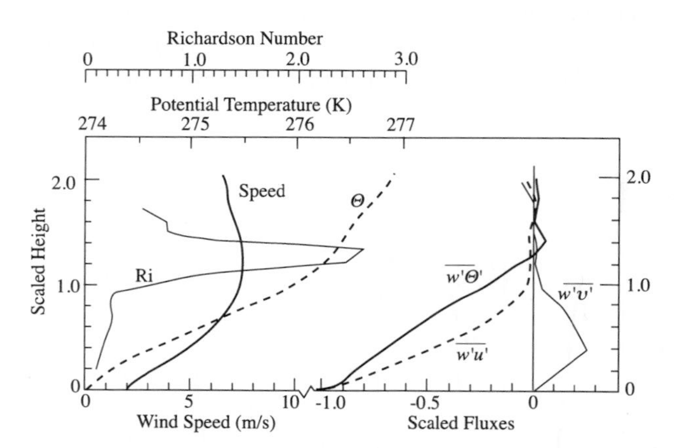

Fig. 4.7 Averaged profiles of wind speed, potential temperature, Richardson number and vertical fluxes of potential temperature $\left(\overline{w'\Theta'}\right)$, and horizontal momentum $\left(\overline{w'u'} \text{ and } \overline{w'v'}\right)$ from nocturnal observations at Haswell, Colorado, on March 24, 1974. The height is scaled by the depth in which turbulence is observed to occur, which on average is about 100 m in this case. Vertical eddy fluxes are scaled by their surface values. [From Mahrt *et al.* (1979). Reprinted with permission from Kluwer Academic Publishers.]

downward against buoyancy. The energy for mixing less dense air downward toward the surface is provided by the mean wind speed shear, which tends to be quite strong under these conditions, often with a low-level wind maximum near the top of the boundary layer and weak winds near the surface. The minimum surface air temperature achieved on a clear night is thus generally lower when the wind speed in the free atmosphere is weak and provides little energy for mixing warm air downward to the radiatively cooled surface.

The atmospheric boundary layer can contain clouds that play an important role in boundary-layer physics and vertical transports. The release of latent heat in clouds can provide buoyancy to drive vertical motions in the boundary layer. Boundary-layer clouds that are important for climate include the fair weather cumulus clouds and stratocumulus clouds. Though less widespread, fog is also an important boundary-layer cloud. The boundary layer also interacts in important ways with deep convective clouds, since the high potential temperature air that drives deep convection is produced in the boundary layer. Except when fog is present, the tops of boundary-layer clouds generally occur near the top of the boundary layer and their bases are some distance above the surface, so that a cloud and a subcloud layer exist within the boundary layer. Stratocumulus clouds modify the boundary-layer physics both through their convective heat and moisture fluxes and through their radiative effects. Because stratocumulus cloud tops are relatively warm and emit longwave radiation efficiently, longwave cooling from cloud tops can be an important mechanism for generating buoyancy within the boundary layer since it cools the air at the top of the boundary layer, which then tends to sink and be replaced by warmer parcels of air rising from below.

Figure 4.8 shows the diurnal variation of temperature in the lowest 1500 m of the atmosphere over Nebraska during a relatively clear summer day. At sunrise the surface is colder than the air a kilometer above the surface. This temperature inversion quickly disappears after sunrise, as insolation warms the surface and this heat is transferred to a shallow layer of air near the ground. Near the middle of the day the surface reaches its maximum temperature and a lapse rate near the dry adiabatic value of 9.8 K km^{-1} is observed near the surface. At this time buoyancy raises warm parcels of air near the surface, and turbulent convection efficiently moves sensible heat upward in the boundary layer. Even before sunset the surface begins to cool in response to efficient upward transport of energy by turbulent motions. After sunset the surface cools rapidly, so that by 10 PM the surface temperature has reached its nighttime value, leaving a very sharp inversion near the ground.

When mean wind speeds are light to moderate, diurnal variations in the temperature profile will affect the exchange of heat, moisture, and momentum between the atmosphere and the surface. The strong density stratification associated with a nocturnal temperature inversion suppresses turbulent transport of momentum from the free atmosphere. Sensible and latent heat fluxes are also suppressed by the strong density stratification, so that the surface temperature responds primarily to the radiative

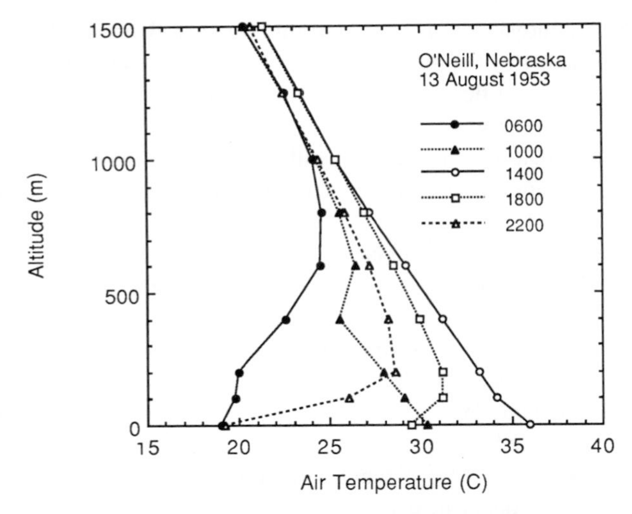

Fig. 4.8 Plot of air temperature at various local times in the lowest 1500 m of the atmosphere at O'Neill, Nebraska on August 13, 1953. Times are given using a 24-hour clock so that 1800 = 6 PM, etc. [Data from Lettau and Davidson (1957).]

forcing, which acts to cool the surface at night. Temperature inversions often develop when a high-pressure system dominates the local weather pattern. The weak surface winds, clear skies, and downward mean motion in the free atmosphere normally associated with high-pressure systems encourage the development of a strong inversion. Pollutants released under a temperature inversion can easily build up to unhealthy levels, because turbulent mixing into the free atmosphere is suppressed by the strong density gradient.

4.5.1 The Neutral Boundary Layer

When the static stability in the boundary layer is near neutral, buoyancy does not play an important role in the turbulent kinetic energy budget. Under neutral conditions the source of energy for boundary-layer turbulence is the kinetic energy of the mean wind of the free atmosphere. The turbulence in the boundary layer produces a strong flux of momentum to the surface. The vertical flux of horizontal momentum at the surface, τ_0, constitutes a drag on the atmospheric flow. In the surface layer the vertical gradient of wind speed (U) under conditions of neutral stability should depend only on the height (z), density (ρ), and surface drag. The characteristic wind speed for use in dimensional analysis is the friction velocity, u_*.

$$u_* = \left(\frac{\tau_0}{\rho} \right)^{1/2} \tag{4.15}$$

Using the friction velocity and the height to scale the wind shear, dimensional analysis suggests that the scaled wind shear should be a constant.

$$\left(\frac{z}{u_*}\right)\frac{\partial U}{\partial z}=\frac{1}{\kappa} \tag{4.16}$$

The von Karman constant, κ, is the same for all neutral boundary layers regardless of the surface characteristics, and has a measured value of approximately 0.4. Equation (4.16) can be integrated with respect to height to obtain the logarithmic velocity profile.

$$U(z)=\left(\frac{u_*}{\kappa}\right)\ln\left(\frac{z}{z_0}\right) \tag{4.17}$$

An additional constant, z_0, is introduced during the integration to obtain (4.17). It is called the *roughness height;* the height at which the wind speed reaches zero. For most natural surfaces the irregularities of the surface are larger than the 1 mm depth of the layer where molecular diffusion dominates, and this roughness can be characterized by the height z_0. Roughness heights for natural surfaces range from about a millimeter for average seas to more than a meter for cities with tall buildings. Roughness heights are estimated by measuring the wind speed profile under neutral conditions and solving (4.17) for z_0. The logarithmic velocity profile has been shown to be a good approximation for many laboratory boundary layers and also for the planetary boundary layer under conditions of neutral stratification. It is valid for heights much greater than the roughness height, $z >> z_0$, and so does not describe the mean wind velocity profile within the plant canopy or very close to a rough surface.

The logarithmic velocity profile law is useful for expressing the momentum flux at the surface in terms of the wind speed at some height in the surface layer. Substituting (4.15) into (4.17) yields an expression for the surface drag in terms of the wind speed, U_r, at some reference height z_r.

$$\tau_o=\rho\, C_D\, U_r^2 \tag{4.18}$$

where,

$$C_D=\kappa^2\ \ln\left(\frac{z_r}{z_0}\right)^{-2} \tag{4.19}$$

The drag coefficient, C_D, depends on the ratio of the reference height to the roughness height. The reference layer can be taken at any level within the surface layer where measurements can be conveniently acquired and where the logarithmic profile is a good approximation of the actual flow. The aerodynamic drag formula (4.18) and related formulas for the sensible and latent heat fluxes at the surface form the basis for empirical estimates of surface fluxes and the specification of surface fluxes in climate models. They allow the calculation of turbulent fluxes using only mean wind speed at a reference height and a few external parameters.

4.5.2 Stratified Boundary Layers

The dimensional analysis for neutral boundary layers can be extended to stratified boundary layers.[2] This theory adds heat flux and buoyancy variables to the dimensional analysis. Characteristic vertical profiles for both wind and temperature are derived in which the vertical coordinate is scaled by a dimensionless combination of the friction velocity, the heat flux, and the buoyancy. From these profiles, bulk aerodynamic formulas can be derived that describe the turbulent heat and momentum fluxes at the surface in terms of mean variables. The coefficients in these formulas now depend on the vertical stability of the atmosphere as well as the roughness height. This theory applies only to the surface layer.

The vertical stability can be characterized with the *Richardson number*. In differential form the Richardson number depends on the vertical derivatives of potential temperature, Θ (see Appendix C), and wind speed, U

$$\mathrm{Ri} = \frac{g}{T_0} \frac{(\partial \Theta / \partial z)}{(\partial U / \partial z)^2} \tag{4.20}$$

where g is the gravitational acceleration and T_0 is the reference temperature. The bulk Richardson number for the boundary layer may be written

$$\mathrm{Ri}_B = \left(\frac{g}{T_0} \right) \left[\frac{z_r \left(\Theta(z_r) - \Theta(z_0) \right)}{U(z_r)^2} \right] \tag{4.21}$$

The Richardson number is large when the potential temperature of the near-surface air is high compared to the potential temperature at the surface. Under such conditions the air is stably stratified, which inhibits vertical mixing. Parcels that are raised up from the surface become negatively buoyant and will be forced back downward by the gravity force. This stabilizing effect can be overcome by the kinetic energy available in the mean wind shear near the ground, which can generate turbulent velocities sufficient to mix stably stratified air. This influence of the kinetic energy is represented by the square of the wind speed in the denominator of (4.21). If the potential temperature decreases with height near the surface, then the boundary layer is unstable and small vertical displacements of parcels will be accelerated by the buoyancy force. The Richardson number for a buoyantly unstable boundary layer is negative. Under these conditions vertical transfer of energy and moisture is relatively efficient because of the free exchange of parcels across level surfaces.

Over land areas it is common for the boundary layer to become unstable during summer days as insolation heats the surface. At night the surface cools faster than the overlying air so that a temperature inversion can develop (Fig. 4.8). The resulting strong density stratification can suppress the nighttime exchanges of heat, momentum, and moisture between the surface and the free atmosphere. The

[2]Monin and Obukhov (1954); Arya (1988).

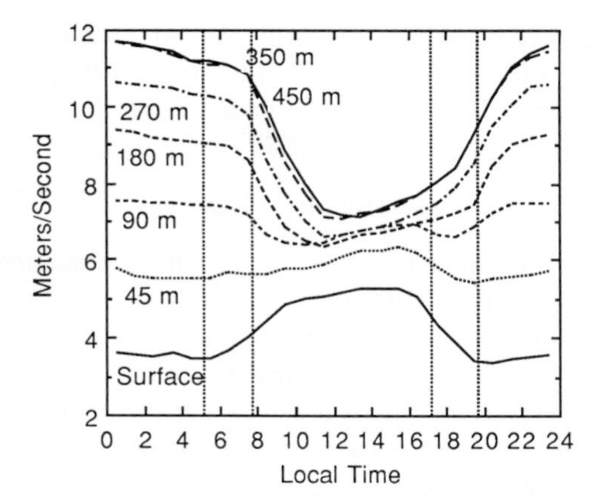

Fig. 4.9 Diurnal cycle of wind speed as a function of height measured from a tower in Oklahoma City and averaged over the period June 1966 to May 1967. [Adapted from Crawford and Hudson (1973). Reprinted with permission from the American Meteorological Society.]

effect of changes in the stability of the boundary layer on momentum fluxes can be seen in diurnal changes in the wind profile. Figure 4.9 shows the diurnal variations in wind speed measured at various heights on a TV tower in Oklahoma. At night wind speeds measured very near the surface decrease, because the downward mixing of momentum from the free atmosphere is reduced by the greater static stability at night. The winds higher in the boundary layer increase at night because the drag from the surface is reduced. During the day, efficient mixing of momentum through a relatively unstable boundary layer causes the wind speed near the surface to increase at the expense of the wind speed higher in the boundary layer.

4.6 Sensible and Latent Heat Fluxes in the Boundary Layer

Sensible and latent heat fluxes from the surface are produced by the turbulent fluid motions in the boundary layer. Transport by molecular diffusion is negligible compared to turbulent transport, except within about a millimeter of the surface. Turbulence is characterized by rapid chaotic fluctuations in wind velocity. Where mean vertical gradients in temperature or humidity exist, the turbulent fluctuations of wind velocity will be accompanied by fluctuations of scalar properties such as temperature and humidity. Vertical fluxes of mass, momentum, and energy are produced by turbulence when the parcels of air moving upward have different properties than parcels of air moving downward. The flux can therefore be measured by the spatial or temporal average of the product of vertical velocity and the property of interest.

For example, if we have measurements of the temperature, T, and vertical velocity, w, at a point near the surface, we may obtain the vertical flux of sensible heat from the time average of the product of vertical velocity and temperature, multiplied by the specific heat and average density of the air.

$$\text{Upward sensible heat flux} = c_p \, \rho \, \overline{wT} \tag{4.22}$$

For this estimate to be accurate, temperature and wind measurements must be taken at frequent enough intervals to define the turbulent fluctuations that produce the vertical transport. Since turbulent fluctuations are very rapid, measurements taken more frequently than every second can be required to directly measure turbulent fluxes. Inhomogeneities in surface conditions may also cause difficulty in obtaining representative fluxes from measurements taken at a single point.

By dividing the variables into a time mean and a deviation from the time mean, or eddy, we may write the upward sensible heat flux as a sum of time mean and eddy contributions.

$$w = \overline{w} + w', \qquad T = \overline{T} + T' \tag{4.23}$$

Here an overbar indicates a time average of the quantity under the overbar and a prime indicates a deviation from that time average. Substituting (4.23) into (4.22) and performing the averages, we obtain the mean and eddy contributions to the vertical flux of temperature.

$$\begin{aligned} \overline{wT} &= \overline{w}\,\overline{T} + \overline{w'T'} \\ \text{Total} &= \text{mean} + \text{eddy} \end{aligned} \tag{4.24}$$

Near the surface the mean vertical velocity is very small compared to the typical eddy or turbulent vertical velocities, and the eddy contribution to the vertical heat flux is dominant. We can then define the latent and sensible heat fluxes as the eddy fluxes of heat and moisture at some level in the atmospheric boundary layer.

$$\text{SH} = c_p \rho \overline{w'T'}, \qquad \text{LE} = L\rho \, \overline{w'q'} \tag{4.25}$$

where ρ is air density, c_p is the specific heat of air at constant pressure, and L is the latent heat of vaporization. Measurements of the turbulent velocity, temperature, and moisture fluctuations necessary to calculate the sensible and latent energy fluxes are not routinely taken, and these rapid, small-scale fluctuations are not simulated in a global climate model. In most cases one must estimate the turbulent fluxes by using variables averaged over larger spatial and temporal scales than those of the turbulent motions in the boundary layer.

Several methods are available for estimating surface fluxes with observations of mean variables. The most common method is through the use of bulk aerodynamic formulas, which relate the turbulent fluxes to observable spatial or temporal averages. One might hypothesize that the sensible heat flux is proportional to the temperature difference between the surface and the air at some standard altitude, z_r, where mean variables are known. Since some of the kinetic energy of boundary-layer turbulence comes from the mean winds blowing over the surface, we might assume that the turbulent fluxes are also proportional to the mean wind speed, U_r, at the standard height. These basic assumptions are consistent with the results of the similarity theory described above, with which we obtain an expression that relates the sensible heat flux to the mean wind speed and temperatures.

$$\mathrm{SH} = c_p \, \rho \, C_{\mathrm{DH}} \, U_r \left(T_s - T_a(z_r)\right) \tag{4.26}$$

The latent heat flux can be related to the difference of specific humidity, q, between the surface and the atmosphere at the reference height.

$$\mathrm{LE} = L \rho \, C_{\mathrm{DE}} \, U_r \, \left(q_s - q_a(z_r)\right) \tag{4.27}$$

In the bulk aerodynamic formulas (4.26) and (4.27) ρ is the air density, c_p is the specific heat at constant pressure, and C_{DH} and C_{DE} are aerodynamic transfer coefficients for temperature and humidity, respectively. Subscripts s and a indicate values for the surface and the air at the reference level, respectively.

The aerodynamic transfer coefficients depend on the surface roughness, the bulk Richardson number, and the reference height. Under ordinary circumstances the values of the transfer coefficients for heat, moisture, and momentum would be nearly equal, and typical values for neutral stability and for a 10-m height above the surface would range from 1×10^{-3} over the ocean to 4×10^{-3} over moderately rough land. If the wind speed at 10 m is 5 m s^{-1} and $C_D = 3 \times 10^{-3}$, then from (4.26) the flux of sensible heat across the surface layer of atmosphere is about 15 W m^{-2} for each degree of temperature difference between the surface and the air at 10 m.

4.6.1 Equilibrium Bowen Ratio for Saturated Conditions

The latent heat flux depends sensitively on the temperature through the dependence of saturation vapor pressure on temperature. Over water or wet land surfaces we may assume that the mixing ratio of water vapor at the surface is equal to the saturation mixing ratio, q^*, at the temperature of the surface.

$$q_s = q^*(T_s) \tag{4.28}$$

The vapor mixing ratio of saturated air at the reference height can be approximated with a first-order Taylor series.

$$q_a^* = q_s^*(T_s) + \frac{\partial q^*}{\partial T} \bullet \left(T_a - T_s\right) + \cdots \tag{4.29}$$

The actual vapor mixing ratio of the air at the reference height can be expressed in terms of the relative humidity at that level.

$$\mathrm{RH} = \frac{q}{q^*} \tag{4.30}$$

$$q_a \cong \mathrm{RH} \bullet \left(q_s^*(T_s) + \frac{\partial q^*}{\partial T} \bullet (T_a - T_s) \right) \tag{4.31}$$

Substituting (4.31) into (4.27) yields an expression for the heat loss from the surface through evaporation in terms of the temperature difference and the relative humidity,

$$\mathrm{LE} \cong \rho L\, C_{\mathrm{DE}}\, U \left(q_s^* (1 - \mathrm{RH}) + \mathrm{RH}\; B_e^{-1} \frac{c_p}{L} (T_s - T_a) \right) \tag{4.32}$$

where

$$B_e^{-1} \equiv \frac{L}{c_p} \frac{\partial q^*}{\partial T} \tag{4.33}$$

The *Bowen ratio* is the ratio of the sensible cooling to the latent cooling of the surface. Comparing (4.32) and (4.26) we see that, when the surface is wet and the air is saturated, RH = 1, the Bowen ratio takes a special value:

$$B_o \equiv \frac{\mathrm{SH}}{\mathrm{LE}} = B_e \tag{4.34}$$

When the surface and the air at the reference level are saturated, the Bowen ratio approaches the value B_e given by (4.33), which can be called the equilibrium Bowen ratio. We presume that the flux of moisture from the boundary layer to the free atmosphere is sufficient to just balance the upward flux of moisture from the surface so that the humidity at the reference height is in equilibrium at the saturation value. The Bowen ratio in such an equilibrium is inversely proportional to the rate of change of the saturation mixing ratio of water vapor with temperature (4.33). The rate of change of the saturation mixing ratio with temperature is very sensitive to the temperature itself. Using the approximate formula (B.3) from Appendix B, it can be shown that

$$\frac{\partial q^*}{\partial T} \approx q^*(T) \bullet \left(\frac{L}{R_v T^2} \right) \tag{4.35}$$

The exponential dependence of the saturation mixing ratio on temperature far outweighs the inverse square of temperature in (4.35), so that the equilibrium Bowen ratio decreases exponentially with temperature. The temperature dependencies of the saturation mixing ratio and the equilibrium Bowen ratio are shown graphically in a log–linear plot in Fig. 4.10. The equilibrium Bowen ratio is unity at about 0°C, and

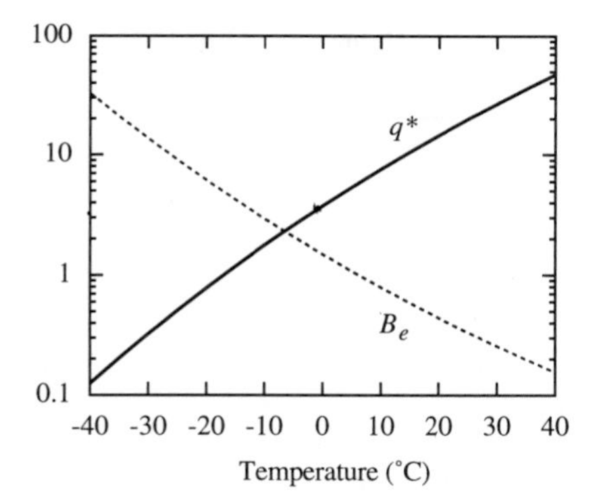

Fig. 4.10 Saturation specific humidity q^* (g kg^{-1}) and equilibrium Bowen ratio B_e, as functions of temperature.

decreases to about 0.2 at 30°C. As the relative humidity in (4.32) is decreased from 1.0 to smaller values, the evaporative cooling increases, so that the equilibrium Bowen ratio is the maximum possible Bowen ratio for a wet surface. The actual Bowen ratio over a wet surface will generally be smaller than the equilibrium Bowen ratio, because the air at the reference height is usually not saturated. As a result of the strong temperature dependence of saturation vapor pressure, latent cooling of the surface dominates sensible cooling from a wet surface at temperatures like those in the tropics, but in high latitudes during winter sensible heat transport can be of greatest importance.

The preceding discussion strictly applies only to conditions where the surface is wet, so that evaporative cooling is not constrained by lack of surface moisture. Over land areas the evaporative cooling may be greatly reduced when moisture cannot be supplied from below the surface rapidly enough to keep the air in contact with the surface saturated. In desert areas the surface is typically so dry that evaporative cooling is small regardless of the temperature, so that sensible cooling and longwave emission must balance solar heating. For vegetated terrain, cooling by evaporation and transpiration through leaves is controlled by the physical and biological condition of the plant canopy and the water content of the soil. The role of soil and vegetation in the surface water and energy balances is discussed further in Chapter 5.

4.7 Variation of Energy Balance Components with Latitude

Some intuition about the relative importance of the various components of the surface energy budget as a function of the climatic regime can be gained from their dependence on latitude (Table 4.5 and Fig. 4.11). The net radiation at the surface

Table 4.5

Mean Latitudinal Values of the Components of the Energy Balance Equation for Earth's Surface

Latitude zone	Oceans				Land			Earth			
	R_s	LE	SH	ΔF_{eo}	R_s	LE	SH	R_s	LE	SH	ΔF_{eo}
80–90N	•	•	•	•	•	•	•	–12	4	–13	–3
70–80N	•	•	•	•	•	•	•	1	12	–1	–9
60–70N	31	44	21	–35	27	19	8	28	27	13	–12
50–60N	39	52	21	–35	40	25	15	40	37	19	–16
40–50N	68	70	19	–21	60	32	28	64	50	23	–9
30–40N	110	114	17	–21	80	31	49	97	78	32	–13
20–30N	150	139	12	–1	92	27	65	127	97	32	–1
10–20N	158	131	8	19	94	39	56	141	108	21	12
0–10N	153	106	5	41	96	64	32	139	96	15	29
0–10S	153	112	5	36	96	66	29	139	101	13	25
10–20S	150	138	7	5	97	54	42	138	119	15	4
20–30S	134	133	9	–8	93	37	56	125	110	21	–7
30–40S	109	106	11	–8	82	37	45	106	98	15	–7
40–50S	76	73	12	–9	54	28	27	74	70	13	–9
50–60S	37	41	13	–17	41	27	15	37	41	15	–19
60–70S	•	•	•	•	•	•	•	17	13	15	–11
70–80S	•	•	•	•	•	•	•	–3	4	–5	–1
80–90S	•	•	•	•	•	•	•	–15	0	–15	0
0–90N								96	73	21	1
0–90S								96	82	15	–1
Globe	109	98	11	0	65	33	32	96	78	18	0

Values in W m^{-2}. [Data from Sellers (1965). Reprinted with permission from the University of Chicago Press.]

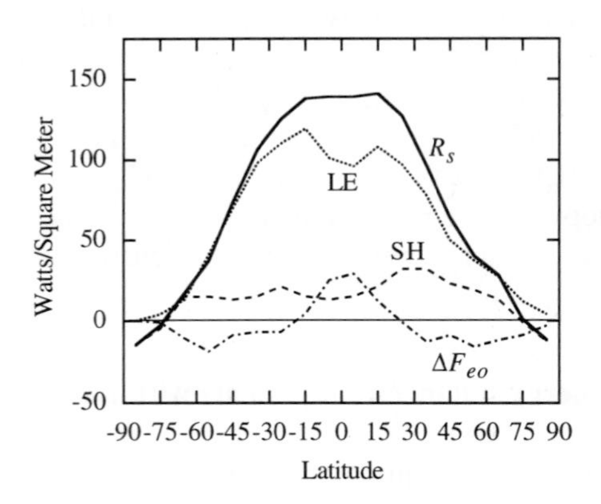

Fig. 4.11 Components of the annual-average surface energy balance plotted against latitude. [Data from Sellers (1965). Reprinted with permission from the University of Chicago Press.]

peaks in the tropics, following the general pattern of insolation at the top of the atmosphere. The net radiation over the tropical ocean is about 60% greater than that over the land at the same latitudes. Most land areas have higher surface albedos than those of water surfaces. Low surface albedo is primarily responsible for the greater net radiation over the oceans, but differences of surface temperature and water vapor abundance between land and ocean areas also contribute.

In polar regions, net radiative energy loss from the surface is balanced by a downward flux of sensible heat. The downward flux of sensible heat is associated with the temperature inversion in high latitudes shown in Fig. 1.4. The polar temperature inversion is supported by poleward heat transport in the atmosphere, which keeps polar regions warmer than they would be in the absence of atmospheric transport.

Ninety percent of the radiative heating of the global ocean surface is balanced by evaporation. Cooling of the surface by evaporation is greatest in the subtropics over the oceans, because surface solar heating and surface water temperature are greatest there. Also, mean downward motion in the free troposphere of the subtropics brings dry air into the boundary layer, encouraging evaporation, and the tradewinds provide a source of kinetic energy for turbulence in the boundary layer. Over land, evaporation peaks near the equator and is reduced in the latitude belt from 10 to 30 degrees because of the dryness of the continents. Energy is available for evaporating water over the subtropical land masses, but little surface water is present to be evaporated. The sensible heat loss from the continents peaks in the subtropics where hot, dry desert conditions are most prevalent. The Bowen ratio, the ratio of sensible to latent cooling of the surface, increases with latitude over both the oceans and the land areas. This reflects the strong dependence of saturation vapor pressure on temperature, which results in an increase of the equilibrium Bowen ratio with decreasing temperature, as explained in the previous section.

In middle and high latitudes the evaporative cooling from the surface is equal to or greater than the energy supplied by radiative heating. Ocean currents move a substantial amount of heat from the tropics to middle and high latitudes. This lateral transport of energy in the ocean supports the very large evaporative heat loss in middle latitudes.

The annual heat budgets for the major continents and oceans are given in Table 4.6. The continents can be easily classified according to their Bowen ratio. Continents with high Bowen ratios are largely "desert" continents, while those with low Bowen ratios are "wet" continents. The driest continent is clearly Australia, with a Bowen ratio of about 2, followed by Africa, which contains the Sahara and Kalahari deserts. The wettest continents are South America and Europe, each with Bowen ratios of about 0.6. North America is also a relatively wet continent, with a Bowen ratio significantly less than 1.0. The sensible and latent surface cooling rates are about equal when averaged over all land areas, while over the global ocean the sensible cooling is only about one-tenth of the latent cooling.

Table 4.6

Annual Energy Balance of the Oceans and Continents

Area	R_s	LE	SH	ΔF_{eo}	SH/LH
Europe	52	32	20	0	0.62
Asia	62	29	33	0	1.14
North America	53	30	22	0	0.74
South America	93	60	33	0	0.56
Africa	90	34	56	0	1.61
Australia	93	29	64	0	2.18
Antarctica	−16	0	−15	0	—
All land	65	33	32	0	0.96
Atlantic Ocean	109	95	11	3	0.11
Indian Ocean	113	102	9	1	0.09
Pacific Ocean	114	103	11	0	0.10
Arctic Ocean	−5	7	−7	−5	−1.00
All oceans	109	98	11	0	0.11

Values in W m^{-2}. [Data from Budyko (1963).]

4.8 Diurnal Variation of the Surface Energy Balance

The surface energy balance at all but polar locations is strongly influenced by the diurnal variation of insolation. Figure 4.12 shows the diurnal variation of the surface radiation balance for grassland in Saskatchewan during a clear summer day with av-

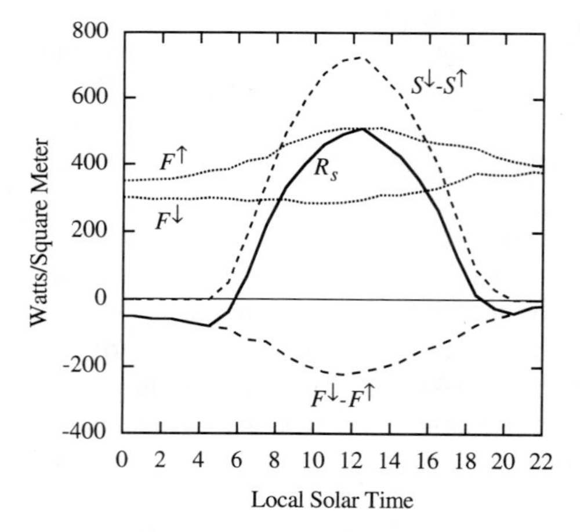

Fig. 4.12 Components of the radiative energy balance for a grass field in Matador, Saskatchewan on July 30, 1971. $F^{\downarrow}$ = downward longwave, $F^{\uparrow}$ = upward longwave, $S^{\downarrow} - S^{\uparrow}$ = net solar, $F^{\downarrow} - F^{\uparrow}$ = net longwave, R_s = net radiation. (After Ripley and Redmann (1976).

erage winds. A dense mat of living and dead grass covers the surface and the surface albedo is about 16%. The average net radiation for the 24-hour period shown was 155 W m^{-2}, with 263 W m^{-2} gained from the sun and 108 W m^{-2} net loss through infrared fluxes. The net downward solar radiation at the ground peaks near local solar noon at about 700 W m^{-2}. The daytime solar heating is large because of the strong insolation, lack of cloudiness, and relatively low surface albedo. Downward longwave radiation is about 300 W m^{-2} and does not change much during the day. The downward longwave radiation has almost no diurnal variation because of the small diurnal variation of air temperature in the free atmosphere. The surface upward emission is about 350 W m^{-2} before sunrise and increases to about 500 W m^{-2} at midday, in response to the warmer daytime surface temperatures. The surface temperature varies from about 10°C before sunrise to about 40°C at midday. The net longwave loss from the surface thus increases from about 50 W m^{-2} at night to about 200 W m^{-2} at midday. The longwave loss is relatively large because the skies are clear and the air humidity is low. The net radiation that results is an almost uniform 50 W m^{-2} loss through longwave cooling during the night, and a strong, solar-driven gain peaking near 500 W m^{-2} at midday.

The diurnal cycle in surface energy balance components for a dry, barren lake bed is shown in Fig. 4.13. The surface soil moisture was measured at about 2% at the time the measurements were taken, so evaporation plays virtually no role in the energy balance. At night the net radiation balance is negative as the surface emits radiation into the dry desert air. The radiative cooling of the surface is balanced by energy lost from the soil as it cools through the night. Sensible heat flux is nearly zero at night because the winds are weak and the cold air near the surface is more dense than the warmer air above. The high vertical stability associated with the nighttime temperature inversion

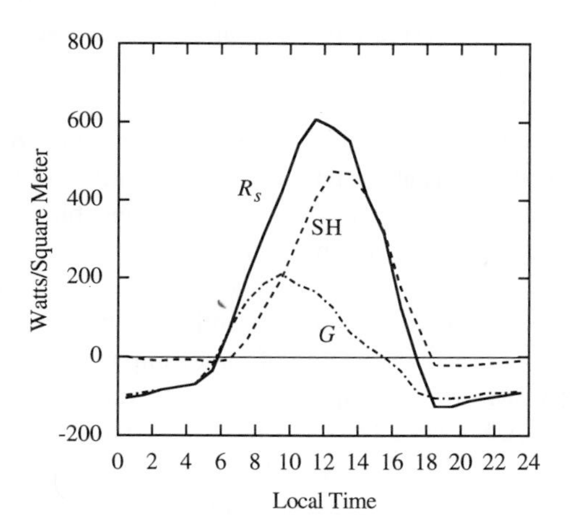

Fig. 4.13 Heat budget for a dry lake bed at El Mirage, California on June 10, 1950. [Data from Vehrencamp (1953). © American Geophysical Union.]

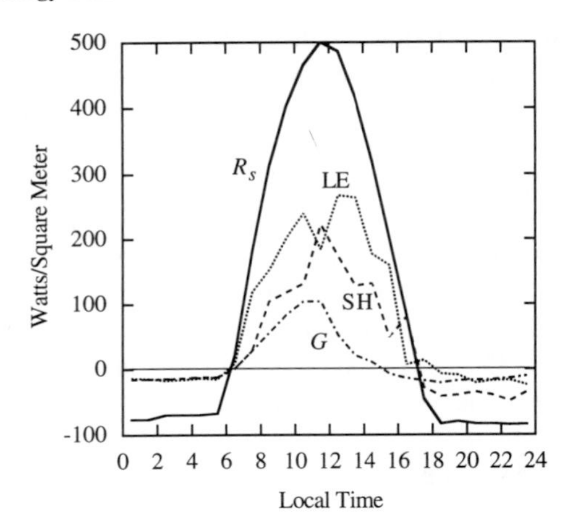

Fig. 4.14 Heat budget for a field of mature corn in Madison, Wisconsin, on September 4, 1952. [Data from Tanner (1960). Reprinted with permission from the Soil Science Society of America.]

suppresses the turbulent exchange of energy. Sunrise occurs at about 5 AM and the surface cooling decreases rapidly. In the early morning the net radiative heating is largely balanced by storage of heat in the ground. After midmorning the strong daytime radiative heating is balanced primarily by upward sensible heat flux by atmospheric turbulence. The sensible heat flux peaks shortly after noon, and by midafternoon the surface has begun to cool because the sensible cooling exceeds the net radiation.

The heat budget for a mature cornfield during a clear day in late summer is shown in Fig. 4.14. Again the net radiation is weakly negative at night and goes through a strong positive maximum during the day. The nighttime radiation loss is balanced about equally by release of stored surface heat, downward sensible heat flux, and dewfall. Although the corn is nearly three meters tall, a substantial amount of heat reaches the soil so that storage in the soil and corn stalks is an important part of the energy balance during the day. The surface soil is dry, but the corn roots have sufficient water. Sensible cooling is about half of the evaporative cooling, when averaged over the day. The peculiar change in the latent and sensible cooling near noon is thought to be related to the north–south orientation of the corn rows.

When a surface is wet, or when growing vegetation has ample soil water for evapotranspiration, the net radiation may be almost entirely used for evaporation, with the soil storage and sensible heat fluxes being of minor importance. When warm and dry air is advected over a surface with ample water, the evaporative cooling may actually exceed the net radiation. Figure 4.15 shows the energy balance on a day when winds carry warm dry air over a well-irrigated alfalfa field. Evapotranspiration equals or exceeds the net radiation at every hour of the day, and evaporation continues even through the night. The excess of the evaporative cooling over the net radiation is provided by downward sensible heat flux, with turbulent motions carrying heat downward to the

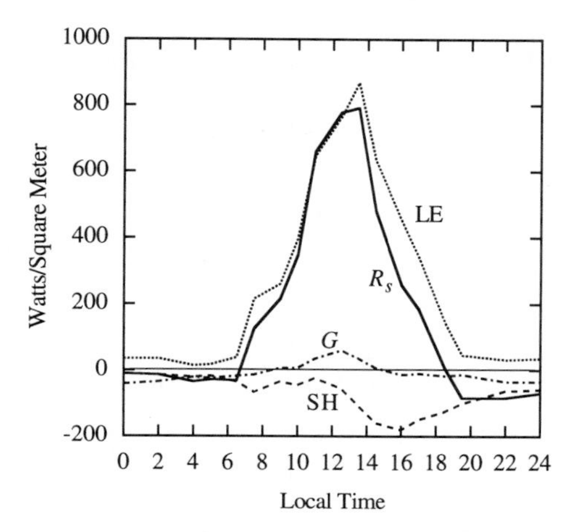

Fig. 4.15 Heat budget for a well-irrigated alfalfa field in Hancock, Wisconsin on July 9, 1956 when the air was advected to the field from a warm dry area. [Data from Tanner (1960). Reprinted with permission from the Soil Science Society of America.]

evaporatively cooled surface. For small irrigated plots in hot, arid regions, the sensible heating of evaporatively cooled surfaces can substantially add to the water demand.

4.9 Seasonal Variation of the Energy Balance of Land Areas

The energy balance of the surface changes with season, especially in middle and high latitudes, where substantial seasonal variations of insolation and temperature occur. In tropical regions, the energy balance may change because of seasonal changes in precipitation, even when the temperature and insolation remain relatively constant. The seasonal variation of the components of the surface energy balance at several locations in middle latitudes is shown in Fig. 4.16. The annual variation of net radiation generally follows that of insolation, with peak values in summer approaching 200 W m^{-2} in land areas, depending on the latitude, sky conditions, and surface albedo. Water surfaces in relatively cloudless areas can have summertime net radiation near 300 W m^{-2}. The mechanism for balancing this net radiation depends on the local surface conditions. Over land areas, it is primarily a question of whether the mechanism is latent or sensible cooling, since storage is small and transport is zero. The apportionment between sensible and latent cooling depends on the availability of surface moisture, the temperature, and the humidity of the air.

In regions with significant precipitation during the summer season, where the surface remains relatively moist, the latent cooling is generally limited by and follows the annual cycle of radiative heating. Where the climate is exceptionally dry, such as in Yuma, Arizona, the latent cooling is negligible, except during months with precipitation. At Yuma significant evaporation occurs in the springtime and during September.

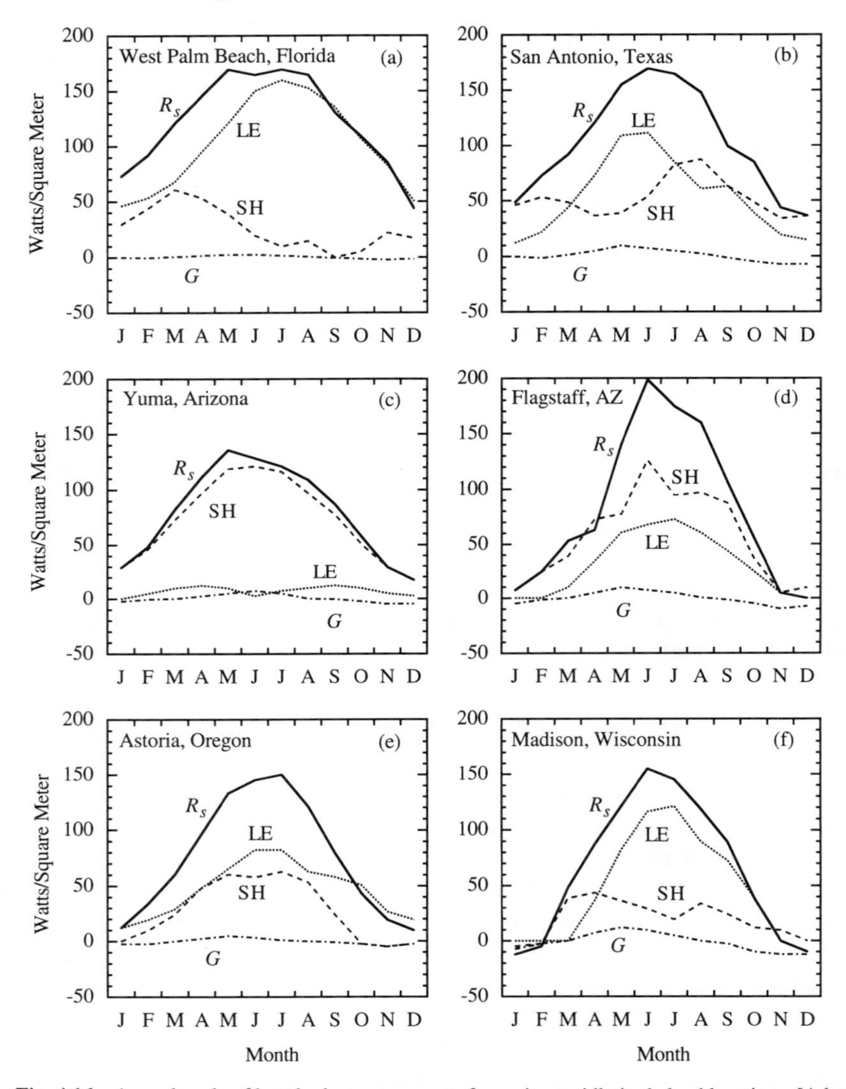

Fig. 4.16 Annual cycle of heat budget components for various midlatitude land locations. [Adapted from Sellers (1965). Reprinted with permission from the University of Chicago Press.]

Flagstaff, Arizona, is at a higher elevation than Yuma and the mountains receive significant summertime precipitation. As a result the evaporative cooling during summer is greater than at Yuma.

At West Palm Beach, Florida, the winters are relatively dry, so that the springtime insolation increase is initially balanced by equal contributions from latent and sensible cooling of the surface. As the summertime convective precipitation begins to wet the surface, the evaporative cooling takes over. At San Antonio, Texas, the surface dries out during spring and summer, so that a gradual increase in the importance of sensible heating occurs over the course of the summer months. At Astoria, Oregon, sensible cooling is important in the summer, even though Astoria gets more annual precipitation than Madison, Wisconsin, where evaporative cooling dominates during summer. This is because the precipitation at Astoria peaks in the winter, when little energy is available for evaporation, so that about 70% of the annual precipitation runs off before being evaporated. During the summer relatively little precipitation falls at Astoria. At Madison precipitation peaks in the summer when the large net radiation provides the energy to evaporate large amounts of water that later is released in convective weather systems such as thunderstorms.

4.10 Surface Energy Flux Components over the Oceans

In ocean areas the heat capacity of the water is sufficient that the energy for evaporation may be derived from the energy of the water itself. The evaporative loss may be less correlated with net radiation than with factors such as the wind speed or the temperature and humidity contrast between the surface and the air above it. Much of the high-latitude evaporation over the oceans takes place in winter over the Kuroshio and Gulf Stream currents. These currents carry warm water poleward along the western margins on the Pacific and Atlantic Oceans where it comes into contact with cold, dry air coming off the continents. The combination produces evaporation rates near 400 W m^{-2} in these regions during winter. Figure 4.17 shows estimates of the

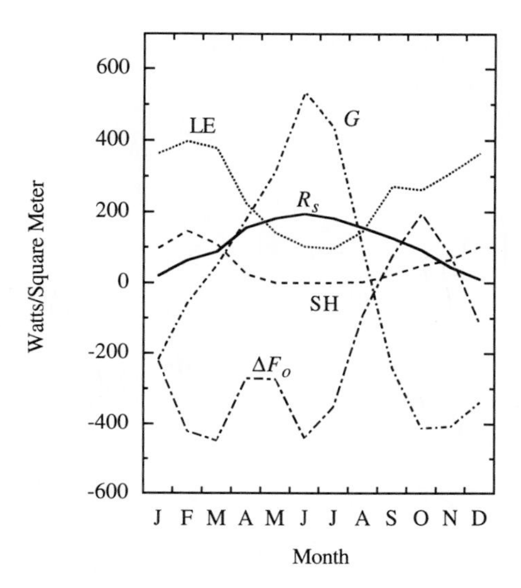

Fig. 4.17 Annual cycle of heat budget components for the Gulf Stream at 38°N, 71°W. (Adapted from Sellers, 1965. Reprinted with permission from the University of Chicago Press.)

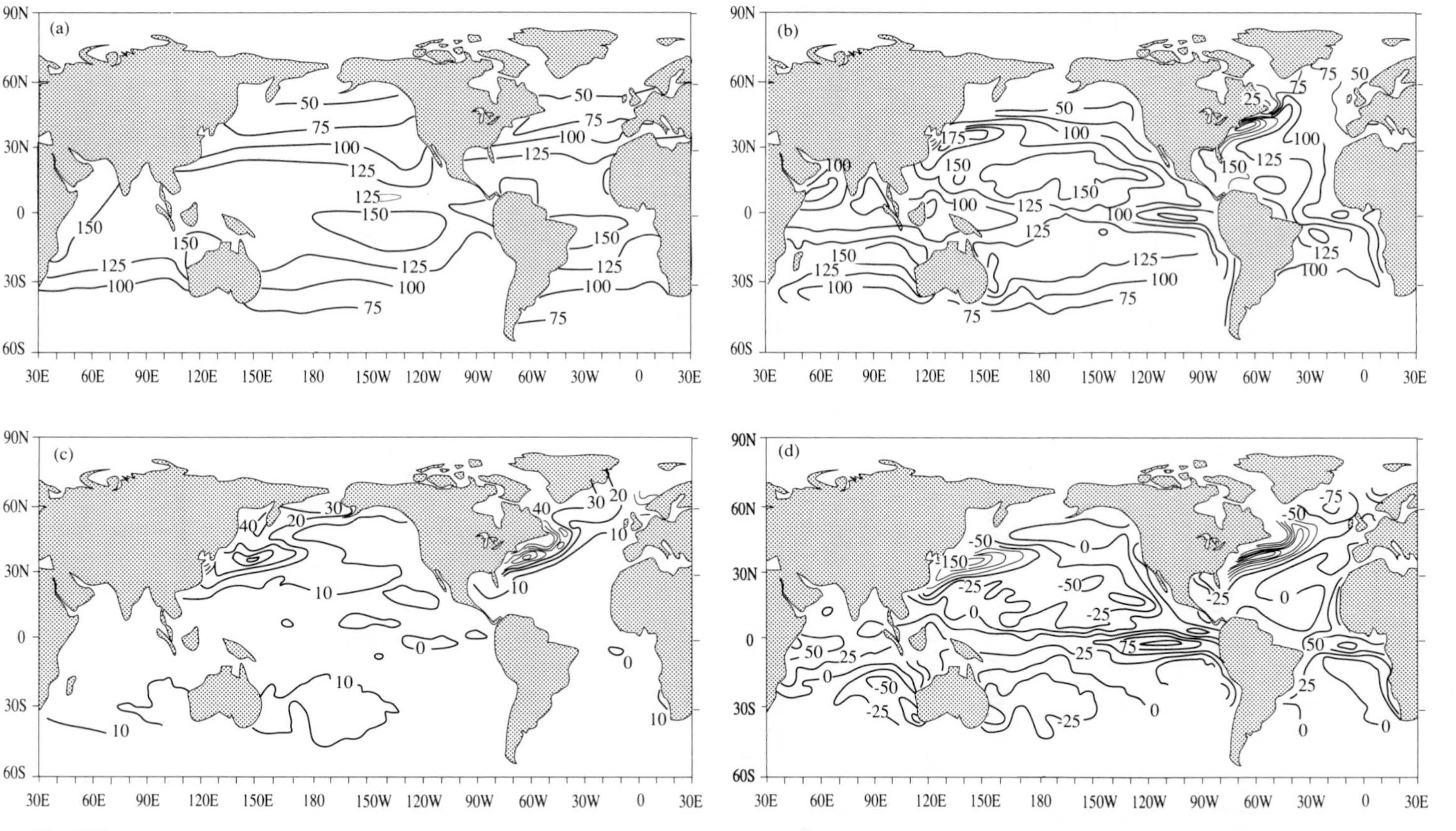

Fig. 4.18 Maps of the annual average energy budget components over the oceans: (a) net radiation; (b) latent heat flux; (c) sensible heat flux; (d) net downward heat flux into the ocean. [From Oberhuber (1988).]

annual cycle of surface energy fluxes over the Gulf Stream at 38°N. The annual variations in most terms are much larger than that of the net radiation. Enormous latent cooling rates in winter are balanced primarily by the release of energy stored in the water temperature. The horizontal energy transport and storage terms are each nearly four times as large as the net radiation.

The surface heat balance components for the oceans vary systematically with latitude and surface conditions (Fig. 4.18). Net radiation is greatest over the tropical oceans, where the surface albedo is low and the surface temperature is moderate [Fig. 4.18(a)]. In these regions it can exceed 150 W m^{-2}. Most of the variation in net radiation comes from the latitudinal decrease of insolation and from cloudiness variations and their effect on surface solar heating. The net longwave radiation typically cools the surface and varies between about 25 and 50 W m^{-2}. In the tropics solar radiation provides about 200 W m^{-2} to the ocean surface and longwave radiation removes about 50 W m^{-2}. In high-latitude ocean regions the longwave component of the surface radiative energy balance contributes a net cooling of about 30 W m^{-2}.

The evaporative heat loss from the surface has its greatest values over the midlatitude, warm, western boundary currents, the Kuroshio and the Gulf Stream [Fig. 4.18(b)]. In these regions at the western edges of the Pacific and Atlantic Oceans, the evaporative heat loss may exceed 200 W m^{-2} and is much greater than the local net radiative heating of the ocean surface. The evaporative cooling of the western boundary currents is greatest in the winter season. The evaporative heat loss is also large over the subtropical oceans, where it consumes most of the energy provided to the surface by net radiation. Evaporation driven by insolation over the tropical and subtropical oceans is the boiler that drives the circulation of the atmosphere and the hydrologic cycle of Earth.

The sensible heat loss from the ocean surface is small, except over the warm western boundary currents of the midlatitude oceans [Fig. 4.18(c)]. These large sensible heat fluxes occur when cold air from the continents flows over the warm ocean currents during winter. In these regions sensible heat fluxes may exceed 50 W m^{-2} in the annual mean, but they are still much less important than evaporative cooling.

The divergence of the heat flux in the ocean, or alternatively, the flux of heat from the surface into the ocean, is large and negative over the western boundary currents [Fig. 4.18(d)]. In these regions the ocean is supplying heat by horizontal transport in the oceans, which is then lost to the atmosphere. The regions where the air is heating the water are found along the equator and along the eastern margins of the oceans where upwelling brings cold water to the surface. In these regions of cold water upwelling, latent heat loss is reduced and more of the net radiation is used to heat the ocean water. The large transfers of energy from the tropical and eastern oceans to the midlatitude western oceans in the Northern Hemisphere play a critical role in determining the climate of maritime areas. The ocean currents that produce these important energy transports are described in Chapter 7.

Exercises

1. If the top 100 m of ocean warms by 5°C during a 3-month summer period, what is the average rate of net energy flow into the ocean during this period in units of W m^{-2}? If the atmosphere warms by 20°C during the same period, what is the average rate of net energy flow into the atmosphere?
2. Derive (4.18) from (4.15) and (4.17).
3. The blackbody emission from the surface can be linearized about some reference temperature T_0.

$$\sigma T_s^4 \approx \sigma T_0^4 + 4\sigma T_0^3 \left(T_s - T_0\right) + \cdots$$

   And the sensible cooling of the surface can be written as

$$\mathrm{SH} \approx c_p \, \rho C_D U \left(T_s - T_a\right)$$

   Calculate and compare the rates at which longwave emission and sensible heat flux vary with surface temperature, T_s. In other words, if the surface temperature rises by 1°C, by how much will the longwave and sensible cooling increase? Assume that $T_0 = 288$ K, T_a is fixed, $\rho = 1.2$ kg m^{-3}, $c_p = 1004$ J kg^{-1} K^{-1}, $C_D = 2 \times 10^{-3}$, and $U = 5$ m s^{-1}.
4. Air with a temperature of 27°C moves across a dry parking lot at a speed of 5 m s^{-1}. The insolation at the surface is 600 W m^{-2} and the downward longwave radiation at the ground is 300 W m^{-2}. The longwave emissivity of the surface is 0.85, and the albedo of the asphalt surface is 0.10. What is the surface temperature in equilibrium? What is the surface temperature if the asphalt is replaced with concrete with an albedo of 0.3 and the same emissivity? The air density and drag coefficient are as in problem 3. *Hint:* Linearize the blackbody emission around the air temperature and use the surface energy equation to show that

$$T_s - T_a = \frac{S^{\downarrow}(0) \cdot \left(1 - \alpha_s\right) + \varepsilon\left(F^{\downarrow}(0) - \sigma T_a^4\right)}{c_p \, \rho C_D U + \varepsilon 4 \sigma T_a^3}$$

5. Do problem 4 for the case in which the parking lot is wet and the air is maintained just at saturation, and include the effect of latent cooling of the surface. Ignore any effects of surface water on the albedo. Compare the surface temperature for wet and dry surfaces. How would the results differ if the air was not saturated? *Hint:* Make use of (4.34).
6. Give the reasons why the net radiation at the surface at Flagstaff is greater than the net radiation at Yuma during summer [Fig. 4.16(c,d)].

Chapter 5 The Hydrologic Cycle

5.1 Water, Essential to Climate and Life

Water continually moves between the oceans, the atmosphere, the cryosphere, and the land. The total amount of water on Earth remains effectively constant on time scales of thousands of years, but it changes state between its liquid, solid, and gaseous forms as it moves through the hydrologic system. The movement of water among the reservoirs of ocean, atmosphere, and land is called the hydrologic cycle. The amount of water moved through the hydrologic cycle every year is equivalent to about a 1-m depth of liquid water spread uniformly over the surface of Earth. This amount of water annually enters the atmosphere through evaporation and returns to the surface as precipitation. To evaporate 1 m of water in a year requires an average energy input of 80 W m^{-2}. The sun provides the energy necessary to evaporate water from the surface. Once within the atmosphere, water vapor can be transported horizontally for great distances and moved upward. This horizontal and vertical movement of water vapor is critical to the water balance of land areas, since about one-third of the precipitation that falls on the land areas of Earth is water that was evaporated from ocean areas and then transported to the land in the atmosphere (Fig. 5.1). The excess of precipitation over evaporation in land areas supports the return of water from the land to the ocean in rivers.

The atmosphere contains a relatively small amount of water (Tables 5.1 and 1.2). If all the water vapor in the atmosphere were condensed to liquid and spread evenly over the surface of Earth, it would be only about 2.5 cm deep. Since 100 cm of water is evaporated and condensed per annum, the atmospheric water is removed by precipitation about 40 times a year, or every 9 days. Because net evaporation is a small residual of a more rapid two-way exchange of water molecules across the air–water interface, the actual residence time of water molecules in the atmosphere is about 3 days. Since nearly a 3-km depth of water is present near the surface of Earth, most of which is in the oceans, and only 2.5 cm can reside in the atmosphere, an average water molecule must wait a very long time in the ocean, in an ice sheet, or in an aquifer, between brief excursions into the atmosphere.

In earlier chapters we saw the important role of water in many aspects of the climate system. Water is crucial to life, and the existence of oceans on Earth has dramatically influenced the character and evolution of Earth's atmosphere. Chemical

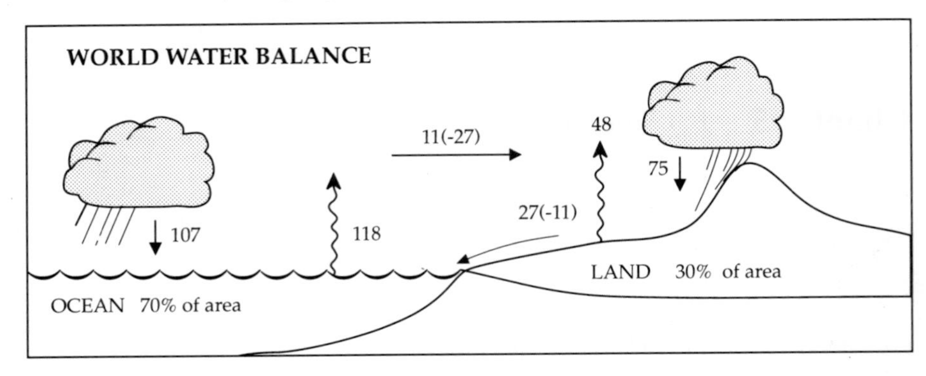

Fig. 5.1 Schematic diagram showing the basic fluxes of water in the global hydrologic cycle. Units are centimeters per year spread over the area of the land or ocean. Since the areas of land and ocean are different, the land–ocean water exchanges by atmospheric transport and river runoff have different values depending on the reference area, as indicated by the parentheses. The smaller values are those referenced to the larger oceanic area.

Table 5.1

Water Volumes of Earth

Category	Volume (10^6 km^3)	Percent
Oceans	1348.0	97.39
Polar ice caps, icebergs, glaciers	227.8	2.010
Groundwater, soil moisture	8.062	0.580[a]
Lakes and rivers	0.225	0.020
Atmosphere	0.013	0.001
Total water amount	1384.0	100.0
Freshwater	36.00	2.60
Freshwater reservoirs as a percent of total freshwater		
Polar ice caps, icebergs, glaciers		77.2
Groundwater to 800-m depth		9.8[a]
Groundwater 800–4000-m depth		12.3[a]
Soil moisture		0.17[a]
Lakes (freshwater)		0.35
Rivers		0.003
Hydrated earth minerals		0.001
Plants, animals, humans		0.003
Atmosphere		0.040
Sum		100.000

[From Baumgartner and Reichel (1975).]

[a]Numbers uncertain.

and biological processes that take place in the oceans continue to regulate atmospheric composition. In the current atmosphere, water vapor is the most important gaseous absorber of solar and terrestrial radiation and accounts for about half of the atmosphere's natural greenhouse effect. Clouds of liquid water and ice contribute about 30% of the atmosphere's natural opacity to thermal radiation and contribute about half of Earth's reflectivity for solar radiation. The evaporation of water from Earth's surface accounts for about half of the cooling of the surface that balances the heating by absorption of solar radiation. As the water vapor rises into the atmosphere it eventually condenses and precipitates, but the energy released during the condensation of atmospheric water vapor helps to drive the circulation systems of the atmosphere. Water can alter the surface albedo of Earth through the deposition of snow and ice and by fostering the development of vegetative cover on land surfaces.

5.2 The Water Balance

To understand how local climates are maintained, it is instructive to consider the water budget for the surface. In order to model the climate, the surface water balance must be accurately represented. The surface water balance may be written

$$g_w = P + D - E - \Delta f \tag{5.1}$$

where g_w is the storage of water at and below the surface, P is the precipitation by rain and snow, D is the surface condensation (dewfall or frost), E is the evapotranspiration, and Δf is the runoff.

Averaged over a long period of time, the storage term is small. Also, dewfall is usually small, or can be incorporated into a generalized precipitation. The resulting hydrologic balance for a long-term average is

$$\Delta f = P - E \tag{5.2}$$

A complementary balance for the atmosphere must also be satisfied. Precipitation minus evaporation is the net flux of water from the atmosphere to the surface and occurs with opposite sign in the atmospheric water balance.

$$g_{\mathrm{wa}} = -(P + D - E) - \Delta f_a \tag{5.3}$$

The terms have the same meaning as in (5.1), except that g_{wa} indicates storage of water in the atmosphere and Δf_a indicates horizontal export of water by atmospheric motions, primarily in the form of water vapor. Adding the budgets for the surface (5.1) and the atmosphere (5.3), we obtain a water balance for the Earth–atmosphere system in which the exchange of water across the surface does not appear.

$$g_w + g_{\mathrm{wa}} = -\Delta f - \Delta f_a \tag{5.4}$$

When averaged over a year, the storage terms on the left of (5.4) are generally small, and the horizontal transport of water out of a region by the atmosphere must be equal

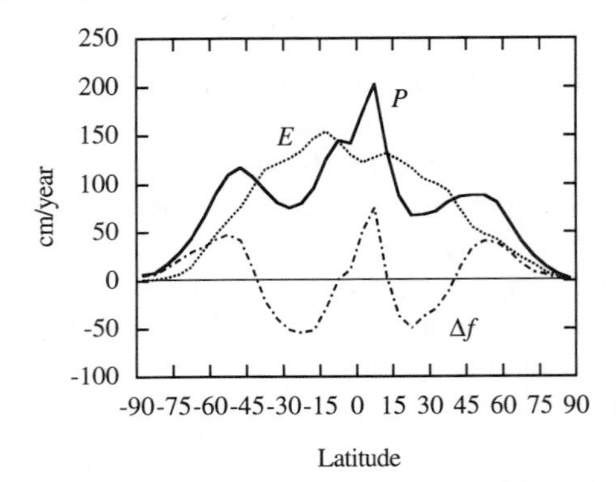

Fig. 5.2 Latitudinal distribution of the surface hydrologic balance, showing evaporation E, precipitation P, and runoff Δf. [Data from Baumgartner and Reichel (1975).]

and opposite to the net horizontal transport below the surface. This means that water carried to continents by atmospheric transport must equal the runoff from rivers.

The distributions with latitude of the terms in the annually averaged surface water balance are shown in Fig. 5.2. Precipitation peaks near the equator, with secondary maxima in middle latitudes of each hemisphere. The equatorial maximum is associated with heavy precipitation in the intertropical convergence zone. Moisture-laden air near the surface flows toward the equator from both hemispheres and converges near the equator, where it is released in thunderstorms, tropical cyclones, and other precipitation-producing weather systems (Fig. 5.3). The secondary maxima in midlatitudes of each hemisphere are associated with the weather systems of that region. In these latitudes cyclonic disturbances with strong winds drive vertical motions that release water. Evaporation varies more smoothly than precipitation, with a broad maximum in the tropics. Precipitation exceeds evaporation in the equatorial belt and again in middle to high latitudes. Evaporation exceeds precipitation in the belt from 15 to 40 degrees of latitude, and these regions export water vapor to be condensed in the latitudes where the precipitation maxima occur. The runoff distribution shown in Fig. 5.2 implies transport of water vapor in the atmosphere from the subtropics to the equatorial and high latitude zones. A return flow in the oceans or rivers carries water back toward the subtropics.

The water balances of the continents and oceans are closely related to their climates and the processes that maintain climate (Table 5.2). The Atlantic and Indian Oceans are net exporters of water vapor, whereas the Pacific and Arctic Oceans receive more water in the form of precipitation than they give up to the atmosphere through evaporation. Comparison of the surface salinity of the Atlantic and Pacific Oceans shows a much higher salinity in the north Atlantic than in the north Pacific. The surface hydrologic balance of the oceans plays an important role in determining their salinity and thereby the deep circulation of the oceans. The saline surface water of the Atlantic is a key factor for allowing surface water to sink to the bottom of the

Fig. 5.3 Hurricane Bonnie located about 500 miles from Bermuda on 19 September 1992. (Photo credit: NASA.)

Table 5.2

Water Balance of the Continents and Oceans (in mm/year)

Region	E	P	Δf	$\Delta f/P$
Land				
Europe	375	657	282	0.43
Asia	420	696	276	0.40
Africa	582	696	114	0.16
Australia	534	803	269	0.33
North America	403	645	242	0.37
South America	946	1564	618	0.39
Antarctica	28	169	141	0.83
All land	480	746	266	0.36
Ocean				
Arctic Ocean	53	97	44	0.45
Atlantic Ocean	1133	761	−372	−0.49
Indian Ocean	1294	1043	−251	−0.24
Pacific Ocean	1202	1292	90	0.07
All ocean	1176	1066	−110	−0.10
Globe	973	973	0	

[From Baumgartner and Reichel (1975).]

ocean, since salinity is an important variable in determining seawater density, especially in high latitudes.

The *runoff ratio*, $\Delta f/P$, is a measure of the wetness of a continent. If it is large, then a significant fraction of the precipitation that falls on that continent flows into the ocean, rather than being evaporated over the land. The dry continents of Africa and Australia have relatively low runoff ratios. Typically, about 40% of the precipitation on a continent runs back to the global ocean in rivers. The evaporation from the surface of a continent is typically 60% of the precipitation that falls on that continent.

5.3 Surface Water Storage and Runoff

The storage term in (5.1) accounts for changes in the amount of water that is retained in the surface. Over land areas this includes the water in the near surface soil and also water that flows deeper and becomes part of an underground water system. An additional important form of water storage is surface snowcover. Distinct seasons of precipitation and drying are a prominent feature of the climate in many regions. For such regions, storage of water in the soil and in snowpack is critical for determining the nature of the environment that develops during the dry season. In many midlatitude regions mountain snowpack is essential for spring and summer river flow, and at lower elevations spring snowmelt helps to replenish soil moisture and groundwater for the summer dry season. The combination of moist soil in springtime followed by summer warmth and sunshine makes many midlatitude land areas agriculturally productive. Storage of precipitated water in snowpack depends only on the surface thermodynamics and physical structure. Storage of water that arrives at the surface as rainfall depends on the frequency and intensity of the precipitation and the characteristics of the soil, its vegetative cover, and the topography of the surface.

Climate interacts only with water that is on the surface or in the soil water zone. The soil water zone extends downward to the depth penetrated by the roots of the vegetation. Plants can draw water from this depth relatively quickly and release it to the atmosphere by *transpiration* through leaves. Because roots of plants can draw moisture from the soil more quickly than water is brought to the surface by nonbiological processes, vegetated surfaces normally release water more quickly to the atmosphere than does bare soil with the same water content. Depending on the conditions, one may need to consider a layer deeper than the root zone in order to predict surface moisture and evapotranspiration. Moisture stored deeper in the soil than the root zone must be brought upward by diffusion or capillary action. Transport through the soil in both liquid and vapor form is possible.

Water is suspended in the soil by adherence to soil particles in thin films. The amount of water that can be held in this manner is called the *field capacity* of the soil. If the soil water content increases above the field capacity, then gravitational forces carry the water downward to the water table, where it becomes part of the groundwater. If the water encounters an impermeable obstacle such as bedrock, then it may flow laterally, seeking lower pressure. Gradual collection of water in subsurface reservoirs results in the formation of aquifers, from which freshwater can be extracted.

The moisture balance of the soil layer and the average soil moisture content are critical to the local climate of land areas. The water in this zone is available for use by plants and can be transpired or evaporated. The soil layer and associated vegetation determine the fate of precipitated water, which may be quickly reevaporated, absorbed by the soil, or run off in stream flow. The transfer of surface water to the soil is called infiltration. The fraction of precipitation that is retained by the soil is determined by soil and vegetation properties and by the rate and frequency of precipitation.

If the surface soil is saturated and the precipitation or snowmelt is more rapid than can be balanced by infiltration and evaporation, then surface ponding will occur. Once the surface depressions in the soil are filled with water, the surface water will begin to flow laterally toward streams and drainage systems. Water runoff from land areas in streams and rivers is important for navigation, fisheries, hydroelectric power generation, irrigation of dry land areas, and municipal water supplies.

5.4 Precipitation and Dewfall

Precipitation is produced when air parcels become supersaturated with water vapor, condensation and droplet formation occur, and the droplets or particles reach the surface without reevaporation. Supersaturation is normally caused by cooling of air parcels during ascent. Ascent of air parcels can be forced by atmospheric motions associated with midlatitude frontal and synoptic weather systems. In the tropics and over continents during summer, ascent, condensation, and precipitation are often associated with convective instability, where parcels of air are forced upward by buoyancy in cumulonimbus clouds. In stratiform cloud systems, light but steady precipitation is generated through the radiative cooling of the tops of the clouds and steady overturning of moist air from beneath. Heavy and persistent precipitation may result when moist air is forced over mountain ranges by prevailing winds.

The geographic distribution of precipitation is shown in Fig. 5.4. The general features of the zonal-average precipitation in Fig. 5.2 are apparent, with the largest precipitation near the equator, where the average water content of the air is high and tropical convective systems are responsible for much of the rainfall. In the subtropics convection and precipitation are suppressed by the downward mean air motion that characterizes this region. In these latitudes precipitation is at a minimum, but high surface insolation and subsidence of dry air give rise to very strong evaporation. In midlatitudes precipitation increases again because of midlatitude synoptic storm systems. The forced ascent of moist surface air in midlatitude weather systems and the westerly flow over obstacles such as the Rocky Mountains give rise to heavy precipitation. In the polar regions precipitation decreases. The entire hydrologic cycle is slowed down in polar regions because of the low temperatures and consequently low water-carrying capacity of the atmosphere.

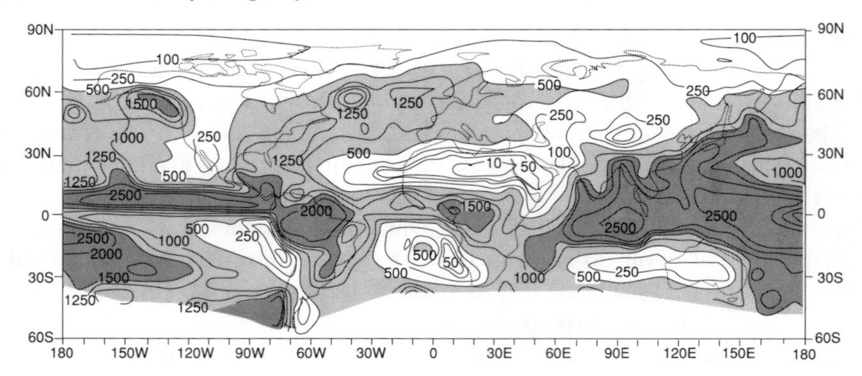

Fig. 5.4 Geographic distribution of annual mean precipitation in mm. [After Shea (1986). Reprinted with permission from the National Center of Atmospheric Research.]

When air comes into contact with a cold surface, usually on relatively clear nights, water vapor may condense directly onto the surface and form dew. Vapor flux from the soil may also be an important contributor to the accumulation of dew, especially at night when the underlying soil may be warmer than the surface. Dewfall is a significant contributor to the surface water balances in some arid climates, but is generally small and lumped together with the precipitation. Fog droplets that are too small to precipitate can be collected from the air by the leaves or needles of plants. In some climates such "combing" of liquid water from the air is an important mechanism whereby plants obtain moisture.

5.5 Evaporation and Transpiration

Evapotranspiration is the removal of water from the surface to the air with an accompanying change in phase from the liquid to the vapor form. It is the sum of evaporation and transpiration. Evaporation refers to direct evaporation of water from the surface itself. *Transpiration* is the passage of water from plants to the atmosphere through leaf pores called *stomata,* which also serve as the point of entry for carbon dioxide required for photosynthesis. Water is absorbed from the soil and carried through the roots and stems of plants to the leaves, where it escapes as water vapor. Stomata normally close at night and open during the day, but they may also close at midday in response to high temperatures, temporary water deficit, or high carbon dioxide concentrations. The differences between evaporation and transpiration are important, but it is difficult to separate the effects of the two processes in practice, so they are generally added to form a single term in water budgets. Evapotranspiration may also include *sublimation,* which refers to the direct conversion of snow and ice to water vapor, without an intermediate liquid phase.

Evaporation from a wet surface is determined by the surface tension at the air–water interface and the rate of decrease of water vapor concentration between the water surface and the adjacent air. The rate at which the water vapor concentration changes with distance from a water surface depends on the molecular diffusivity and the ventilation of the air near the water surface by air motions. Normally, turbulent air motions are of primary importance for carrying water vapor away from a surface and dominate in determining gradients on scales larger than a few millimeters. The interaction of surface water waves with atmospheric turbulence can influence the rate of evaporation over the oceans. Over land the structure of the surface and the vegetation covering it can have a substantial effect on the rate of evaporation. The collection of vegetable matter covering the land surface is called the *plant canopy,* which may be as thin as a layer of moss or as thick as a tall forest.

Plant canopies have important effects on the water and energy balances of the surface. Some of these effects are illustrated in Fig. 5.5. Precipitation that falls on a plant canopy can be intercepted by leaves and stems or fall directly onto the soil.

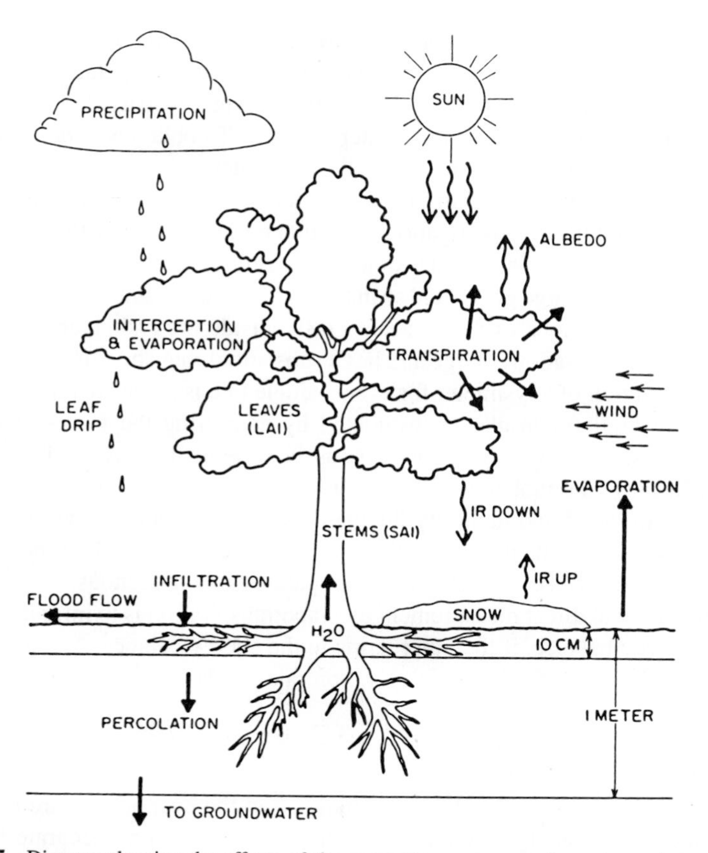

Fig. 5.5 Diagram showing the effects of the vegetation canopy on the water and energy fluxes. [From Dickinson (1984). © American Geophysical Union.]

Water that falls on the leaves can be evaporated from the leaves or drip to the surface. Interception of precipitation by leaves and evaporation from leaves can greatly decrease runoff if the rainfall rate is not too intense and the air is relatively dry. The leaf structure of a plant presents a much larger surface area on which evaporation can take place than the ground surface alone. The energy balance of the leaves is also of importance, since it determines how rapidly water can be evaporated. The structure and arrangement of leaves and branches affect the absorption of solar radiation, the emission of longwave radiation, and the ventilation of the surface by air motions. A parameter often used to characterize plant canopies is the *leaf area index* (LAI). It is defined as the ratio of the area of the top sides of all the leaves in the canopy projected onto a flat surface to the area of the surface under the canopy. It is equal to the number of leaves that would be crossed by a vertical line passing through the canopy, on average.

5.5.1 Measurement of Evapotranspiration

Evapotranspiration can be estimated in a variety of ways. One of the most accurate methods is by weighing the moisture change in the soil and its vegetative cover with a device called a *lysimeter*. A lysimeter is a container of soil set on a balance or provided with some other means of measuring water content.[1] To obtain accurate results, the lysimeter must be large enough to contain the soil water zone and associated vegetation. The lysimeter should be set in a larger environment where the surface conditions are similar to those under investigation, if results representative of the natural environment are desired. For example, a lysimeter containing a depth of growing grass and soil can be buried in a large grass field so that the grass in the lysimeter experiences exactly the same conditions as the adjacent grass outside the lysimeter. For environments where the vegetation is large and has a substantial root structure, such as a forest, the construction of a lysimeter for even a single tree is a considerable challenge.

Evapotranspiration can also be estimated by measuring the fluxes of moisture away from the surface by taking simultaneous measurements of vertical velocity and humidity. Because the moisture is carried upward by turbulent motions, the device used to measure wind and humidity fluctuations must respond on the time scale of seconds. It is also a challenge to obtain representative spatial and temporal means, particularly if the surface characteristics are spatially inhomogenous.

An alternative to direct measurement of evapotranspiration is to infer it as a residual in the energy balance, if the other terms in the energy balance can be measured. Rearranging (4.1) to solve for the evaporation rate yields

$$E = \frac{1}{L}\left(R_s - \mathrm{SH} - \Delta F_{\mathrm{eo}} - G\right) \tag{5.5}$$

When the surface is moist, the net radiation and the evaporation are the largest terms in the surface energy balance (see Chapter 4), so that an accurate measure-

[1]Brutsaert (1982).

ment of the net radiation and approximations to the other terms in the surface energy balance will provide a good estimate of evapotranspiration. Radiation can be measured very accurately and over long periods with relatively inexpensive instrumentation, and most weather stations have a device for measuring insolation. Sensible heat loss can be estimated from bulk aerodynamic formulas if measurements of mean wind speed and temperature at two levels are available. Measurements of temperature profiles in the soil or water can be used to estimate energy storage below the surface.

5.5.2 Evaporation from a Wet Surface

Penman (1948) derived a method of calculating the evaporation from wet surfaces with minimal input data. The Bowen ratio is the ratio of sensible to latent surface energy flux. It may be estimated by using the bulk aerodynamic formulas (4.26) and (4.27).

$$B_o = \frac{\mathrm{SH}}{\mathrm{LE}} \cong \frac{c_p}{L} \frac{(T_s - T_a)}{(q_s - q_a)} \tag{5.6}$$

Here we have assumed that the aerodynamic transfer coefficients for heat and moisture are equal. If the surface air is saturated, and the surface and reference-level air temperatures are not too different, we may make the following approximation:

$$\frac{\left(q_s^* - q_a^*\right)}{(T_s - T_a)} \cong \frac{dq^*}{dT} \tag{5.7}$$

where q^* is the saturation mixing ratio of water vapor. Using (5.7) in (5.6), and the assumption that the surface air is saturated, we obtain

$$B_o = B_e \left(1 - \frac{\left(q_a^* - q_a\right)}{\left(q_s^* - q_a\right)} \right) \tag{5.8}$$

where B_e is the equilibrium Bowen ratio defined in (4.33). It should be noted that the use of the Bowen ratio can be problematical in the presence of temperature inversions if the denominator in (5.7) is near zero.

The surface energy balance (5.5) may be rewritten as

$$E\,(1 + B_o) = E_{\mathrm{en}} \tag{5.9}$$

where

$$E_{\mathrm{en}} = \frac{1}{L}\left(R_s - \Delta F_{\mathrm{eo}} - G\right) \tag{5.10}$$

E_{en} is the evaporation rate necessary to balance the energy supply to the surface by radiation, horizontal flux below the surface, and storage. Substituting (5.8) for the Bowen ratio in (5.9) yields

$$E\left(1+B_e\right)=E_{\mathrm{en}}+E\ B_e\ \frac{\left(q_a^*-q_a\right)}{\left(q_s^*-q_a\right)}. \tag{5.11}$$

Using the bulk aerodynamic formulas, the evaporation may be eliminated from the second term on the right in (5.11) to yield an expression for the evaporation from a wet surface, which is often called Penman's equation.

$$E=\frac{1}{\left(1+B_e\right)}E_{\mathrm{en}}+\frac{B_e}{\left(1+B_e\right)}E_{\mathrm{air}} \tag{5.12}$$

The evaporating capacity of the air is defined by

$$E_{\mathrm{air}}=\rho\ C_{\mathrm{DE}}\ U\left(q_a^*-q_a\right)=\rho\ C_{\mathrm{DE}}\ U\ q_a^*\ (1-\mathrm{RH}) \tag{5.13}$$

and depends on the relative humidity of the air, RH, as well as the air temperature and wind speed.

The advantage of (5.12) is that measurements of atmospheric variables at only one level are required. Over land surfaces the horizontal transport term is zero, and for time scales of a month or longer the storage term can also be ignored, so that only measurements of the net radiation, air temperature, specific humidity, and wind speed at one level are required to evaluate evaporation. The Penman equation (5.12) also shows the relative roles of air humidity and available radiation in driving evaporation over a wet surface. At high temperatures the equilibrium Bowen ratio is small and evaporation is mostly dependent on available energy. As the equilibrium Bowen ratio becomes small, the evaporation rate approaches a value necessary to balance the energy input to the surface. This occurs at temperatures greater than about 25°C. At lower temperatures, and consequently higher equilibrium Bowen ratios, the evaporation rate is more dependent on the supply of unsaturated atmospheric air. At temperatures near or below freezing, the equilibrium Bowen ratio is large and the evaporation is dependent primarily on the drying capacity of the air.

5.5.3 Potential Evaporation

Evapotranspiration is constrained by the surface water supply, the energy available to provide the latent heat of vaporization, and the ability of the surface air to accommodate water vapor. The potential evaporation is defined as the rate of evapotranspiration that would occur if the surface was wet, and is therefore the maximum possible evapotranspiration for the prevailing atmospheric conditions. It measures the effect of energy supply and air humidity on the evapotranspiration rate and avoids the more difficult issue of soil moisture availability and the physiological processes in plants that bring moisture from the soil to the atmosphere. If the potential evaporation exceeds the actual evapotranspiration, then a moisture deficit exists, and one may infer a dry surface. Potential evaporation can be calculated using a variety of theoretical and empirical techniques. One method would be to calculate it from Penman's equation, which relates the evaporation from a wet surface to net radiative heating and mean air temperature, humidity, and wind speed at one level.

5.6 Modeling the Land Surface Water Balance

The water balance of the surface is intimately coupled with the surface energy balance. Over water surfaces, the joint energy–water balance problem is simplified because the air at the surface can always be assumed to be saturated, and the storage and retrieval of water from below the surface is not an issue. Over land surfaces, the heat and water balances are very sensitive to the amount of water below the surface and the rate at which it can be brought to the surface and evaporated or transpired through plants. Transpiration through plants is dependent not simply on the soil moisture and atmospheric conditions, but also on the physiological state of the plant cover.

5.6.1 The Bucket Model of Land Hydrology

The simplest model for the soil water budget is the bucket model. The soil is assumed to have a fixed capacity to store water that is available for evapotranspiration. The rate of change of the mass of water in the soil per unit area w_w, is determined by the rainfall rate P_r, the evapotranspiration rate E, the melting of snow M_s, and the runoff rate Δf.

$$\frac{\partial w_w}{\partial t} = \rho_w \frac{\partial h_w}{\partial t} = P_r - E + M_s - \Delta f \tag{5.14}$$

The amount of available water in the soil can be expressed as an equivalent depth h_w, using a standard water density ρ_w. In the bucket model, the soil is assumed to have a fixed capacity to store moisture, corresponding to an equivalent water depth, h_c, which would typically be about 15 cm. If the soil moisture equals the capacity of the soil, then the soil is assumed to be saturated. If the sum of rainfall plus snowmelt exceeds evaporation when the soil is saturated, then runoff at a rate just sufficient to keep the soil saturated is predicted.

To complete the soil moisture balance model for regions with snowfall, a separate budget for snowcover must be retained. If precipitation occurs when the surface temperature is below freezing P_s, it can be assumed to result in surface snowcover. The snowcover can be measured in terms of its water mass per unit area w_s, or an equivalent depth of water h_s.

$$\frac{\partial w_s}{\partial t} = \rho_w \frac{\partial h_s}{\partial t} = P_s - E_{\text{sub}} - M_s \tag{5.15}$$

The maximum carrying capacity of the surface for snow or ice is determined by the lateral flow of ice sheets and does not become a factor until the ice is several hundred meters thick. Snowcover is removed by sublimation, E_{sub}, or melting. The snowcover lies on top of the soil and does not enter into the soil moisture balance unless it melts. Melting occurs when the surface temperature rises to the freezing point of water. The latent heat of fusion must be supplied to the surface energy balance when melting occurs. Melting continues at the rate necessary to keep the surface temperature from rising above 0°C until the temperature falls below freezing or the snowcover is completely removed.

The rate of evaporation depends on the soil moisture. The soil moisture can be used to relate the actual evaporation to the potential evaporation: the evaporation that would occur if the surface were wet. If measurements of air humidity, air temperature, wind speed, and surface temperature are available, the bulk aerodynamic formula can be used to calculate potential evaporation.

$$\mathrm{PE} = \rho_a \, C_{\mathrm{DE}} \, U \left(q^*(T_s) - q_a \right) \tag{5.16}$$

If insufficient data to evaluate (5.16) are available, then another approximate formula can be used to estimate PE.

The actual evapotranspiration may be related to the potential evaporation and the soil moisture content.

$$E = \beta_E \bullet \mathrm{PE} \tag{5.17}$$

Healthy vegetation may transpire at the rate of potential evaporation, even when the soil is not saturated. When the soil moisture falls below a certain level h_v, the vegetation will no longer transpire at the potential rate. For soil moisture availability less than h_v, it is simplest to assume that β_E varies linearly between zero and one.

$$\beta_E = \begin{cases} 1.0, & h_w \geq h_v \\ \left(\dfrac{h_w}{h_v} \right), & 0 < h_w < h_v \end{cases} \tag{5.18}$$

The simple bucket model can easily be elaborated by adding a deep layer that exchanges water with the upper layer at a slow rate depending on the relative saturation of the two layers. This allows the soil water zone to be replenished with moisture from below without the occurrence of precipitation. In this case an additional budget equation for the deep layer is required, and a term describing the exchange with the deep layer must be added to the soil moisture equation (5.14). A thin layer near the surface can also be added to allow better treatment of short time scales associated with rainstorms or diurnal variations.

5.6.2 More Elaborate Models of Land Surface Processes

To improve significantly on the bucket model of land surface hydrology, much more complex models must be introduced that describe the interactions of the atmosphere with vegetation and soil. Such models must fully couple the momentum, heat, and moisture budgets near the surface and describe each with compatible levels of sophistication and detail. The processes that must be considered include those illustrated in Fig. 5.5. Plants play a central role in the momentum, energy, and moisture transfers at the surface, and must be included in a model. Plants have effects on boundary processes through their physical properties and biological processes, and they have the ability to move water through their leaves at a rapid rate to facilitate photosynthesis when water is available. However, in times of water stress, plants can reduce their transpiration rate by closing their stomata.

The rate at which plants transpire water depends on the availability of photosynthetically active radiation, temperature, air humidity, the availability of water within the plant, and the physiological state of the plant. The rate of water movement through plants is limited by the vapor phase in the leaves, rather than by the uptake of liquid water in the roots. The vapor pressure gradients that drive transpiration are strongly related to leaf temperature. For this and other reasons, leaf temperature is important to model, which requires calculation of the energy transfer through the plant canopy and a heat budget for the leaf structure of the plant. The transfer of solar radiation through the plant canopy is important, since it determines the distribution of heat input throughout the plant canopy and at the soil surface. Much of the insolation will be absorbed by the vegetative cover, rather than by the underlying soil. Because plants and other natural surfaces have very different albedos for visible and near-infrared radiation, these two frequency bands of solar radiation must be treated separately in accurate calculations. The obstacle to free airflow presented by the physical structure of the plant canopy affects the ventilation within the canopy, which is important to the turbulent fluxes of momentum, heat, and moisture. The similarity hypotheses and resulting velocity profiles that lead to the bulk aerodynamic formulas are not valid within the plant canopy. Mean wind speeds and turbulent kinetic energy are much smaller within the canopy than just above it. The air properties within the plant canopy and at the soil surface can be very different from those near the top of the canopy, which is in direct contact with the atmosphere.

In addition to drawing moisture from the soil, plants can also intercept a substantial amount of precipitated water and store it on leaves and stems. The effectiveness of a plant in intercepting rainfall is dependent on the leaf structure, leaf orientation, the leaf area per unit of surface area, and on the frequency and intensity of precipitation. Tall vegetation with a large leaf area index can greatly decrease the supply of moisture to the soil by intercepting precipitation and facilitating its reevaporation before it reaches the ground. Interception losses of this nature can range from 15 to 40% of precipitation for coniferous forests and 10–25% for deciduous forests in midlatitudes. Interception loss is greater if the precipitation rate is low or is intermittent, and less if the precipitation rate is high or continuous. To model interception and storage on leaves, budgets must be calculated for the amount of water stored on the surfaces of leaves. The removal of this surface water by evaporation depends on the supply of energy and unsaturated air at each level within the plant canopy.

The soil is the main reservoir from which evapotranspiration is drawn. Three layers within the soil may be distinguished by their interaction with the atmosphere. A thin layer very near the surface determines the interaction of the atmosphere with the bare ground surface on the time scale of individual precipitation episodes. If this thin layer becomes saturated during a rain shower, then runoff may occur. If this layer becomes dry, then surface evaporation is very small and transpiration by plants becomes the only mechanism whereby water can be efficiently removed from the soil. Below the surface layer is a deeper layer in which the roots of the vegetation reside and draw moisture from the soil. Below the root zone is a deeper layer to which moisture is carried by gravity if the soil is saturated, and from which moisture can be drawn by capillary action.[2]

The ability of soil to hold moisture may be measured by its field capacity, which is defined as the maximum volume fraction of water that the soil can retain against gravity. Typical values for loam are about 30%, but field capacities range from 10% for sand to 55% for peat. If the volumetric water content of the soil falls below a certain level, plants are unable to draw moisture from the soil and will remain wilted at all times of day. This permanent wilting threshold is typically one-third to one-half of the field capacity. The soil moisture available to plants is the difference between the volumetric soil water fraction and the wilting threshold. The maximum available soil moisture is the difference between the field capacity and the wilting threshold, and is 15–20% for typical soils.

The hydraulic conductivities of most soils decrease rapidly as the soil dries out, so that conduction of water becomes very slow if the water content is much below field capacity. If the surface layer of the soil becomes very dry, then infiltration of subsequent rainfall may be inhibited. Similarly, the soil in the root zone may become very dry, while the soil moisture several centimeters below the deepest roots remains near field capacity. The amount of water that is available to the vegetation is thus approximately equal to the available volumetric soil water fraction times the rooting depth. If the roots extend down about one meter, and the available field capacity is about 15%, then the total amount of water available to plants when the soil is saturated is equivalent to about 15-cm depth of water. This may be greater or less depending on the type of soil and vegetation. Plants that live in sandy soil tend to develop deep roots, which offsets the low volumetric fraction of available water in sand.

5.7 Annual Variation of the Terrestrial Water Balance

The annual variation of the surface water balance at a location is intimately related to the local climate and its potential for human habitation and agriculture. The natural vegetation is adapted to the normal cycle of water surplus and water shortage that a region experiences. The annual variation of the water balance can be used as one

[2]Bras (1990).

means of classifying climates.[3] The water balance depends on the annual variation of precipitation and evaporation, which together largely determine the soil moisture.

The accuracy with which the water balance can be determined depends on the amount and quality of the data available. Ideally one would require good measurements of evaporation, precipitation, and soil moisture. Evaporation and soil moisture are not routinely measured at most locations, however. Most climatological stations report surface air temperature and humidity, precipitation, and wind speed. With these variables potential evaporation can be estimated from semiempirical formulas like (5.12), (5.16), or their simplified forms. If a soil moisture balance is calculated using the bucket model, then the actual evapotranspiration can be estimated from the potential evapotranspiration using (5.17). Figure 5.6 shows results of such a water balance analysis for a variety of locations.

The west coast of North America in middle latitudes experiences a wintertime maximum in precipitation associated with winter storms [Fig. 5.6(a–c)]. The wintertime peak of precipitation is smaller and occurs later in southern California than on the Pacific Coast of Canada. In Juneau, Alaska the monthly precipitation peaks in October, whereas in Los Angeles it does not peak until January or February. The summertime minimum of precipitation becomes deeper and of longer duration toward the south. Because the summer minimum of precipitation corresponds with the season of strongest insolation and warmest temperatures, the soil moisture can become depleted in the summer months. At Seattle, evapotranspiration exceeds precipitation beginning in May, and by July the soil is sufficiently dry that the potential evaporation exceeds the actual evapotranspiration. This continues until October, when the precipitation again exceeds the evapotranspiration. In San Francisco the dry season begins in June and extends until November. In Los Angeles the soil is nearly always dry, and only during the months of December through February does the actual evapotranspiration approach the potential evaporation. Acapulco, Mexico is in the tropics (~17°N), and its precipitation maximum comes in the summer months, when tropical convection reaches northward from the equatorial region [Fig. (5.6(d)]. During the heavy precipitation months of June through October, the precipitation exceeds the potential evaporation. The remainder of the year is very dry. Because of the relatively constant temperature and humidity in the tropics, the potential evaporation varies little during the year.

In much of the interior of North America the precipitation peaks in the spring or early summer when the soil is moist and temperatures are high enough to allow the air to carry large amounts of water vapor [Fig. 5.6(e–h)]. As the summer season progresses, the soil dries out and the precipitation amount decreases. During middle and late summer the potential evaporation generally exceeds the actual evaporation, indicating relatively dry soil. At Sioux City, Iowa the precipitation peaks in June and continues into the middle and late summer. The combination of warmth, insolation, and precipitation during the summer in this part of the American Midwest makes it

[3]Thornthwaite (1948).

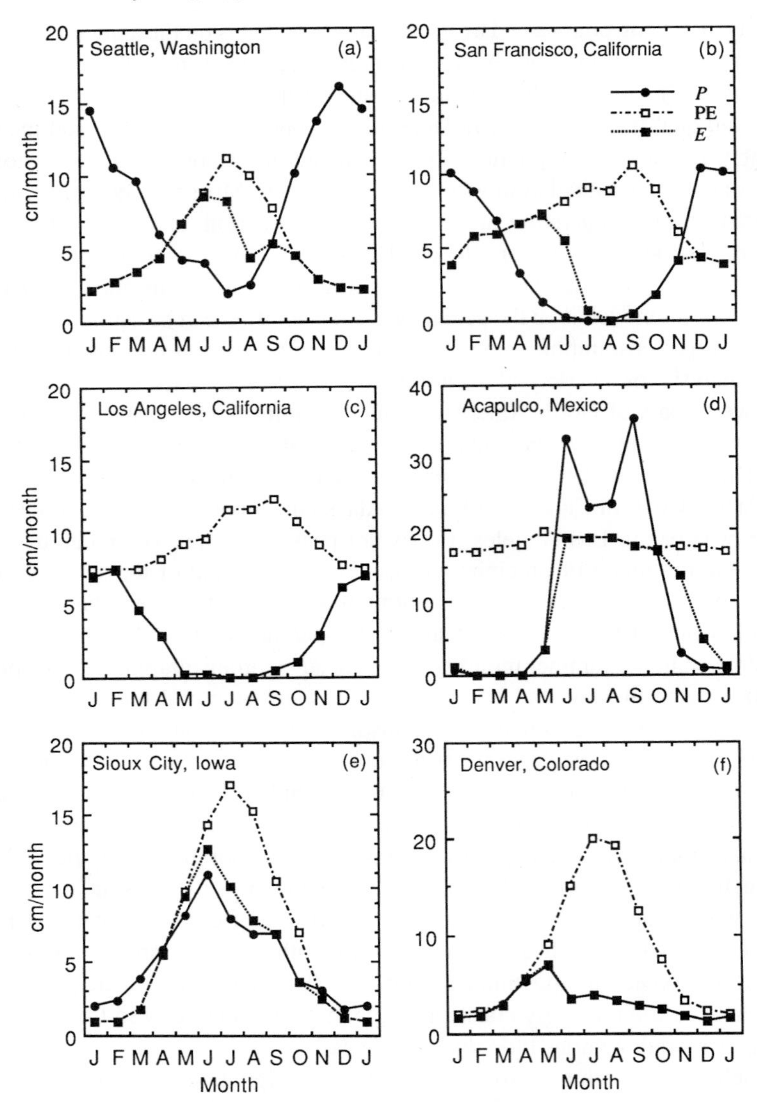

Fig. 5.6 Annual cycle of the water balance at various locations. Where precipitation does not appear, it is equal to evaporation (e.g., Los Angeles, Phoenix). Where potential evaporation does not appear, it is equal to evaporation (e.g., Churchill). [Data from Eagleman (1976).]

well suited for agriculture. Farther west and at the higher elevation of Denver, the precipitation is smaller and peaks earlier in the year. The air is dry and warm in the summer months and the potential evaporation greatly exceeds precipitation. At Phoenix, Arizona the precipitation is small at all times of year compared to the high

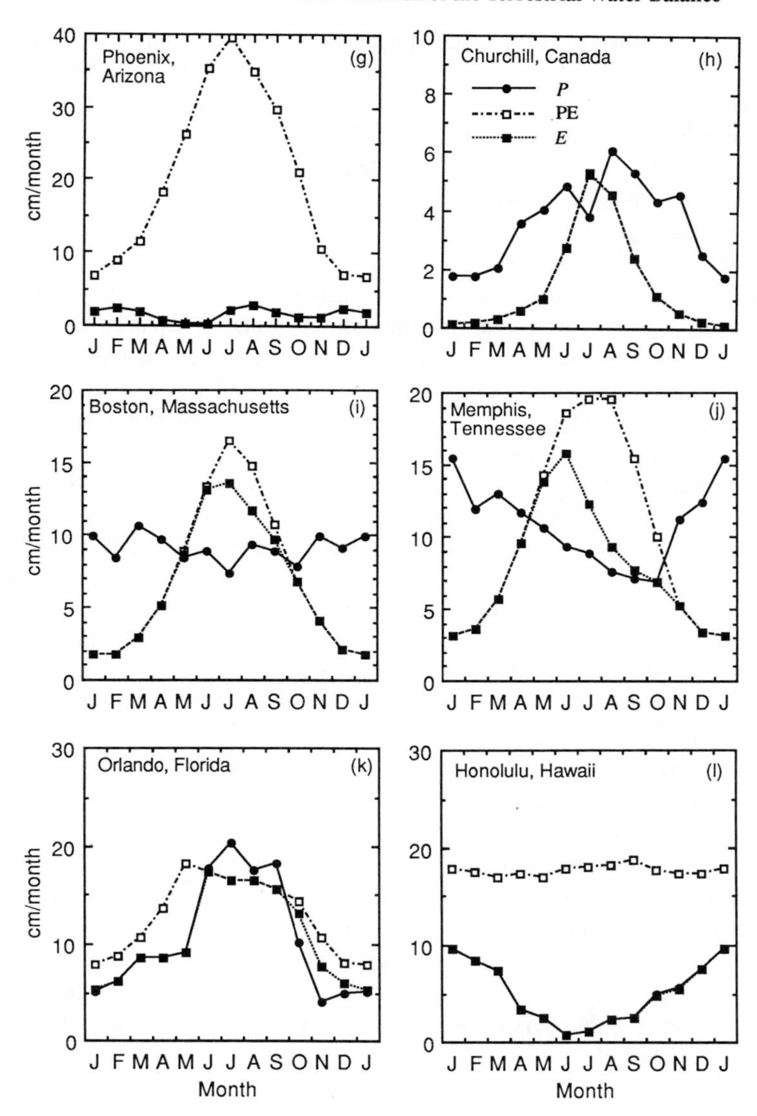

Fig. 5.6—*Continued*

drying capacity of the warm, dry air there. The potential evaporation greatly exceeds the precipitation at all seasons, indicating a desert climate.

In high latitudes the low saturation vapor pressure associated with the relatively cold temperature constrains the rates of evaporation and precipitation. At many locations the precipitation is greater than the potential evaporation during most of the year. Water evaporated at warmer latitudes is transported into high latitudes by atmospheric motions and is precipitated when large-scale motions drive saturated air upward. The energy available at the surface is insufficient to allow the evaporation of this water. The growing season is short, so that the vegetation is not especially effective in bringing water from the soil to the atmosphere. Soils in high latitudes therefore typically have a high water content. At very high latitudes this water is mostly frozen. Churchill, Canada (~59°N) is an example of a cold continental climate where the evapotranspiration is always energy limited and exceeds the precipitation only during July [Fig. 5.6(h)].

In the northeastern United States the monthly precipitation amount is almost independent of season. The frontal precipitation of winter is replaced in summer by more convective precipitation, such that the total precipitation remains almost constant. The potential evapotranspiration follows the insolation and temperature and peaks in the summer. During the summer months the evaporation exceeds the precipitation, causing the soil to dry out somewhat. In Boston, Massachusetts the potential evaporation exceeds the actual evapotranspiration from July to September [Fig. 5.6(i)]. In the southeastern United States, as characterized by Memphis, Tennessee, a modest wintertime peak in precipitation is observed. Farther south on the Florida peninsula, Orlando shows a summertime maximum in precipitation, with very heavy precipitation from June to September. This precipitation is mostly associated with thunderstorms, which are driven by solar heating of the land and begin during the hottest part of the day. Honolulu is in the subtropics where potential evaporation exceeds precipitation at all seasons [Fig. 5.6(l)]. Precipitation peaks in winter and is balanced by evaporation in all seasons.

In general there are four basic types of water balance cycles, although many locations show a combination of several types. The west coast of North America in middle latitudes has a winter precipitation maximum and summer dry period. The interior of the continent experiences a spring or summer rainfall maximum, followed by a drying period of varying intensity in the late summer. On the east coast of North America in midlatitudes the precipitation amount is almost independent of season, but the potential evaporation peaks in the summer, producing some reduction in soil moisture. In tropical or subtropical latitudes the precipitation is very small in winter, but thunderstorms yield large amounts of rain during the warmest part of the summer.

Exercises

1. The approximate volume of water retained in soil moisture and groundwater is given in Table 5.1. Use the data in Fig. 5.1 to calculate the time it would take for precipitation over land to deliver an amount of water equal to the soil water and groundwater. How long would it take to replace the groundwater and soil moisture if only 10% of the runoff could be redirected to replenishing the groundwater?

2. Use the bulk aerodynamic formula (4.32) to calculate the evaporation rate from the ocean, assuming that $C_{DE} = 10^{-3}$, $U = 5$ m s^{-1}, and that the reference-level air temperature is always 2°C less than the sea surface temperature. Calculate the evaporation rate for (a) $T_s = 0$°C, $q_s{}^* = 3.75$ g kg^{-1}, RH = 50%; (b) $T_s = 0$°C, $q_s{}^* = 3.75$ g kg^{-1}, RH = 100%; (c) $T_s = 30$°C, $q_s{}^* = 27$ g kg^{-1}, RH = 50%; and (d) $T_s = 30$°C, $q_s{}^* = 27$ g kg^{-1}, RH = 100%. Assume a fixed density of 1.2 kg m^{-3}. How would you evaluate the importance of relative humidity versus the importance of surface temperature for determining the evaporation rate?
3. Calculate the Bowen ratio using the bulk aerodynamic formulas for surface temperatures of 0, 15, and 30°C, if the relative humidity of the air at the reference level is 70% and the air – sea temperature difference is 2°C. Assume that the transfer coefficients for heat and moisture are equal.
4. Use the results of problem 3 to explain why high-latitude land areas often have high surface moisture content.
5. Why is local winter and spring snow accumulation important for the summer soil moisture of midlatitude continental land areas? How do you think the August climate would change if the winter and spring snowfall were replaced by rainshowers?
6. What are some of the shortcomings of the bucket model of land hydrology? How are these limitations addressed by more sophisticated models for land surface processes?
7. Derive (5.12) using the method outlined in the text.

Chapter 6 Atmospheric General Circulation and Climate

6.1 The Great Communicator

The movement of air in the atmosphere is of critical importance for climate. Atmospheric motions carry heat from the tropics to the polar regions and thereby reduce the extremes of temperature that would otherwise result. Water from the oceans is evaporated and carried in the air to land, where rainfall supports plant and animal life. Winds supply momentum to ocean surface currents that transport heat and oceanic trace constituents such as salt and nutrients. The circulation of the atmosphere is a key component of the climate, since it both responds to temperature and humidity gradients and helps to determine them by transporting energy and moisture. The atmosphere provides the most rapid communication between geographic regions within the climate system.

The global system of atmospheric motions that is generated by the uneven heating of Earth's surface area by the sun is called the *general circulation*. A complete description of the atmosphere's general circulation includes mean winds, temperature and humidity, the variability of these quantities, and the covariances between wind components and other variables that are associated with large-scale weather systems. A statistical description of the general circulation of the atmosphere has been constructed from the ensemble of daily global flow patterns estimated with data from balloons and satellites. The general circulation of the atmosphere can be simulated by solving the equations of motion on a computer, and such general circulation models form a part of global climate models.

6.2 Energy Balance of the Atmosphere

Atmospheric motions are generated by geographic variations in heating of the surface caused by meridional gradients of insolation, albedo variations, and other factors. By transporting energy, winds generally act to offset the effects of these heating variations on atmospheric temperature. The local energy balance of an atmospheric column of unit horizontal area includes the effects of radiation, sensible heat exchange with the surface, condensation heating, and the horizontal flux of energy in the atmosphere.

$$\frac{\partial E_a}{\partial t} = R_a + \mathrm{LP} + \mathrm{SH} - \Delta F_a \tag{6.1}$$

where $\partial E_a / \partial t$ is the time rate of change of the energy content of an atmospheric column of unit horizontal area extending from the surface to the top of the atmosphere, R_a is the net radiative heating of the atmospheric column, LP is the heating of the atmospheric column by latent heat release during precipitation, SH is the sensible heat transfer from the surface to the atmosphere, and ΔF_a is the horizontal divergence of energy out of the column by transport in the atmosphere.

The net radiative heating of the atmosphere is the difference between the net radiative heating at the top of the atmosphere and the net radiation at the ground.

$$R_a = R_{\mathrm{TOA}} - R_s \tag{6.2}$$

The storage of energy in the atmosphere is negligible, particularly when averaged over a year, so that the atmospheric energy balance is the sum of radiative heating, sensible heating, and latent heating, balanced against the export of energy by atmospheric motions.

$$R_a + \mathrm{LP} + \mathrm{SH} = \Delta F_a \tag{6.3}$$

The annually and zonally averaged net effect of radiative transfer on the atmosphere is a cooling of about -90 W m^{-2}, which is nearly independent of latitude (Fig. 6.1). The radiative cooling corresponds to an atmospheric temperature decrease of about 1.5°C per day (see Chapter 3). The energy lost from the atmosphere in one

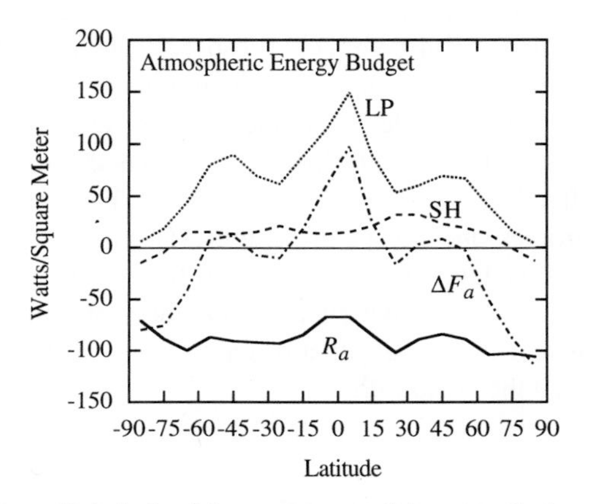

Fig. 6.1 Distribution with latitude of the components of the atmospheric energy balance averaged over longitude and over the annual cycle. Units are W m^{-2}. [Data from Sellers (1965). Used with permission from the University of Chicago Press.]

week through radiative transfer equals about 2.5% of the atmosphere's total energy content. If only the atmosphere's thermal capacity were considered, this cooling rate, acting alone, would bring the global mean surface air temperature to below freezing in about 2 weeks. Under normal circumstances the radiative cooling is balanced in the global mean by condensation heating and sensible heat transfer from the surface.

Heating of the atmosphere by the transfer of sensible heat from the surface is relatively small. The largest contribution to balancing the radiative loss from the atmosphere is the release of latent heat of vaporization during precipitation. In contrast to the radiative cooling, the condensation heating has a very distinctive structure with latitude corresponding to that of precipitation. It peaks at about 150 W m^{-2} near the equator, drops to near 50 W m^{-2} in the subtropics, peaks again near 80 W m^{-2} in midlatitudes, and then decreases sharply to near zero at the poles. The latitudinal structure of the precipitation is reflected in the latitudinal structure of the atmospheric energy flux divergence. Atmospheric motions export close to 100 W m^{-2} from the equatorial region, have a relatively small net effect on the energy balance between about 20 and 60 degrees of latitude, and import about 100 W m^{-2} into the polar regions. The poleward transport of energy by the atmosphere has a broad, flat maximum between the equatorial and polar regions. This poleward transport of energy by the atmosphere is one of the important climatic effects of the general circulation of the atmosphere.

6.3 Atmospheric Motions and the Meridional Transport of Energy

Motions in the atmosphere can be associated with many physical phenomena, which have a wide variety of space and time scales (Fig. 6.2). Small-scale phenomena such as turbulence and organized mesoscale phenomena such as thunderstorms are effective primarily at transporting momentum, moisture, and energy vertically. Only very large-scale phenomena such as extratropical cyclones, planetary-scale waves, and slow meridional circulations that extend over thousands of kilometers are effective at transporting momentum, heat, and moisture horizontally between the tropics and the polar regions. The upward flux of energy and moisture in the boundary layer and the poleward flux of energy by planetary-scale circulations in the atmosphere have equal importance for climate. These phenomena have characteristic spatial scales that differ by nearly 10 orders of magnitude: from millimeters to 10 thousand kilometers.

6.3.1 Wind Components on a Spherical Earth

Wind velocities in the atmosphere are measured in terms of a local Cartesian coordinate system inscribed on a sphere. At each latitude (ϕ) and longitude (λ) on a sphere

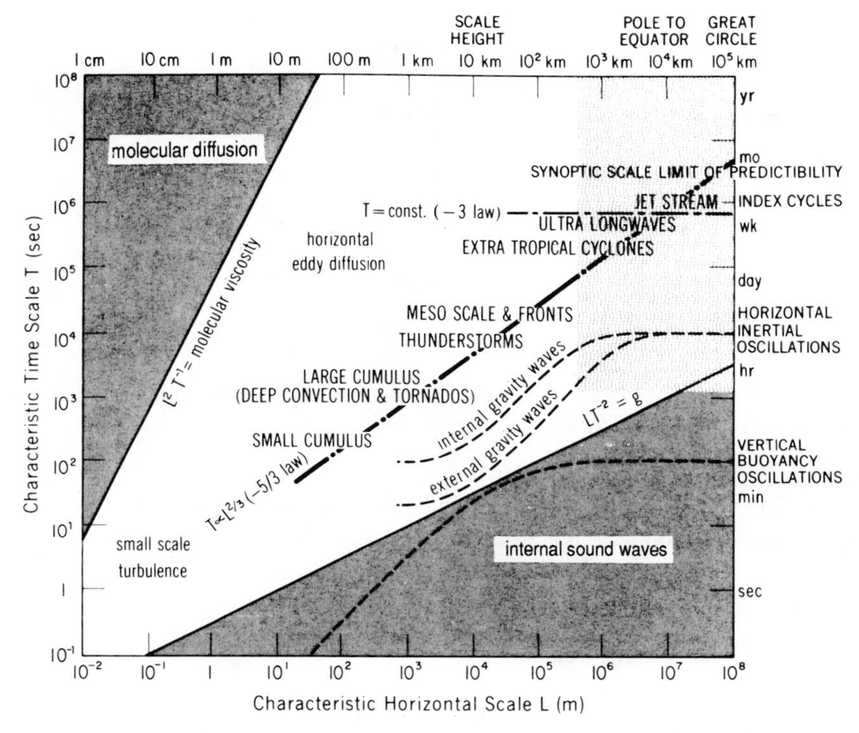

Fig. 6.2 Diagram showing the space and time scales of phenomena in the atmosphere. Light shading represents approximately scales that can be resolved in climate models. [From Smagorinsky (1974).]

of radius a, the zonal and meridional components of horizontal velocity are defined in the following way (Fig. 6.3):

$$
\begin{aligned}
u &= a\cos\phi\,\frac{D\lambda}{Dt} = \text{zonal or eastward wind speed} \\
v &= a\,\frac{D\phi}{Dt} = \text{meridional or northward wind speed}
\end{aligned}
\tag{6.4}
$$

Here D/Dt represents the material derivative—the temporal tendency that is experienced by an air parcel moving with the flow. The vertical component of velocity can be measured in terms of the rate of change of altitude, or the rate of change of pressure following the motion of air parcels.

$$
\begin{aligned}
w &= \frac{Dz}{Dt} = \text{rate of change of altitude following an air parcel} \\
\omega &= \frac{Dp}{Dt} = \text{rate of change of pressure following an air parcel}
\end{aligned}
\tag{6.5}
$$

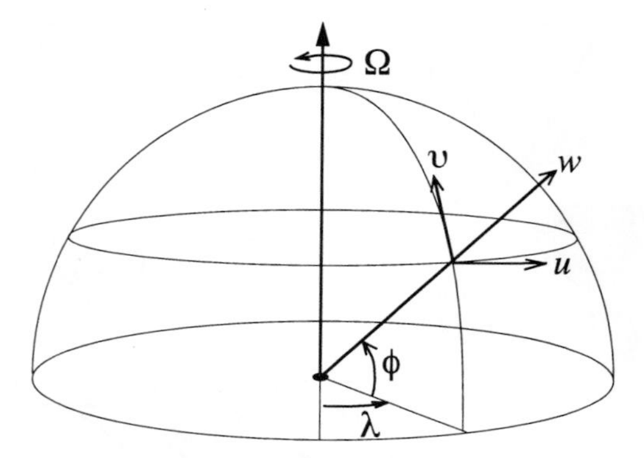

Fig. 6.3 Diagram showing local Cartesian coordinates on a sphere and the zonal (u), meridional (υ), and vertical (w) components of the local vector wind velocity.

The vertical velocity and the pressure velocity are related to each other through an approximate equation, which is valid if a hydrostatic balance is maintained.

$$\omega \cong -\rho g w \tag{6.6}$$

6.3.2 The Zonal-Mean Circulation

In describing the circulations of the atmosphere it is convenient to consider the zonal average, which is the average over longitude, λ, at a particular latitude and pressure, and is represented with square brackets.

$$[x] = \frac{1}{2\pi} \int_0^{2\pi} x \, d\lambda \tag{6.7}$$

Because of the relatively rapid rotation of Earth, and because diurnally averaged insolation is independent of longitude, averaging around a latitude circle captures a physically meaningful subset of the climate. For climatological purposes, we are normally interested in averages over a period of time, Δt, that is long enough to average out most weather variations. This time interval may correspond to a particular month, season, or year, or it may be an average over an ensemble of many months, seasons, or years.

$$\overline{x} = \frac{1}{\Delta t} \int_0^{\Delta t} x \, dt \tag{6.8}$$

Climatological zonal averages are usually obtained by averaging over both longitude and time.

The distribution of the zonal mean of the eastward component of wind, $[u]$, through latitude and height is one of the best known characterizations of the global

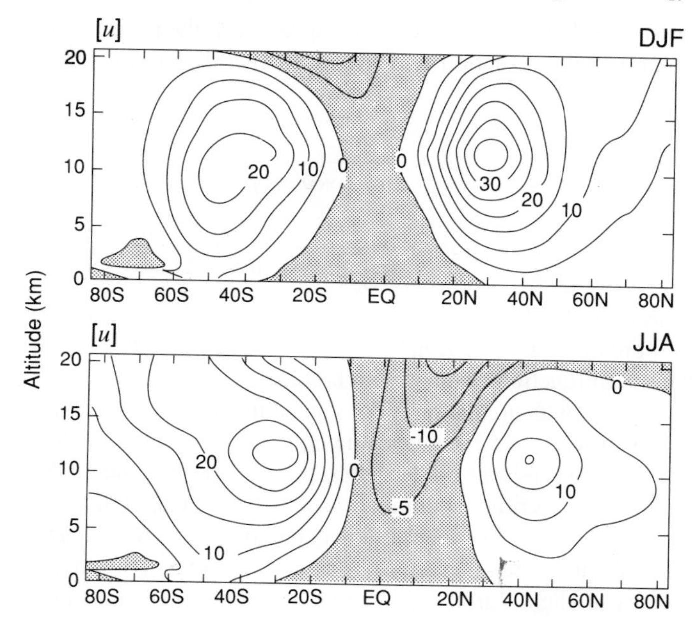

Fig. 6.4 Latitude–height cross section of zonal-average wind speed for DJF and JJA. Contour interval is 5 m s^{-1}; easterly values are shaded. [Data from Oort (1983).]

atmospheric circulation, and is often called the *zonal-mean wind* (Fig. 6.4). In meteorology, winds are called westerly when they flow from west to east and easterly when they flow from east to west. The zonal-mean wind is westerly through most of the troposphere, and peaks at speeds in excess of 30 m s^{-1} in the subtropical jet stream, which is centered near 30 degrees of latitude and at an altitude of about 12 km. The subtropical jet stream is strongest in the winter season. The zonal winds at the surface are westerly at most latitudes between 30 and 70 degrees, but in the belt between 30°N and 30°S zonal-mean easterly surface winds prevail.

The zonal-average meridional and vertical components of wind are much weaker than the zonal wind. Maximum mean meridional winds are only about 1 m s^{-1}, and mean vertical wind speeds are typically a hundred times smaller than the mean meridional wind. The *mean meridional circulation* (MMC), which is composed of the zonal-mean meridional and vertical velocities, can be described by a mass streamfunction, which is defined by calculating the northward mass flux above a particular pressure level, p.

$$\Psi_M = \frac{2\pi a \cos\phi}{g} \int_0^p [v]\, dp \tag{6.9}$$

The mass flow between any two streamlines of the mean meridional streamfunction is equal to the difference in the streamfunction values. The conservation of mass for the zonal-mean flow implies a relationship between the mass streamfunction for the mean meridional circulation and the mean meridional velocity and pressure velocity.

$$[v] = \frac{g}{2\pi a \cos\phi} \frac{\partial \Psi_M}{\partial p} \tag{6.10}$$

$$[\omega] = \frac{-g}{2\pi a^2 \cos\phi} \frac{\partial \Psi_M}{\partial \phi} \tag{6.11}$$

The mean meridional velocity thus depends on the rate at which the streamfunction changes with pressure, and the zonal-average pressure velocity depends on the rate at which the streamfunction changes with latitude.

The mean meridional circulation is dominated in the solstitial seasons by a single circulation cell in which air rises near the equator, flows toward the winter hemisphere at upper levels, and sinks in the subtropical latitudes of the winter hemisphere (Fig. 6.5). This mean meridional circulation cell is often called the *Hadley cell* after George Hadley, who in 1735 proposed it as an explanation for the tradewinds. The mean meridional winds near the surface bring air back toward the equator. The rising branch is displaced slightly into the summer hemisphere. The mean meridional circulations for the equinoctal seasons and for the annual average consist of two smaller cells of about equal strength located on opposite sides of the equator.

In midlatitudes weaker cells called *Ferrel cells* circulate in the opposite direction to the Hadley cell. In these midlatitude mean meridional circulation cells, rising occurs in cold air and sinking in warmer air. These cells are therefore thermodynamically indirect, in that they transport energy from a cold area to a warm area. The mean meridional circulation is a small component of the total flow in midlatitudes, and the Ferrel cells are a byproduct of the very strong poleward transport of energy by eddy circulations. Eddies are the deviations from the time or zonal average, and are a key component of the general circulation of the atmosphere.

6.3.3 Eddy Circulations and Meridional Transport

The cyclones and anticyclones that are responsible for most of the weather variations in midlatitudes produce large meridional transports of momentum, heat, and moisture. These disturbances have large wind and temperature variations on scales of several thousand kilometers, which do not appear in a zonal average, but have a profound effect on the zonal-mean climate. The fluctuations associated with weather appear as deviations from the time average.

$$x' = x - \overline{x} \tag{6.12}$$

In addition to temporal variations associated with midlatitude cyclones, the atmosphere exhibits variations around latitude circles associated with continents and

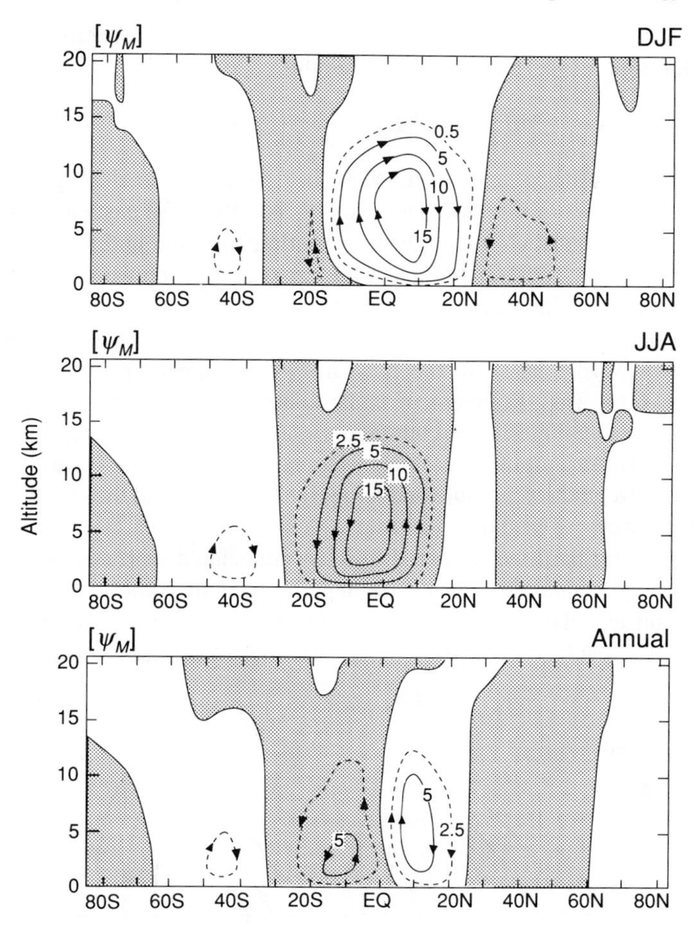

Fig. 6.5 Latitude–height cross section of the mean meridional mass circulation. Shaded values are negative; units are 10^{10} kg s^{-1}. [Data from Oort (1983).]

oceans that are quasistationary and appear clearly in time averages. These are characterized by the deviations of the time mean from its zonal average.

$$\overline{x}^* = \overline{x} - [\overline{x}] \tag{6.13}$$

Northward eddy fluxes of temperature are produced when northward-flowing air is warmer than southward-flowing air, so that, when averaged over longitude, the product of meridional velocity and temperature is positive, even when the mean meridional wind is zero. Using the definitions of the time and zonal averages, the northward transport of temperature averaged around a latitude circle and over time can be written as the sum of contributions from the mean meridional circulation, the stationary eddies, and the transient eddies, which are shown respectively as the three terms on the right of (6.14).

$$\left[\overline{vT}\right] = \left[\overline{v}\right]\left[\overline{T}\right] + \left[\overline{v}^{*}\overline{T}^{*}\right] + \left[\overline{v'T'}\right] \tag{6.14}$$

Transient eddy fluxes are associated with the rapidly developing and decaying weather disturbances of midlatitudes, which generally move eastward with the prevailing flow and contribute much of the variations of wind and temperature, especially during winter. These disturbances are very apparent on weather maps.

The positive correlation between meridional velocity and temperature in large-scale atmospheric waves results from the tendency of the temperature wave to be displaced westward relative to the pressure wave, especially in the lower troposphere (Fig. 6.6). This arrangement is associated with a conversion from energy available in the mean meridional temperature gradient to the energy of waves. Cyclone waves whose amplitude is increasing rapidly with time have a large zonal phase shift between their pressure and temperature waves, and thus produce efficient poleward transports of heat and moisture.

Eddy fluxes by the time-averaged flow are associated with stationary planetary waves. *Stationary planetary waves* are departures of the time average from zonal symmetry and are plainly visible in monthly mean tropospheric pressure patterns (Fig. 6.7). They result from the east–west variations in surface elevation and surface

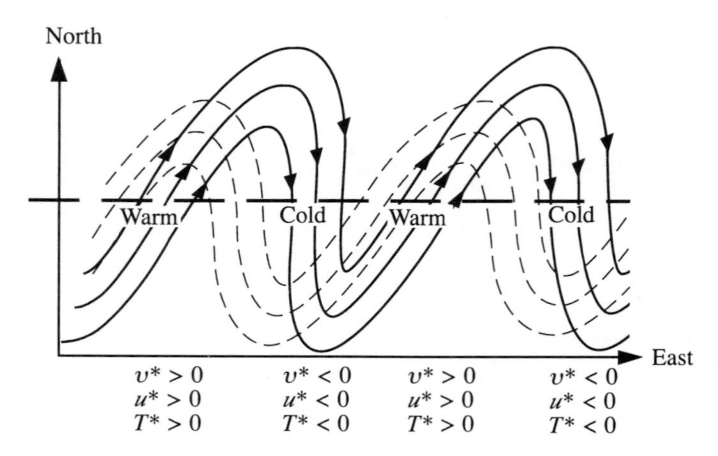

Fig. 6.6 Schematic of the streamlines (solid) and isotherms (dashed) associated with a large-scale atmospheric disturbance in midlatitudes of the Northern Hemisphere. Arrows along the streamline contour indicate the direction of wind velocity. The streamlines correspond approximately to lines of constant pressure, since the winds are nearly geostrophic. The signs of the deviations of the wind components from their zonal-average values are shown to illustrate that the NE–SW tilt of the streamlines indicates a northward zonal momentum transport, and the westward phase shift of the temperature wave relative to the pressure wave gives a northward heat transport.

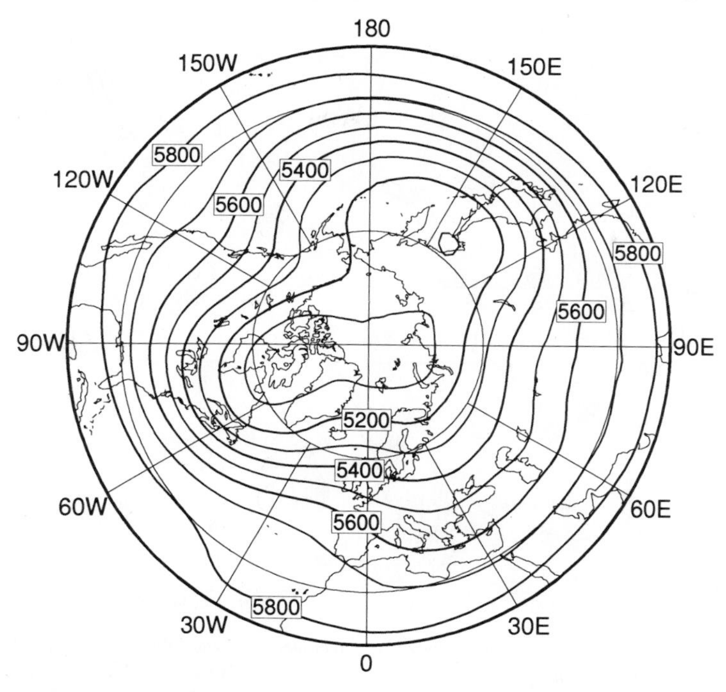

Fig. 6.7 Average height of the 500-mb pressure surface during January in the Northern Hemisphere. Contour interval is 100 m.

temperature associated with the continents and oceans. Stationary eddy fluxes are largest in the Northern Hemisphere where the Himalaya and Rocky mountain ranges provide mechanical forcing of east–west variations in the time-mean winds and temperatures. The thermal contrast between the warm waters of the Kuroshio and Gulf Stream ocean currents and the cold temperatures in the interiors of the continents also provides strong thermal forcing of stationary planetary waves during winter.

The poleward fluxes of temperature by stationary and transient eddies peak at about 50 degrees of latitude in the winter hemisphere in the lower part of the troposphere (Fig. 6.8). The low-level maximum is associated with the structure of growing extratropical cyclones, in which the phase difference between temperature and pressure is largest in the lower troposphere. The fluxes exhibit a minimum near the tropopause and then increase with height into the winter stratosphere. Temperature fluxes have a large seasonal variation in the Northern Hemisphere, with large values in winter and fairly small values during summer. In the Southern Hemisphere the seasonal contrast is less. Transient eddy fluxes dominate the meridional flux of temperature except in the Northern Hemisphere during winter, when stationary eddies contribute up to half of the flux.

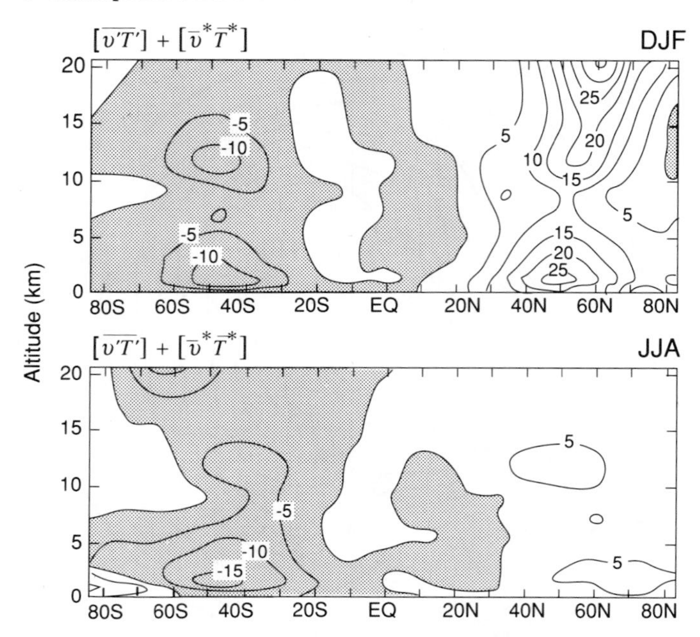

Fig. 6.8 Meridional cross section of the zonally averaged northward flux of temperature by eddies. Note that in the Southern Hemisphere the poleward fluxes are negative as a result of our arbitrarily defining north as the positive direction. Contour interval is 5 K m s^{-1}. [Data from Oort (1983).]

6.3.4 Vertically Averaged Meridional Energy Flux

Four types of atmospheric energy are important for determining the meridional transport of energy (Table 6.1). Internal energy is the energy associated with the temperature of the atmosphere, and potential energy is the energy associated with the gravitational potential of air some distance above the surface. Together internal and potential energy constitute about 97% of the energy of the atmosphere. Although kinetic energy comprises a small fraction of the total energy, it is still very important to understand its gen-

Table 6.1

Kinds and Amounts of Energy in the Global Atmosphere

Name	Symbol	Formula	Amount × 10^6 J m^{-2}	% of total
Internal energy	IE	$c_v T$	1800	70
Potential energy	PE	gz	700	27
Latent energy	LH	Lq	70	2.7
Kinetic energy	KE	$\frac{1}{2}(u^2 + v^2)$	1.3	0.05
Total energy	IE + PE + LH + KE		2571	100

eration and maintenance, because the motions are the means by which energy is transported from equator to pole. Motions are also important in converting one form of energy to another. Furthermore, most of the internal and potential energy is unavailable for conversion into other forms. For example, in a dry, hydrostatic atmosphere without mountains, one can show that the ratio of the potential energy to the internal energy is $R/c_v = 0.4$. This simple relation between internal and potential energy reflects the fact that much of the internal energy of the atmosphere is required simply so that the atmosphere may "hold itself up" against gravity, and is not available for generating motion.

Insolation drives the circulation by heating the tropics more than the polar regions. Winds are driven by the density and pressure gradients generated by this uneven heating. The circulation responds not to the total amount of energy in the atmosphere, but to the temperature gradients on constant pressure surfaces. For this reason the maximum kinetic energy occurs during winter, when the meridional temperature gradients are strongest, and not in the summer, when the total amount of energy in the atmosphere is greatest.

The meridional transport of energy by the atmosphere may be divided into contributions from the mean meridional circulation and the eddy or wave motions that are superimposed on the zonal-mean flow. These may be integrated through the mass of the atmosphere to reveal the total meridional flux of energy in various forms by the mean meridional circulation and the eddies (Fig. 6.9). The meridional energy flux by the mean meridional circulation is mostly confined to the tropics, where the mass flux associated with the Hadley cell is large. The net fluxes of sensible and latent heat by the Hadley cell are both equatorward, and peak near 10 degrees latitude in the winter hemisphere. The equatorward flow near the surface brings warm moist air with it. Heavy precipitation occurs where warm moist air converges in the vicinity of the equator. The release of latent heat and the convergence of sensible heat flux drive strong rising motion in the upward branch of the Hadley cell. In the upward branch of the Hadley cell latent and internal energy are converted into potential energy. The poleward flow of potential energy in the upper branch of the Hadley cell exceeds the sum of the equatorward flow of latent and internal energy in the lower branch, giving a small net poleward flow of energy.

The poleward flow of energy in the Hadley cell can more easily be seen by considering the moist static energy, which is the sum of sensible, potential, and latent energy (Fig. 6.10).

$$\text{Moist static energy} = c_p T + g z + L q = \text{sensible} + \text{potential} + \text{latent} \quad (6.15)$$

where q is the mass-mixing ratio of water vapor and L is the latent heat of vaporization. In a stably stratified atmosphere, moist static energy increases with altitude, so that a mean meridional circulation cell will transport energy in the direction of flow in the upper branch of the cell. The net transport of energy is generally much less than the transport of any individual type of energy. The net transport in the Hadley cell is only about 10% of the potential energy transport. Mean meridional circulation cells are not a particularly efficient means of poleward energy transport, especially in view of the strong constraints the angular-momentum balance places on these circulation cells.

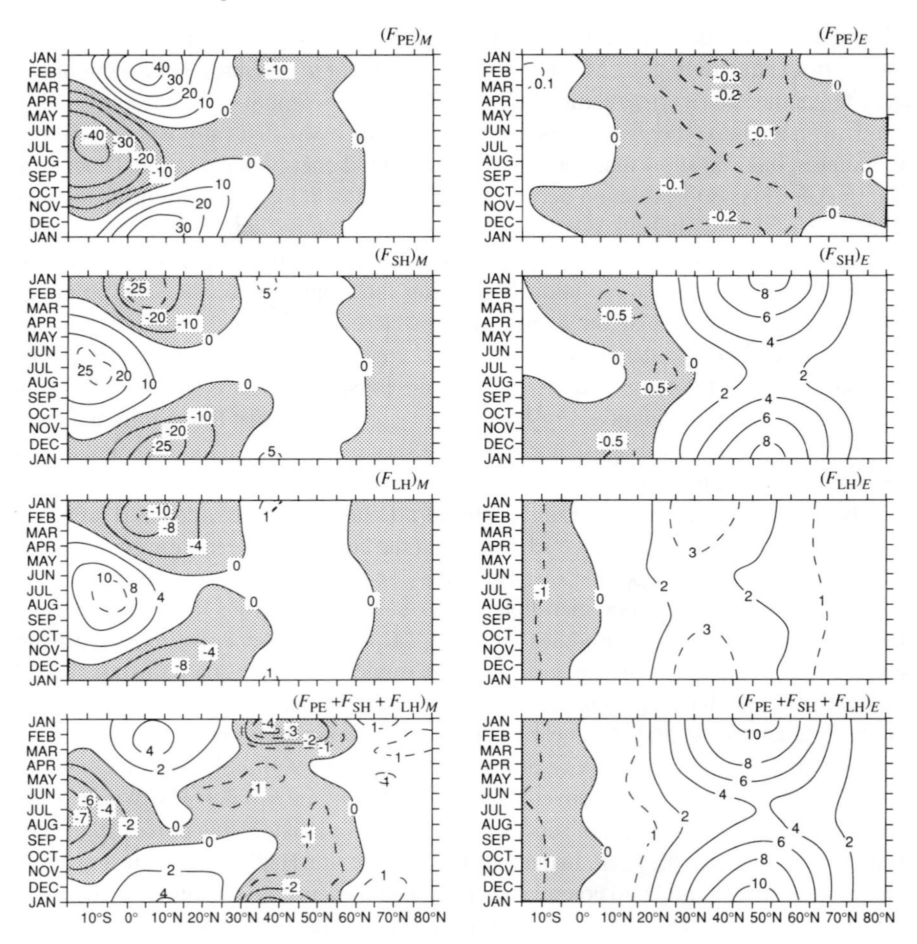

Fig. 6.9 Northward fluxes of potential energy (F_{PE}), sensible heat (F_{SH}), latent heat (F_{LH}), and total energy by the atmosphere as functions of latitude and season. Panels on the left show fluxes by the mean meridional circulation and those on the right by eddy circulations. Units are 10^{19} calories per day and southward fluxes are shaded. [From Oort (1971). Reprinted with permission from the American Meteorological Society.]

Fluxes by eddies are largest in midlatitudes and peak strongly in winter when the radiative forcing of meridional temperature gradients is greatest (Fig. 6.9). Eddy transport of potential energy is small, but large fluxes of latent and sensible heat are both poleward. Latent fluxes peak near 30 degrees latitude, where the temperatures are warmer and the specific humidity is greater than at higher latitudes. Eddy fluxes of sensible heat peak near 50 degrees, where they constitute about 80% of the total poleward energy flux. The meridional flux in midlatitudes is dominated by the transient eddy flux, which is the flux associated with traveling disturbances (Fig. 6.11).

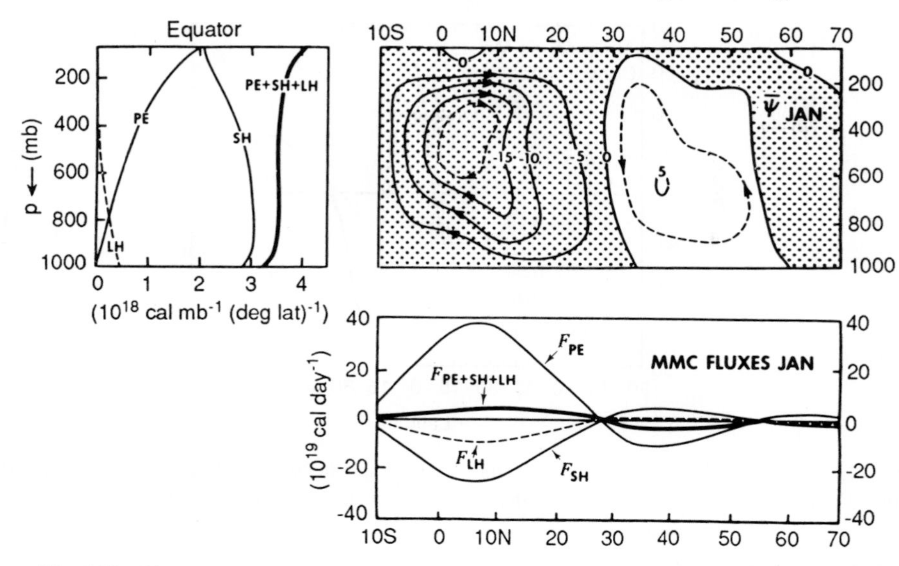

Fig. 6.10 Diagrams explaining the transport of energy by the mean meridional circulation using the month of January. The vertical distribution of potential energy (PE), sensible heat (SH), latent heat (LH), and total moist static energy (PE + SH + LH) near the equator (upper left). The mean meridional mass streamfunction for January contoured in pressure–latitude cross section (upper right). The northward fluxes of various energy types by the mean meridional circulation (lower right). [From Oort (1971). Reprinted with permission from the American Meteorological Society.]

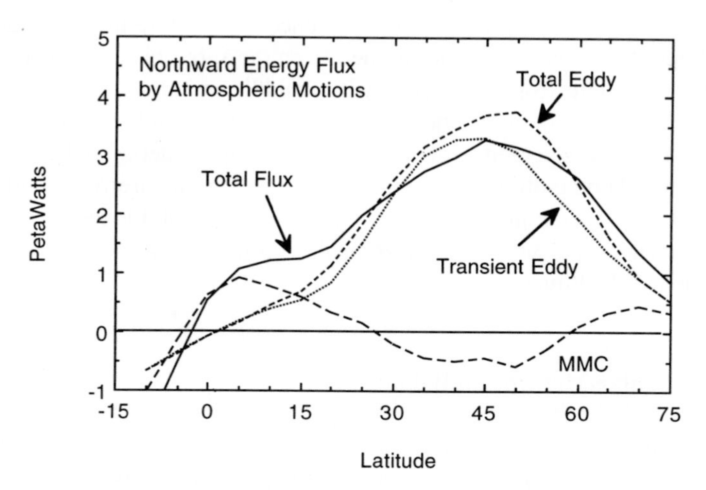

Fig. 6.11 Annual average northward energy flux plotted versus latitude in the Northern Hemisphere. Units are 10^{15} W. Mean meridional circulation (MMC). [Data from Oort (1971). Reprinted with permission from the American Meteorological Society.]

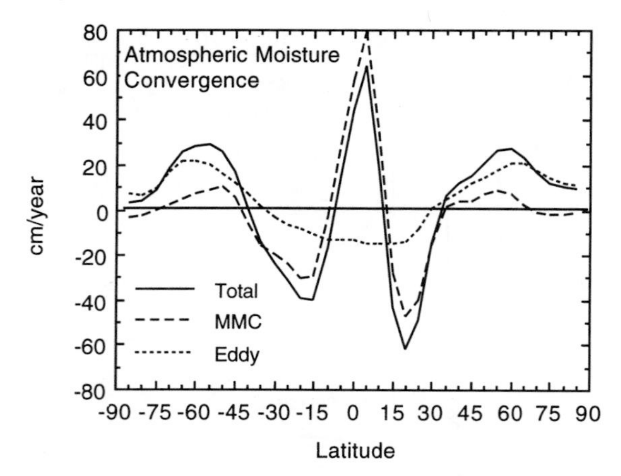

Fig. 6.12 Convergence of the meridional flux of water vapor in the atmosphere in cm year^{-1}. [Data from Peixóto and Oort (1984). Reprinted with permission from the American Physical Society.]

Only in the tropics is the transport of energy by the mean meridional circulation of primary importance.

6.3.5 Meridional Water Flux in the Atmosphere

The mean meridional circulation and eddies transport water and play an important role in determining the nature of the hydrologic cycle. The atmospheric moisture convergence in the tropics is dominated by the transport provided by the mean meridional circulation, which produces a convergence of moisture in the equatorial region of about 75 cm year^{-1} that is supplied from the subtropics (Fig. 6.12). Eddies remove water from the tropics and supply it to middle and high latitudes. According to (5.4) the divergence of atmospheric vapor transport should be equal and opposite to the runoff. The convergence of the zonally averaged meridional flux of water vapor by atmospheric motions shown in Fig. 6.12 has a structure very similar to that of the zonally averaged runoff shown in Fig. 5.2, as it should. Fluxes of moisture are confined very near the surface, because of the strong decrease of the water vapor concentration with altitude (Fig. 6.13).

6.4 The Angular-Momentum Balance

The general circulation of the atmosphere is heavily constrained by the conservation of angular momentum. *Angular momentum* is the product of mass times the perpendicular distance from the axis of rotation times the rotation velocity. The angular momentum about the axis of rotation of Earth can be written as the sum of the angu-

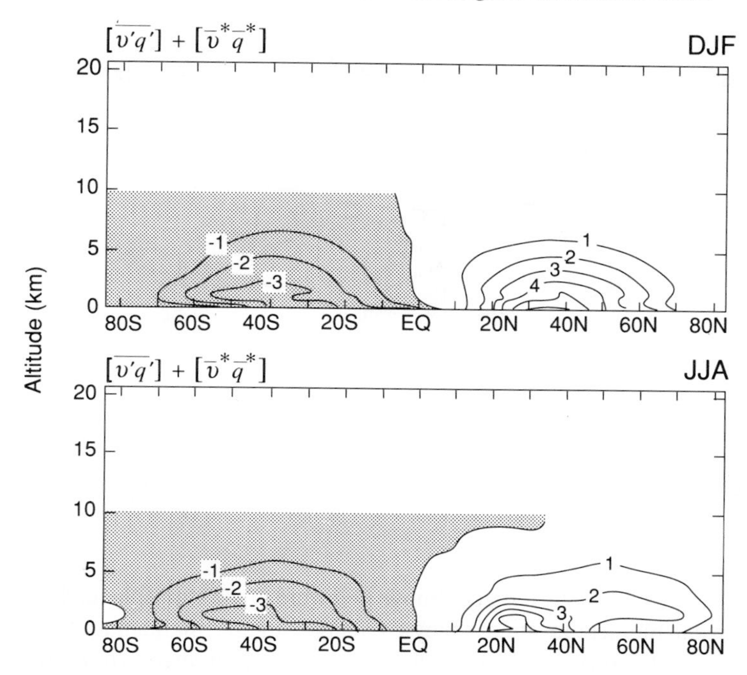

Fig. 6.13 Zonal cross section of the zonally averaged northward flux of specific humidity by eddies. Contour interval is 1 m s^{-1} g kg^{-1}. [Data from Oort (1983).]

lar momentum associated with Earth's rotation, plus the angular momentum of zonal air motion measured relative to the surface of Earth. Because the depth of the atmosphere is so thin compared to the radius of Earth, the altitude in the atmosphere is unimportant for the angular momentum of the air, and we may use a constant radius, a, to describe the distance from the center of Earth. The latitude has a strong influence on the angular momentum, however, because increasing latitude decreases the distance to the axis of rotation (Fig. 6.14). The distance to the axis of rotation is the radius of Earth times cosine of latitude. The zonal velocity consists of the velocity of Earth's surface associated with rotation plus the relative zonal velocity of the wind.

$$M = (\Omega a \cos\phi + u)\, a \cos\phi = (u_{\text{earth}} + u)\, a \cos\phi \tag{6.16}$$

The zonal velocity of Earth's surface at the equator, about 465 m s^{-1}, is very large compared to typical zonal wind speeds in the atmosphere.

$$u_{\text{earth}} = \Omega a \cos\phi = 7.292 \times 10^{-5}\ \text{rad s}^{-1} \bullet 6.37 \times 10^{6}\ \text{m} \bullet \cos\phi$$
$$= 465\ \text{m s}^{-1} \bullet \cos\phi$$

The atmospheric angular momentum associated with Earth's rotation is thus much larger than the angular momentum associated with the zonal winds normally observed

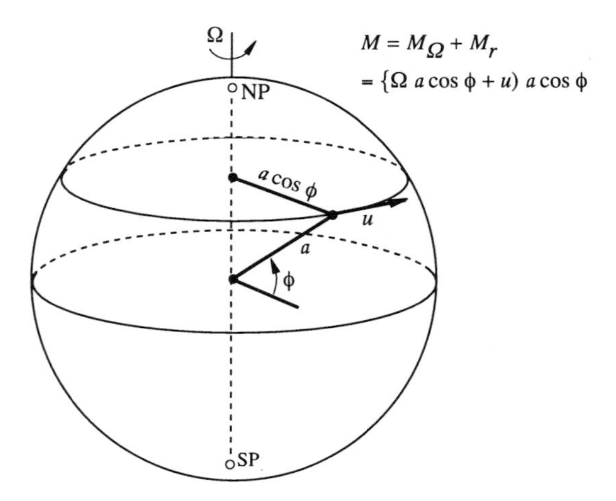

Fig. 6.14 Diagram showing the component of angular momentum about the axis of rotation of Earth. [From Peixóto and Oort (1984). Reprinted with permission from the American Physical Society.]

in the troposphere. When air parcels move poleward in the atmosphere, they retain the same angular momentum unless they exchange angular momentum with other air parcels or with the surface. Since the distance to the axis of rotation of Earth decreases as a parcel moves poleward on a level surface, the relative eastward zonal velocity of the parcel must increase to maintain a constant total angular momentum. Thus, poleward-moving parcels experience an eastward acceleration relative to Earth's surface.

If a parcel of air moves from one latitude to another while conserving angular momentum, then (6.16) implies a relationship between the zonal velocities the parcel will have at any two latitudes.

$$M = \left(\Omega a \cos \phi_1 + u_1\right) a \cos \phi_1 = \left(\Omega a \cos \phi_2 + u_2\right) a \cos \phi_2 \qquad (6.17)$$

If a parcel starts out at the equator with zero relative velocity and moves poleward to another latitude while conserving its angular momentum, we have that

$$M = \Omega a^2 = \left(\Omega a \cos \phi + u_\phi\right) a \cos \phi \qquad (6.18)$$

which can be rearranged to yield an expression for the zonal velocity at any other latitude u_ϕ,

$$u_\phi = \Omega a \frac{\sin^2 \phi}{\cos \phi} \qquad (6.19)$$

By substituting numbers into (6.19) we find that a parcel of air with the angular momentum of Earth's surface at the equator will have a westerly zonal wind speed of 134 m s^{-1} at 30°N or 30°S. This is much greater than the maximum zonally averaged wind speeds in the subtropical jet stream (Fig. 6.4), and we infer that the poleward angular momentum transport in the upper, poleward-flowing branch of the

Hadley cell is more than adequate to explain the existence of a 40 m s^{-1} jet at 30° latitude. The interesting part is explaining why the subtropical jet stream is not stronger than it is. A parcel traveling at a mean meridional velocity of 1 m s^{-1} would require about 30 days to travel from the equator to 30°N, allowing plenty of time for small-scale turbulence or some other slow process to reduce the angular momentum of the parcel. In fact, drag by small-scale eddies is probably not the most important mechanism for reducing the angular momentum of air parcels in the upper branch of the Hadley cell. Rather it is the large-scale eddies in the atmosphere that transport momentum out of the Hadley cell, into midlatitudes, and downward to the surface.

The zonally averaged meridional flux of angular momentum in the atmosphere can be written

$$\left[\overline{vM}\right] = \left[\overline{v}\right]\left(\Omega a \cos\phi + \left[\overline{u}\right]\right) a \cos\phi + \left\{\left[\overline{u'v'}\right] + \left[\overline{u}^{*}\overline{v}^{*}\right]\right\} a \cos\phi \qquad (6.20)$$

and consists of the transport of the zonal-mean angular momentum by the mean meridional velocity and an eddy flux term that depends on the covariance between zonal and meridional velocity fluctuations around a latitude circle. This eddy flux term becomes the dominant one near 30 degrees latitude, where the mean meridional velocity is small. The eddy flux of zonal momentum peaks at about the tropopause level at 35° latitude, where it is poleward in both hemispheres and strongest in the winter season (Fig. 6.15).

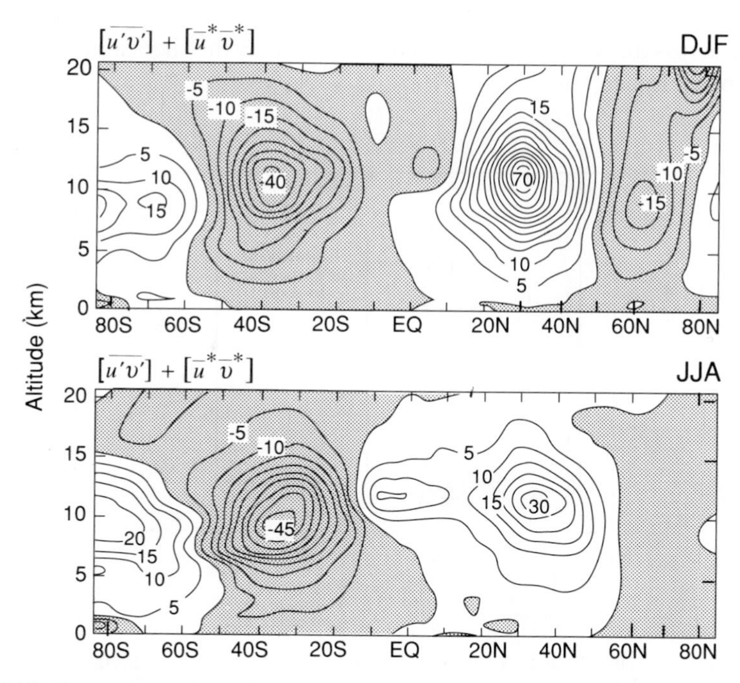

Fig. 6.15 Zonal cross sections of the northward flux of zonal velocity by eddies. Contour interval is 5 m^2 s^{-2}. [Data from Oort (1983).]

Northward zonal momentum flux by large-scale atmospheric disturbances is produced when the streamlines of the flow are oriented such that high and low anomalies tilt from southwest to northeast in the Northern Hemisphere as illustrated in Fig. 6.6. It can be seen that when the streamlines are tilted in this manner, the eastward component of wind is greater when the meridional wind component is poleward, and the eastward component of wind is weaker when the meridional flow is equatorward. Therefore an average over longitude of the product of the deviations of the zonal and meridional components of wind will be positive, indicating a northward flux of zonal angular momentum.

The flow of angular momentum in the atmosphere is shown schematically in Fig. 6.16. In the tropical easterlies, where the atmosphere rotates more slowly than Earth's surface, eastward angular momentum is transferred from Earth to the atmosphere via frictional forces and pressure forces acting on mountains. This westerly angular momentum is transported upward and then poleward in the Hadley cell. Atmospheric eddies transport angular momentum poleward and downward into the midlatitude westerlies. Where the surface winds are westerly, the atmosphere is rotating faster than Earth's surface and the eastward momentum is returned to Earth. It is clear that the surface zonal winds cannot be of the same sign everywhere, since eastward angular momentum must flow into the atmosphere where the surface winds are easterly, and must return to Earth where the surface winds are westerly. Thus the tropical easterly winds and midlatitude westerlies that so regulate the routes of sailing ships across the major oceans are required to satisfy the angular momentum constraint on atmospheric flow.

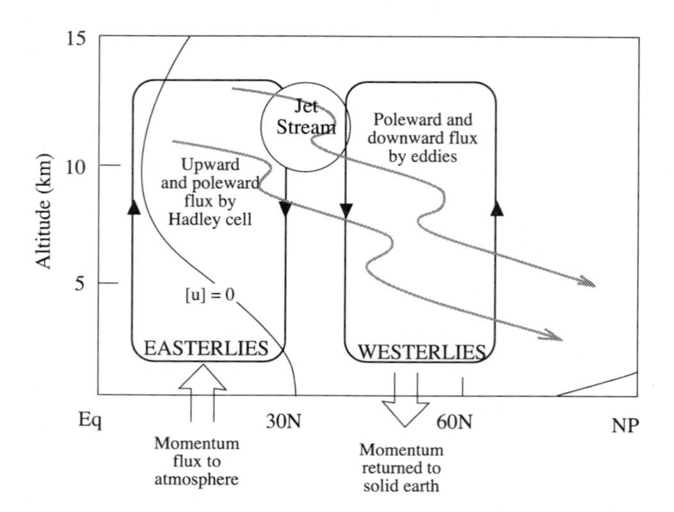

Fig. 6.16 Schematic illustration of the flow of angular momentum from Earth through the atmosphere and back to Earth.

6.5 Large-Scale Circulation Patterns and Climate

The circulation of the atmosphere is not precisely zonally symmetric, and east–west variations of winds and temperature are important for regional climates. For example, the subtropical jet of midlatitudes is not equally strong at all longitudes, but has local maxima associated with the distribution of land and ocean in the Northern Hemisphere (Fig. 6.17). During winter the subtropical jet stream has two local wind-speed maxima downstream of the Tibetan Plateau and Rocky Mountains over the Pacific and Atlantic Oceans, respectively. These maxima in the time-average wind speed have associated with them maxima in the transient eddy activity and eddy fluxes of heat and moisture. They define the so-called *storm tracks,* where vigorous midlatitude cyclones are most frequently observed. The seasonal migration of these storm tracks plays a key role in the annual variation of precipitation along the west coast of North America discussed in Chapter 5.

The distribution of surface air pressure is an important indicator of the general circulation (Fig. 6.18). Near the surface air tends to spiral inward toward low pressure centers, as is clearly indicated for the midlatitude low-pressure centers over the Pacific and Atlantic oceans during winter. By the conservation of mass this converging air must rise over the low pressure center. Conversely, the flow near the surface spirals outward from high-pressure centers, so that high pressure is generally indicative of subsiding motion above the boundary layer and suppressed convection. The surface pressure near 30 degrees latitude is generally higher than the surface pressure at the equator. This is consistent with the surface tradewinds of the tropics, which generally blow toward the equator from both hemispheres and meet in the *Intertropical Convergence Zone* (ITCZ) near the equator, where the surface pressure is low and where deep convection occurs with its attendant latent heat release and large-scale upward motion.

The seasonal variations in sea-level pressure are most apparent in the Northern Hemisphere. During winter the high-latitude oceans are characterized by low-pressure centers with the Aleutian and Icelandic lows centered in the northern margins of the Pacific and Atlantic Oceans, respectively. A high-pressure center lies over Asia. During summer the land–sea pressure contrast is reversed in midlatitudes, with the highest pressures over the oceans and the lowest pressures over the land areas. The oceanic highs are centered over 40°N, compared to their position at about 30°N during winter. The dominant low-pressure feature during Northern Hemisphere summer is centered over Asia at about 30°N, and is associated with the Asian summer monsoon.

The dramatic shifts in land–sea pressure distribution are driven by seasonal changes of insolation and the different responses of the land and ocean to heating. Over the oceans the response of surface temperature to seasonal variations of insolation is smaller because the energy is put into a deep layer of ocean with a large heat capacity and because evaporation consumes much of the heat input. Land surfaces

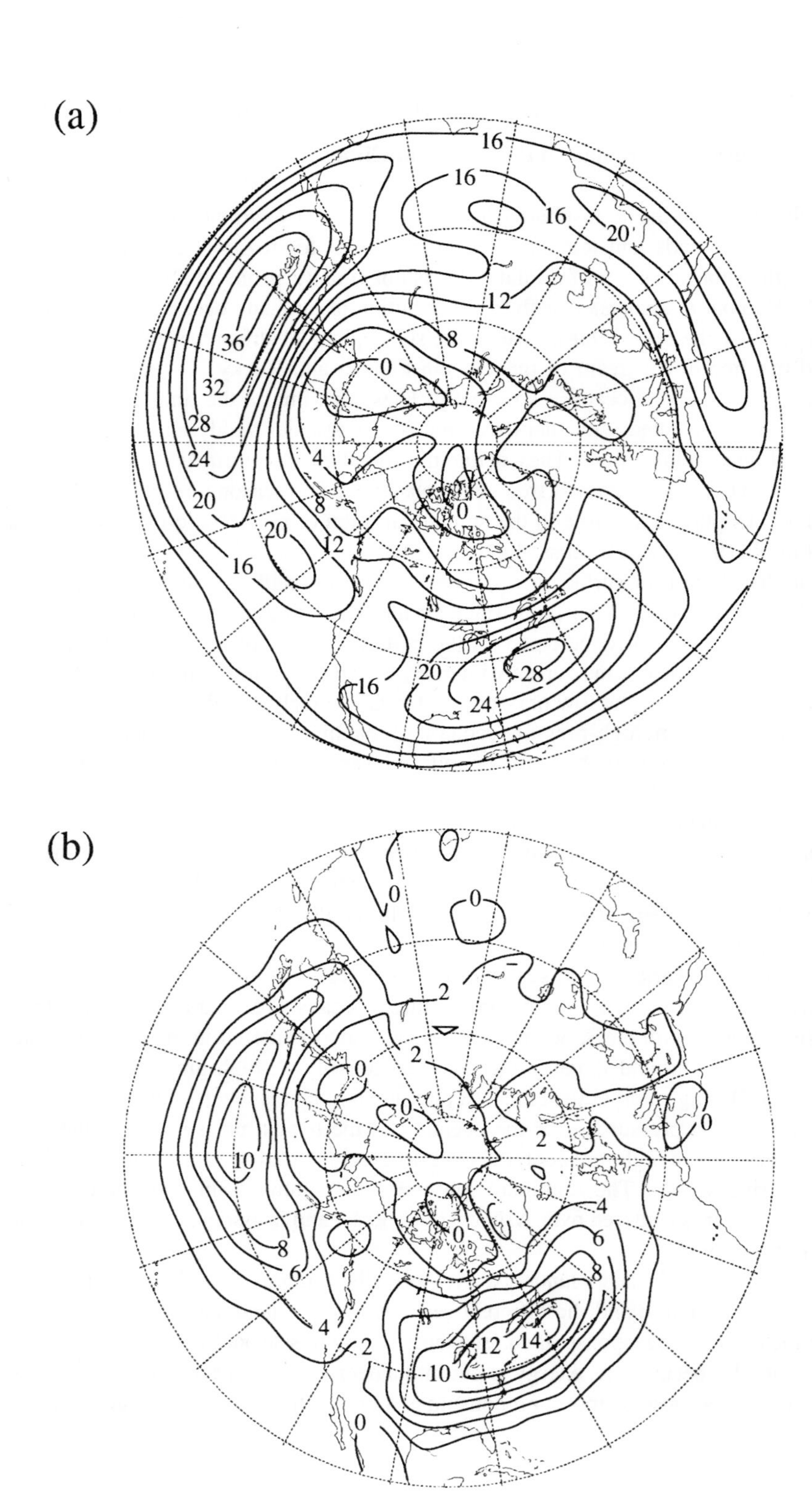

(a)
16
16
16
20
12
8
36
32
28
24
20
0
4
8
0
20
12
16
16
20
28
24
(b)
0
0
2
2
0
0
0
2
10
4
0
8
6
6
8
4
2
12
14
10
0

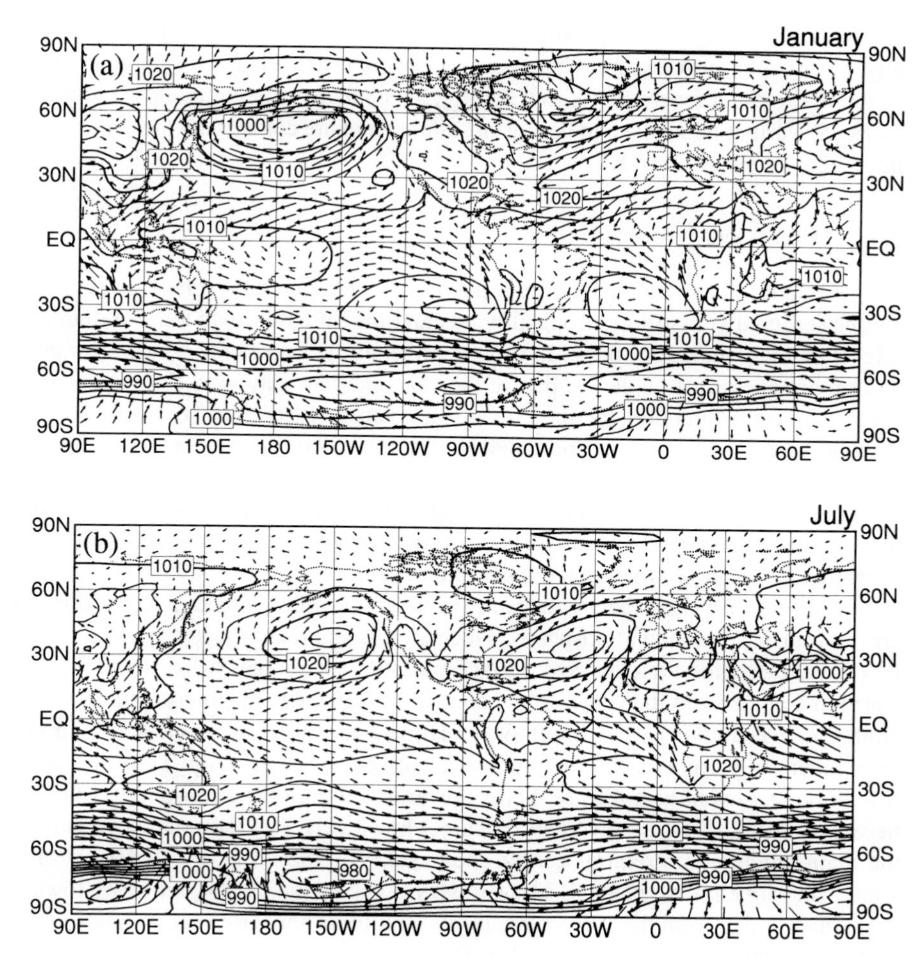

Fig. 6.18 Maps of mean sea-level pressure for (a) January and (b) July. Wind vectors for the 1000-mb level are superimposed. Data are 1980–87 analyses from a forecast model. Contour interval is 5 mb and largest vector represents a wind speed of 12 m s^{-1}.

have a much smaller capacity for storing heat, and often are not sufficiently wet that evaporation can balance large summertime increases in insolation. As a result the land surfaces warm up dramatically in summer and cool in winter. The pressure variations along latitude circles in midlatitudes are associated with the dynamical

---

Fig. 6.17 (a) Time-mean structure of upper-tropospheric wind speed (500 mb) during winter (m s^{-1}), and (b) contours of northward temperature flux ($\overline{v'T'}$) at 850 mb by eddies with periods shorter than 6 days (m s^{-1} K). [From Blackmon (1977). Reprinted with permission from the American Meteorological Society.]

response to the land-sea temperature and heating contrasts. The low pressures generally occupy the warm regions where the atmosphere is heated, and the high pressures occur where the temperature is low and the atmosphere is being cooled. Land surfaces are warmer than adjacent oceans in summer and are colder than the oceans during winter (see Figs. 1.5–1.7).

6.5.1 Monsoonal Climates

The seasonal movement of maximum insolation from one hemisphere to the other causes seasonal shifts in the tropical winds and precipitation, which have dramatic effects on the populations there. In some regions of the tropics the winds blow consistently in one direction during part of the year and then may weaken or blow from a very different direction for the rest of the year. Such regular seasonal changes in wind speed and direction are called monsoons.[1] The word *monsoon* is derived from an Arabic word meaning season. In many parts of Africa, Asia, and Australia, seasonal changes in wind direction are accompanied by dramatic shifts in precipitation regime between very dry and very rainy. Nowhere is this phenomenon more dramatically displayed than in the Asian monsoon.

During summer the Tibetan plateau is heated by insolation, and much of this energy is transferred to the atmosphere, which warms significantly, leading to a reduction in surface pressure. The low surface pressure encourages a low-level flow of warm, moist air from the ocean to the land (Fig. 6.19), which supports intense precipitation over India and the slopes of the Himalayas during summer. A similar encroachment of summertime precipitation occurs over eastern Asia. The heating by insolation and latent heat release drives upward motion in the atmosphere, which is necessary to balance the convergence of air at low levels. The seasonal movement of the main precipitation regions can be seen clearly in the OLR (Fig. 2.10). During winter the situation is reversed. The Himalayas cool dramatically, air flows toward the Indian Ocean from the land at low levels, and India and surrounding lands experience a wintertime drought (Fig. 6.20). The switch from dry to moist conditions occurs abruptly at the time of summer monsoon onset, typically around the middle of June. The date of monsoon onset and the duration and intensity of rainfall during the rainy season vary from year to year. These fluctuations in summer precipitation can produce either flood or drought, and are of great importance for the populace in the regions affected.

6.5.2 Desert Climates

Desert climates occur in land areas where the precipitation is significantly less than the potential evapotranspiration, so that the surface becomes dry. Precipitation will

[1]Fein and Stephens (1987).

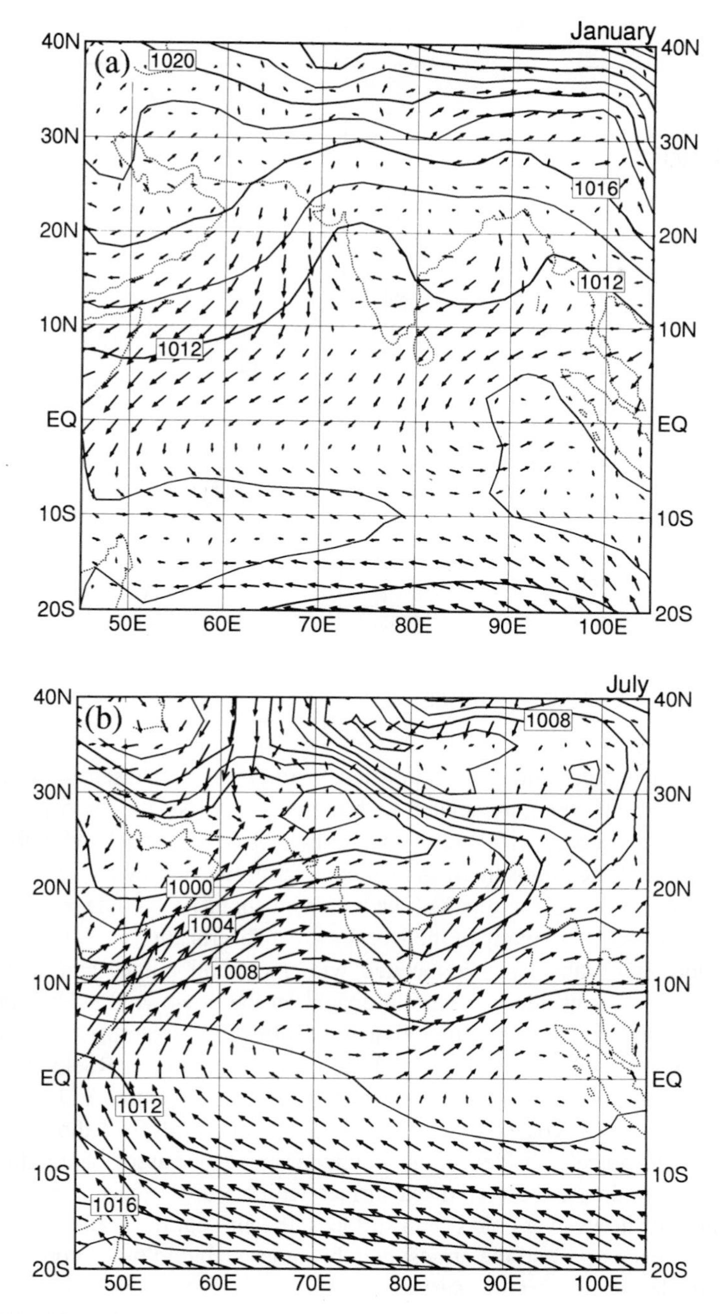

Fig. 6.19 Maps of mean sea-level pressure and 1000-mb winds in the region of the Asian Monsoon during (a) January and (b) July. Contour interval is 2 mb and largest vector represents a wind speed of 17 m s^{-1}.

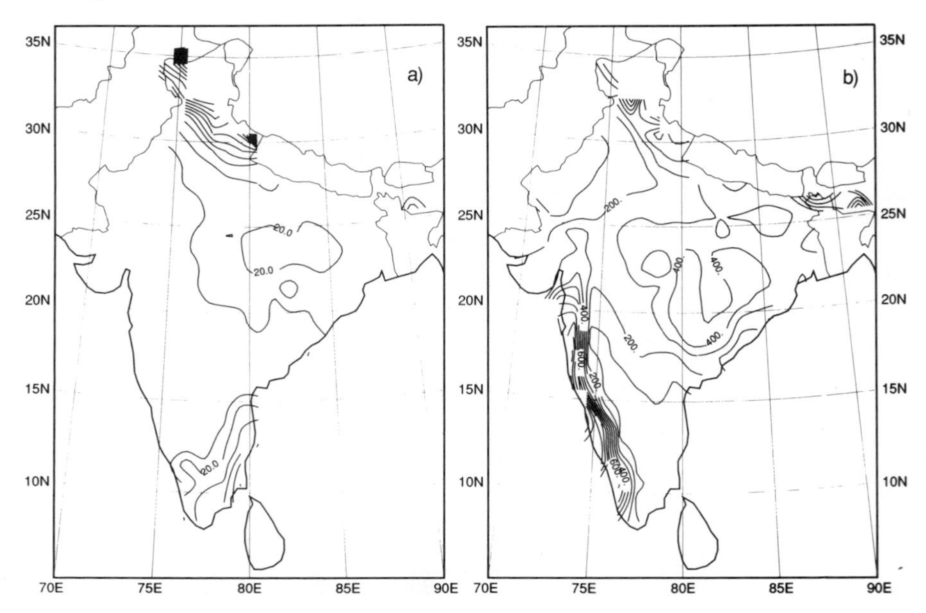

Fig. 6.20 Maps of the mean monthly precipitation over India during (a) January (contour interval 10 mm) and (b) July (contour interval 100 mm).

generally occur where humid air is uplifted. Aridity can be caused by a variety of mechanisms, which are all associated with the circulation of the atmosphere. Rainfall can be suppressed by widespread, persistent subsidence associated with the general circulation of the atmosphere. As we have seen, the Hadley circulation implies downward air motions along much of the belt from about 10 to 40 degrees latitude in both hemispheres, where many of the world's great deserts are found (Fig. 6.21). Regions of localized subsidence associated with mountain ranges can also lead to deserts. Persistent subsidence generally occurs on the downwind side of mountain ranges that stand athwart the prevailing wind direction. Examples are the North American Desert, which is downwind of the Cascade, Sierra, or Rocky Mountains, and the Monte and Patagonian Deserts, which are to the east of the Andes Mountains in South America. The coastal deserts of Peru and Chile result from three causes: the large-scale subtropical subsidence, the localized subsidence associated with easterly flow over the Andes, and the cold sea surface temperatures adjacent to the deserts. When warm air flows over cold ocean surfaces, the air near the surface is cooled by sensible heat exchange and the atmosphere becomes stable to moist convection, because cold, dense air near the surface cannot easily rise through the warmer air above.

Deserts may also be formed in regions that are blocked from a supply of humid air. If air is forced to cross a mountain range, its humidity is greatly reduced as the air

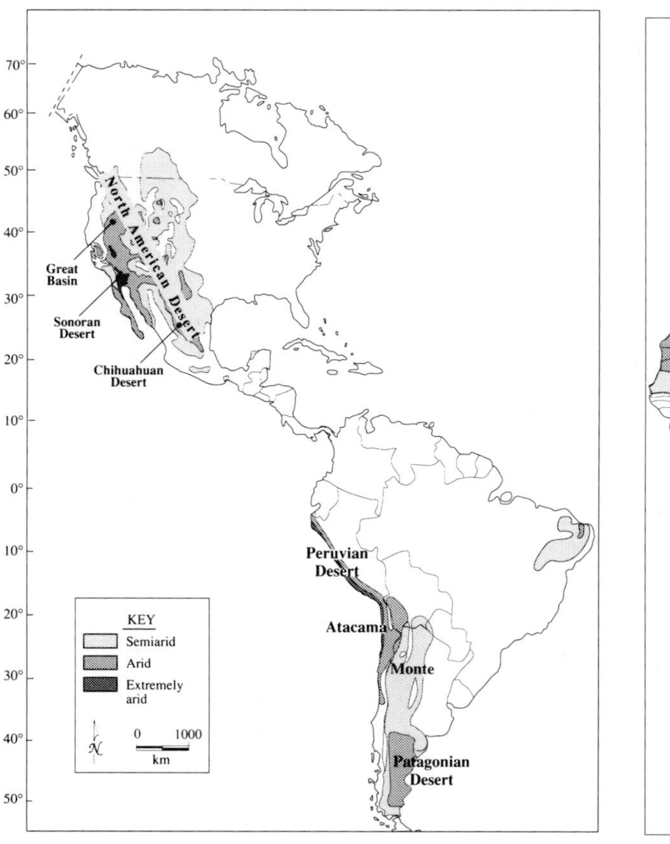

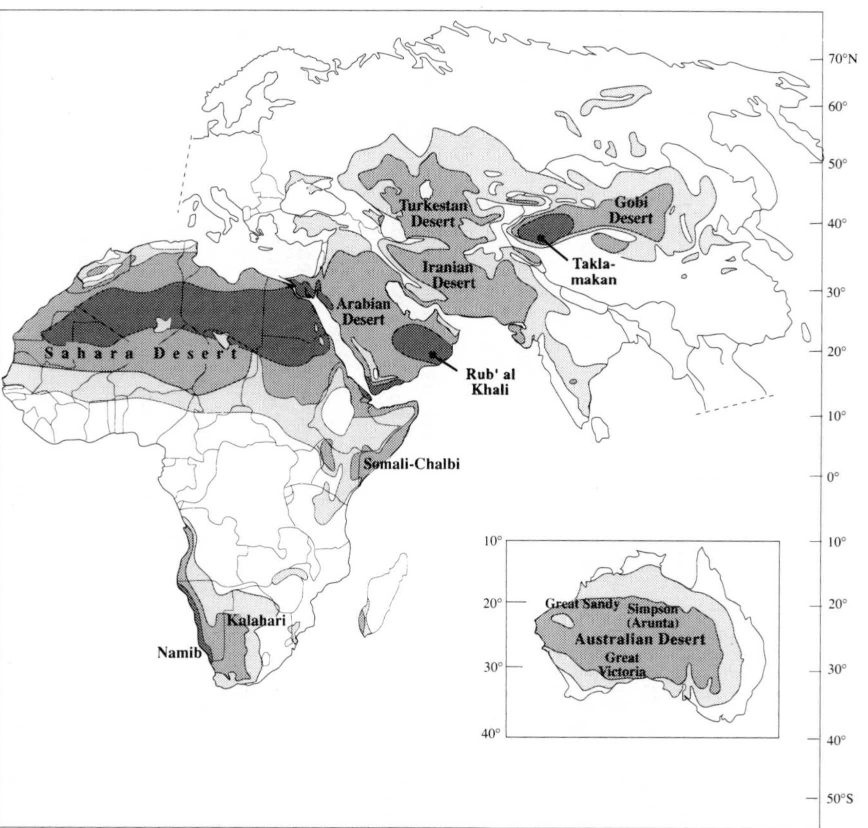

Fig. 6.21 Map showing arid lands around the world. Meigs classification taken from *Mosaic* magazine (Vol. 8, Jan/Feb 1977). [See McGinnies *et al.*, eds., (1968). Permission granted by the Office of Arid Lands Studies.]

stream is uplifted on the windward side. As the air rises it cools, and the water vapor condenses out as heavy precipitation. On the leeward side the air sinks and warms, giving the air an extremely low humidity. The deserts of central Asia, such as the Gobi, occur in regions that are either very distant from the ocean or separated from the ocean by a major mountain range.

The Sahara and Arabian (Rub' al Khali) Deserts constitute one of the largest and most arid regions on Earth. This results from their large land mass and their position in the subtropics where mean air motions are downward and insolation is large. Because of their high surface temperature and high albedo, these deserts are regions of net radiation loss in the annual average (Fig. 2.11). The atmospheric circulation provides energy to these regions by import of high energy air at high levels of the atmosphere. The air then subsides over the deserts as it radiatively cools. The outflow of this air at low levels exports moisture and helps to sustain the great dryness of these regions.

Deserts all share a very dry surface and precipitation much lower than the potential evaporation, but can vary greatly in their mean temperatures. For example, Ouallen, Algeria, a town in the Sahara at 22.8°N, 5.53°E, has a maximum monthly mean temperature of 38°C in July, a minimum of 16°C in January, and an annual precipitation of about 22 mm. In contrast, Antofagasta, Chile in the coastal Atacama Desert at 23.5°S, 70.4°W, has an even lower annual precipitation of 2 mm, but has much colder temperatures. Antofagasta has a monthly mean maximum temperature of 20°C in January and a minimum of 13°C in July. The coastal Peruvian and Atacama Deserts are remarkable because the low temperatures and frequent stratus cloud and fog associated with the cold temperatures of the adjacent ocean are juxtaposed with an extreme lack of precipitation.

Most deserts have a large diurnal variation in surface temperature, because the surface warms substantially during the day in response to insolation, and cools rapidly at night because longwave radiation emitted by the surface exceeds the downward longwave coming from the dry, cloudless atmosphere. Seasonal variations of temperature can also be rather large in midlatitude arid lands, with freezing temperatures in winter and very hot daytime temperatures in summer.

6.5.3 Wet Climates

Wet climates occur where the precipitation is heavy and exceeds the evapotranspiration for much of the year. Tropical wet climates are supported by the natural tendency of the atmospheric circulation to bring warm moist air to the equator. Islands that fall under the Intertropical Convergence Zone are favored by heavy rainfall, and all of the major tropical land masses have a region with a wet climate near the equator. The major regions of convection and precipitation in the tropics can be seen in the seasonal means of outgoing longwave radiation (OLR) in Fig. 2.10. The regions of lowest seasonal mean OLR occur where the most extensive upper-level clouds exist [Fig. 3.21(a)], and these clouds are in turn representative of intense convection and rainfall. Along the equator three minima in the OLR are associated with rainy regions over South America, Africa, and the Indonesian–western Pacific region. These zones of heavy precipitation move north and south with the seasons and tend to occur in the summer hemisphere.

The combination of equatorial location, shallow seas, and relatively small land masses in the region of Malaysia, Indonesia, and New Guinea gives rise to the biggest region of intense precipitation on Earth. Here high sea surface temperatures and strong solar forcing of diurnal land–sea breezes around the many islands foster frequent intense convection and precipitation. This large region of intense convection provides a strong thermal forcing for the large-scale circulation of the tropical atmosphere. Rising motion driven by latent heat release in the Indonesian sector drives a powerful circulation cell along the equator with surface inflow from the east and west and outflow just below the tropopause. The east–west circulation cells along the equator associated with the convective regions over Indonesia, South America, and Africa can be illustrated with a zonal mass streamfunction defined as

$$\Psi_Z = \frac{a}{g} \int_{\phi_S}^{\phi_N} \int_{0}^{p} \overline{u}^* \, dp \, d\phi \tag{6.21}$$

where the integral is taken over latitudes in the Northern and Southern Hemispheres, ϕ_N and ϕ_S, that are equidistant from the equator. $\overline{u}^*$ is the deviation of the time-mean eastward component of wind from its zonal-mean value. Values of the zonal mass streamfunction indicate the eastward mass flow above the level in question. If the range of latitudes over which the average is taken is fairly large, then the zonal gradients in the streamfunction are indicative of vertical motion. Large regions of rising motion are characteristic of the Indonesian, South American, and African regions and subsidence occurs in between (Fig. 6.22). The largest of the circulation cells oriented along the equator extends across the Pacific Ocean and is known as the *Walker circulation,* after the meteorologist Sir Gilbert Walker, who first described it.

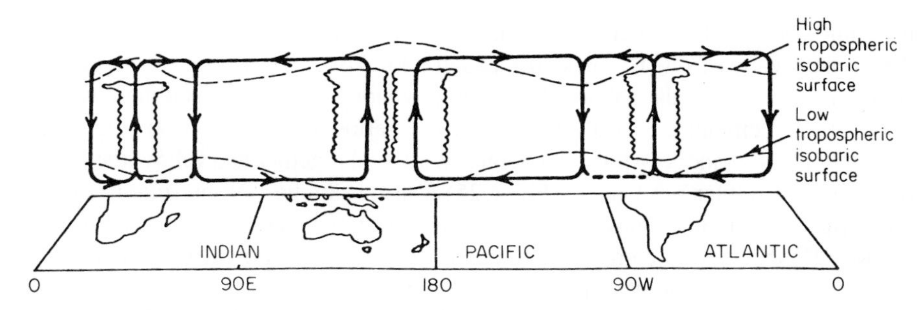

Fig. 6.22 Schematic view of the east-west Walker circulation along the equator indicating low-level convergence in regions of convection where mean upward motion occurs. [From Webster (1983).]

The seasonal movement in the mass of convection centered over Indonesia has both north–south and east–west components. During Southern Hemisphere summer the convection extends southeastward over the south Pacific Ocean and is identifiable as far away as 30°S, 140°W (Fig. 2.10). This feature is called the *South Pacific Convergence Zone* (SPCZ). During Northern Hemisphere summer the SPCZ is retracted back across the dateline, and the Indonesian convection extends northwestward into the Bay of Bengal, where it becomes connected with the convection associated with the Asian summer monsoon.

The largest area of tropical rain forest exists in the Amazon Basin of South America. The Amazon Basin receives more than 2 m of precipitation per year. The heavy precipitation over the vast area of the Amazon Basin is favored by the basin's equatorial location and its orientation in relation to the prevailing easterlies of the tropical atmosphere. Northeasterly winds carry large amounts of water vapor into the Amazon Basin from the Atlantic Ocean. The westward flow of moist air is unimpeded by the very gradual slope of the basin until the barrier formed by the Andes Mountains at the western extremity of the basin prevents the water vapor from simply flowing across South America without precipitation. Once convection is initiated over the Amazon Basin, the resulting latent heat release drives upward motion in the atmosphere that must be supported by further inflow of humid air from the east at low levels. A circulation of this nature is capable of delivering large amounts of precipitation to the surface of the basin and supports a huge volume of surface runoff to the Amazon River, which has the largest flow of any river in the world. The rainfall is seasonal, however, with maximum rainfall during the early months of the calendar year and minimum rainfall centered around the month of August, when the ITCZ moves northward and the heaviest precipitation falls in Central America. The heaviest precipitation follows the solar declination angle into the summer hemisphere, so that the rainy seasons on opposite sides of the equator tend to occur 6 months apart (Fig. 6.23).

6.5.4 Tropical Wet and Dry Climates

Many tropical regions have a wet season and a dry season of varying lengths. An interesting example that covers a very large area is the northern half of Africa, where the climate varies from wet near the equator to extremely arid at 25°N, with the southern boundary of the Sahara Desert roughly along 16°N. Between these two extremes the precipitation is seasonal, with a wet period and a dry period (Fig. 6.24). The precipitation and low-level convergence are tied together and follow the insolation into the summer hemisphere (Fig. 6.25). During summer, strong solar heating of the surface results in a transfer of heat to the air, which can support mean upward motion. This upward motion can draw moist, low-level air from the south that has its origin in the Gulf of Guinea or in the moist land areas of central Africa. As a result precipitation in the semiarid lands at the southern margin of the Sahara Desert occurs

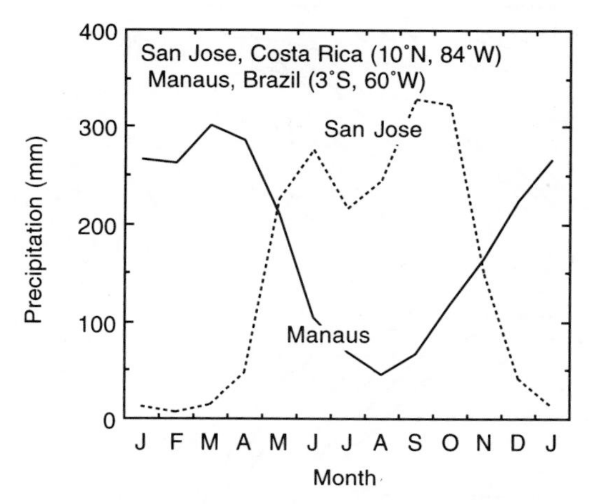

Fig. 6.23 Annual cycle of monthly precipitation at Manaus (Brazil) and San Jose (Costa Rica). Units are mm/month.

only during the summer season, when solar heating of the surface is sufficient to drive upward motion and low-level convergence there. In the tropical western Sahara during July the minimum sea-level pressure and apparent low-level convergence occur near 22°N, 5°W [Fig. 6.25(b)]. The precipitation associated with this

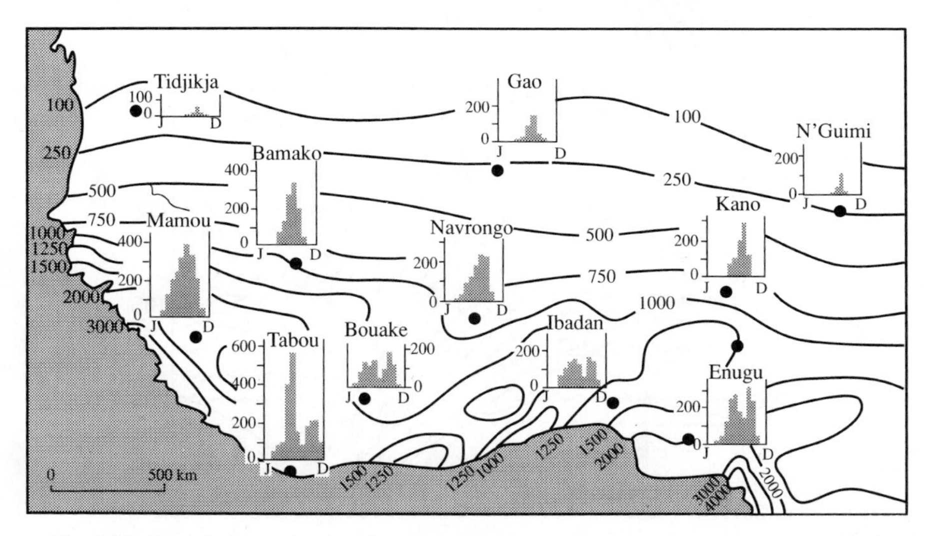

Fig. 6.24 Precipitation and seasonality of precipitation in NW Africa. Units are mm/year for contours of annual mean precipitation and mm/month for bar graphs showing the monthly precipitation from January (J) through December (D). [From Ledger (1969). Permission granted from Methuen and Co.]

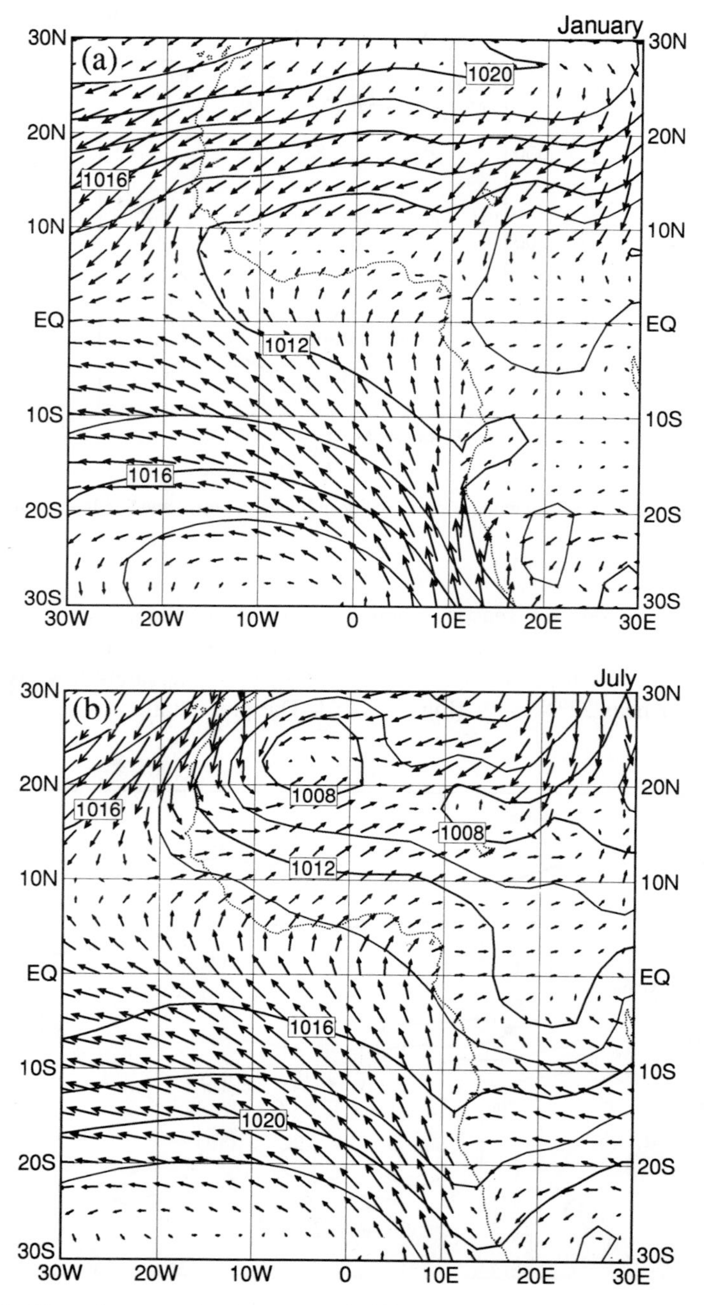

Fig. 6.25 Maps of mean sea-level pressure and 1000-mb winds in the region of the African Monsoon during (a) January and (b) July. Contour interval is 2 mb, and largest wind vector represents a speed of 12 m s^{-1}.

low-level convergence is quite small, however, because the air must follow a long trajectory over dry land areas so that it is quite dry when it reaches the area where the low-level winds converge and support shallow rising motion. The low-level wind convergence during January occurs at about 7°N in western Africa, which is much closer to the moisture source in the Atlantic Ocean [Fig. 6.25(a)], and which therefore produces very heavy precipitation (Fig. 6.24). Along the southern coast of western Africa the precipitation peaks twice a year during early summer and late fall. This double maximum is associated with the sun being exactly over the equator twice a year.

The wet season in Africa decreases in duration and reliability as one moves northward from the equator. As a general rule, the fractional variation of precipitation from year to year increases as the mean annual precipitation decreases, so lands that have low precipitation also have a high probability of unusually dry or wet years. At the southern margin of the Sahara Desert, in a region often called the Sahel, life is particularly sensitive to failures of the summer rainfall, especially when the rains fail for several years in succession as they did in 1910–1913, 1938–1942, and 1969–1973 (Fig. 6.26). Drought periods in the Sahel have been related to decadal changes in sea surface temperature,[2] and may be made worse by the local environmental changes induced by humans and their domestic animals.[3] Domestic animals such as goats and cattle need vegetation to eat, and humans need firewood for cooking. If the density of human and animal populations is such that the land is denuded of vegetation, then the desert may advance into regions that were previously semiarid. The complete removal of the surface vegetation is more likely during a drought episode, since the growth rate of plants cannot match the harvest rate by humans and animals.

Changes in surface environment of arid lands can be very long-lasting. When the rains fail and the surface vegetation is removed, the dominant surface covering may become windblown sand, which has a higher albedo than a vegetated surface. Such a surface will reflect more of the insolation, so that the heating rate necessary to drive low-level convergence may not be attained, and the rains will be more likely to fail. The reverse feedback process is also possible. Random weather events or the remote influence of sea surface temperature changes can lead to a succession of wet years. A few years of enhanced rainfall can foster the development of surface vegetation that will hold down the sand and decrease the surface albedo, thereby increasing the probability of additional wet years. The desert margin is thus a sensitive region, and can be altered by modest forces driving it toward greater or lesser aridity. Paleoclimatic evidence presented in Chapter 8 shows that the Sahara was not always a desert, and that natural changes can transform its climate to a much wetter one.

[2]Folland *et al.* (1986).
[3]Charney (1975); Rasool (1984).

Exercises

1. Discuss why the net energy transport of the Hadley circulation is much less than the poleward transport of potential energy. In what ways are the Hadley circulation and its energy transports related to the climate of the tropics?
2. Draw Fig. 6.6 for the Southern Hemisphere using the same coordinate axes. Make sure that what you draw transports heat and momentum toward the South Pole.
3. Derive (6.20) from (6.16), (6.12), and (6.13).
4. Calculate the zonal velocity of an air parcel at the equator, if it has conserved angular momentum while moving to the equator from 20°S, where it was initially at rest relative to the surface.
5. The tropical easterlies and midlatitude westerlies occupy about the same surface area of Earth. Would you expect the surface westerly winds to be stronger, weaker, or about the same as the surface easterlies? Explain your answer with equations.
6. Estimate the rate at which air must subside over the Sahara to balance the heat loss by radiation shown in Fig. 2.11a. *Hint:* Approximate the vertical gradient of moist static energy with that of potential energy and then use downward advection against this gradient to balance a radiative heat loss, e.g., $w[\partial(gz)/\partial z] = R_{TOA} \bullet (p_s/g)^{-1}$.
7. It has been speculated that you might be able to bring rainfall to the Sahara by decreasing the surface albedo. Estimate how much the albedo of the Sahara would need to decrease in order to convert its current net annual loss of radiation to a net annual gain of radiation of the same magnitude. This net radiative heating could support mean upward motion that would produce precipitation. How can you account for the likely surface temperature increases and associated longwave energy loss increases that will occur if you increase the solar absorption without providing moisture for evaporation?
8. Using the results of problems 6 and 7 and the tropical humidity distribution in Fig. 1.9, estimate how much albedo decrease you would need to generate net

---

Fig. 6.26 Three photographs showing the variation of surface conditions from the subtropical Sahara Desert to the Intertropical Convergence Zone. Top: the Saharan oasis of Ghardaia, Algeria at 32°N. The water table approaches the surface in this depression so that date palms can be grown by irrigation. The annual rainfall is about 75 mm and the surrounding land is barren of vegetation. Middle: Sahel region grassland between Agadez and Tanout, Niger at 16°N. The annual rainfall is about 400 mm. Bottom: equatorial rain forest on the rim of the Congo Basin near Bondo, Zaire at 4°N. Annual rainfall is about 1800 mm. (Photos courtesy of S. G. Warren.)

radiative heating to drive mean vertical motion sufficient to produce runoff from the Sahara equivalent to 1 m of precipitation per year. Hint: Equate the vertical transport of water vapor at the top of the boundary layer with the precipitation rate P, e.g., $(w\rho_a q)_{1\ \mathrm{km}} = P\rho_w$, where ρ_a and ρ_w are the density of air and liquid water, respectively, and q is the mass mixing ratio of water vapor. Compare your result with the net radiation over the Amazon Basin during the rainy season (Fig. 2.11).

Chapter 7 The Ocean General Circulation and Climate

7.1 Cauldron of Climate

The global ocean plays several critical roles in the physical climate system of Earth. These roles are related to key physical properties of the ocean: it is wet, it has a low albedo, it has a large heat capacity, and it is a fluid. Oceans provide a perfectly wet surface, which when unfrozen has a low albedo and is therefore an excellent absorber of solar radiation. The oceans receive more than half of the energy entering the climate system, and evaporative cooling balances much of the solar energy absorbed by the oceans, making them the primary source of water vapor and heat for the atmosphere. The world ocean is thus the boiler that drives the global hydrologic cycle. The world ocean also provides the bulk of the thermal inertia of the climate system on time scales from weeks to centuries. The great capacity of the oceans to store heat reduces the magnitude of the seasonal cycle in surface temperature by storing heat in summer and releasing it in winter. Because seawater is a fluid, currents in the ocean can move water over great distances and carry heat and other ocean properties from one geographic area to another. The equator-to-pole energy transport by the ocean is important in reducing the pole-to-equator temperature gradient. Horizontal and vertical transport of energy by the ocean can also alter the nature of regional climates by controlling the local sea surface temperature.

In addition to its direct physical effects on the climate system, the global ocean can affect the climate indirectly through chemical and biological processes. The ocean is a large reservoir for the chemical elements that form the atmosphere. Exchange of gases across the air–sea interface controls the concentration of trace chemical species containing oxygen, carbon, sulfur, and nitrogen, which are important in determining the radiative and chemical characteristics of the atmosphere. For example, the ocean controls the concentration of carbon dioxide in the atmosphere by exchanging gaseous carbon dioxide across the air–sea interface. Carbon dioxide is converted to organic solids in the ocean, and carbon is then stored by deposition of these solids onto the sea floor. Sulfur-bearing gases released from biological and chemical processes in the ocean enter the atmosphere where they are converted to aerosols that form the nuclei on which cloud droplets form. Evaporation of sea spray also forms salt particles that can form cloud condensation nuclei. The cloud condensation nuclei produced in the ocean can have a substantial influence on the energy balance of Earth, through their effect on the optical properties and extent of clouds.

7.2 Properties of Seawater

Oceanic currents and the resulting heat transports are determined primarily by the physical properties of the ocean. To specify the physical state of seawater requires three variables: pressure, temperature, and salinity. As described in Chapter 1, *salinity* is the mass of dissolved salts in a kilogram of seawater, and is generally measured in parts per thousand, which we denote with the symbol ‰. The average salinity and temperature of the world ocean are approximately 34.7‰ and 3.6°C, respectively. The effect of the ocean on atmospheric composition through biological and chemical processes depends on a more complex mix of physical, chemical, and biological properties. For example, the oxygen and nutrient content of seawater are of critical importance for life in the sea. Trace amounts of key minerals may be very important for local biological productivity.

Only about the first kilometer of ocean between 50°N and 50°S is warmer than 5°C [Fig. 7.1(a)], so that much of the mass of the ocean is between −2°C and 5°C. The thermal structure of the ocean at most locations can be divided into three vertical sections. The top 20–200 m of water in contact with the atmosphere usually has

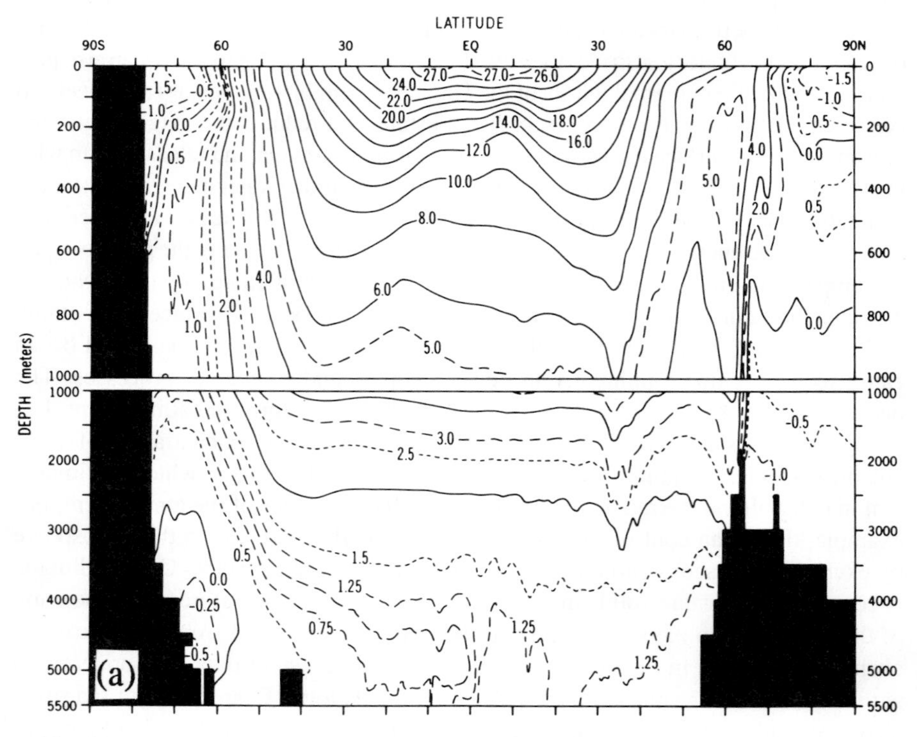

Fig. 7.1 Annual-mean zonal average for the global ocean of (a) potential temperature (°C), and (b) salinity [‰ (‰ = parts per thousand)], and (c) potential density ($\rho_t - 1000$, kg m^{-3}). [From Levitus (1982).]

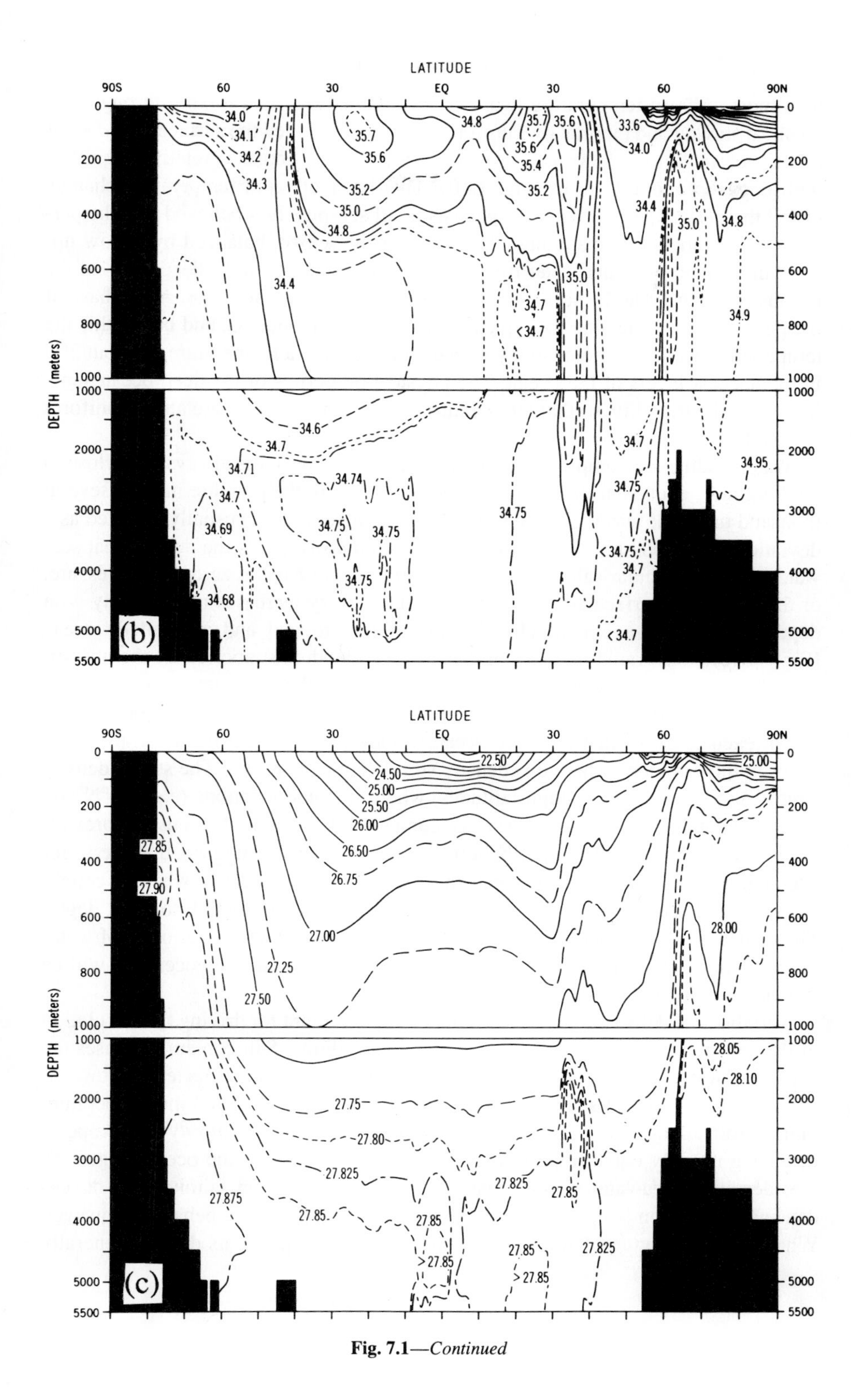

Fig. 7.1—*Continued*

an almost uniform temperature, which is maintained by rapid mixing through mechanical stirring and thermal overturning. This layer is called the surface *mixed layer* of the ocean. Below the mixed layer the temperature decreases relatively quickly with depth to about 1000 m (Fig. 1.10). This layer of rapid temperature change, called the *permanent thermocline,* persists in all seasons. It is believed that the permanent thermocline is maintained by heating from above, balanced by a slow upward movement of colder water from below. The cold water in the deep abyss of the oceans is produced at the surface in a few regions of the polar ocean. At the base of the permanent thermocline the typical temperature is about 5°C, and below this the temperature decreases more slowly with depth, reaching a temperature of about 2°C in the deepest layers of the ocean. The physical properties of the deep ocean show little spatial variability, so that temperature, salinity, and density are almost uniform (Fig. 7.1).

Water is almost incompressible, so the density of seawater is always very close to 1000 kg m^{-3}, even near the bottom of the ocean where the pressure may be several thousand times the surface air pressure. Density of seawater is usually reported as a deviation from 1000 kg m^{-3}, $\rho-1000$. *Potential density,* ρ_t, is the density that seawater with a particular salinity and temperature would have at zero water pressure, or the density at surface air pressure. Potential density increases most rapidly with depth in the first several hundred meters of the tropical and midlatitude ocean [Fig. 7.1(c)]. This rapid increase of density with depth is supported by the absorption of solar radiation near the surface, which sustains the warm temperatures there. The strong density stratification in the upper ocean inhibits vertical motion and turbulent exchanges, so that the deep ocean is somewhat isolated from surface influences in those regions where this density stratification is present. The strong density stratification is reduced in high latitudes, where in some locations (e.g., 65°N and 75°S) the potential density at the surface comes much closer to the densities prevailing in the deep ocean. The distribution of potential density suggests that the water occupying the bulk of the deep ocean came from the polar regions, where at certain locations and seasons surface water becomes dense enough to sink to great depth. The distributions of other tracers also suggest that slow downward motion of water in high latitudes extends downward and equatorward into the deep ocean, as will be discussed in Section 7.6.

Variations of density on pressure surfaces are important for driving the circulation of the ocean, and depend on the temperature and salinity. Salt content increases the density of water, and seawater expands and becomes less dense as its temperature increases. The salinity of seawater ranges from about 25 to 40‰ and the temperature ranges from about −2 to 30°C. By varying within these ranges salinity and temperature have roughly equal importance for density variations in the ocean (Fig. 7.2). The density of seawater is very nearly linearly dependent on salinity. The dependence of density on temperature does not have this simple linear behavior, however. When the temperature of water approaches its freezing point, its density generally

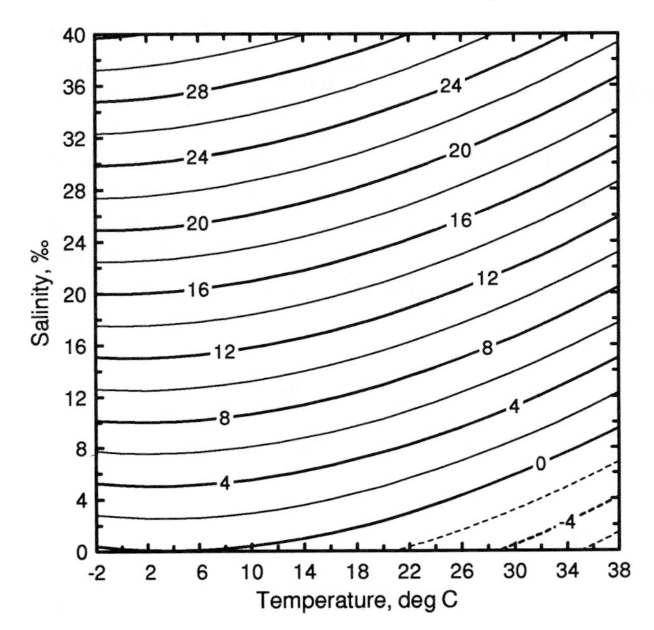

Fig. 7.2 Contours of seawater density anomalies (ρ_t − 1000, kg m^{-3}) plotted against salinity and temperature.

becomes less sensitive to temperature. For pure water, for example, the maximum density occurs at 4°C, and the water then expands slightly as it is cooled further. Therefore, fresh water lakes that are cooled from the top continue to overturn convectively until the entire water column reaches 4°C, because water that is at 4°C will always be more dense than warmer water. When the entire water column is cooled to 4°C, surface water that is cooled further will become less dense than the column and will "float" at the surface. When it reaches 0°C the surface water will freeze and form a layer of surface ice, which provides a layer of insulation between the cold air above and the warmer water below. If the lake is deep enough, the water near the bottom will remain at about 4°C, although the air temperature above the surface ice may fall to many degrees below zero. This fact allows fish in high-latitude or high-altitude lakes to survive the winter in the liquid water beneath the surface ice.

For seawater with salinity greater than 24.7‰, the density continues to increase with decreasing temperature until freezing occurs, although more slowly as the freezing point is approached. Therefore, if the salinity is initially well mixed, the entire water column must be cooled to the freezing point before ice can form. Sea ice is able to form in the high-latitude oceans because the salinity decreases significantly near the surface (Fig. 1.11). Lower salinities near the surface cause a decrease in density that offsets the increase in density associated with colder temperatures near the surface, allowing water near the surface to freeze while warmer water is present below. The low surface salinities result primarily from the excess of precipitation over evaporation in these latitudes. In the Arctic Ocean the supply of freshwater from rivers flowing from the surrounding continents contributes importantly to low surface salinities and therefore to the stable density gradient. Because salinity increases with depth, the surface is able to form ice without bringing the temperature of the entire water column to the freezing point. It has been hypothesized that, if the flow of certain key rivers were diverted from the Arctic Ocean to supply irrigation water to continental interiors farther south, the heat balance of the Arctic could be severely distorted, because the normal configuration of a thin layer of surface ice on a mostly unfrozen Arctic Ocean may no longer be stable. Increased salinity of surface waters in the Arctic might lead to either complete removal of most arctic sea ice or complete freezing of the Arctic Ocean from surface to bottom.

7.3 The Mixed Layer

The primary heat source for the ocean is solar radiation entering through the top surface. Almost all of the solar energy flux into the ocean is absorbed in the top 100 m. Infrared and near-infrared radiation are absorbed in the top centimeter, but blue and green visible radiation can penetrate to more than 100 m if the water is especially clear. The depth to which visible radiation penetrates the ocean depends on the amount and optical properties of suspended organic matter in the water, which vary greatly with location, depending on the currents and the local biological productivity. The principal component of suspended matter in surface water is *plankton,* which are plants and animals that drift in the near-surface waters of the ocean (Fig. 7.3). The solar flux and heating rate in the ocean are greatest at the surface and decrease exponentially with depth, in accord with the Lambert–Bouguet–Beer Law as described in Chapter 3. Under average conditions the solar flux and heating rate are reduced to half of their surface value by a depth of about one meter, but significant heating can still be present at more than 100 m below the surface.

Since the solar heating is deposited over a depth of several tens of meters in the upper layers of the ocean, and cooling by evaporation and sensible heat transfer to the atmosphere occurs at the surface, there must be an upward flux of energy in the upper ocean to maintain an energy balance between surface loss terms and subsurface heating. Molecular diffusion is an important heat transport mechanism only in the top centimeter of the ocean. Elsewhere the heat flux is carried by turbulent mixing, convective overturning, and mean vertical motion, which is called *upwelling* or *downwelling* in the ocean. Turbulent mixing in the surface layer of the ocean is greatly aided by the supply of mechanical energy by the winds and

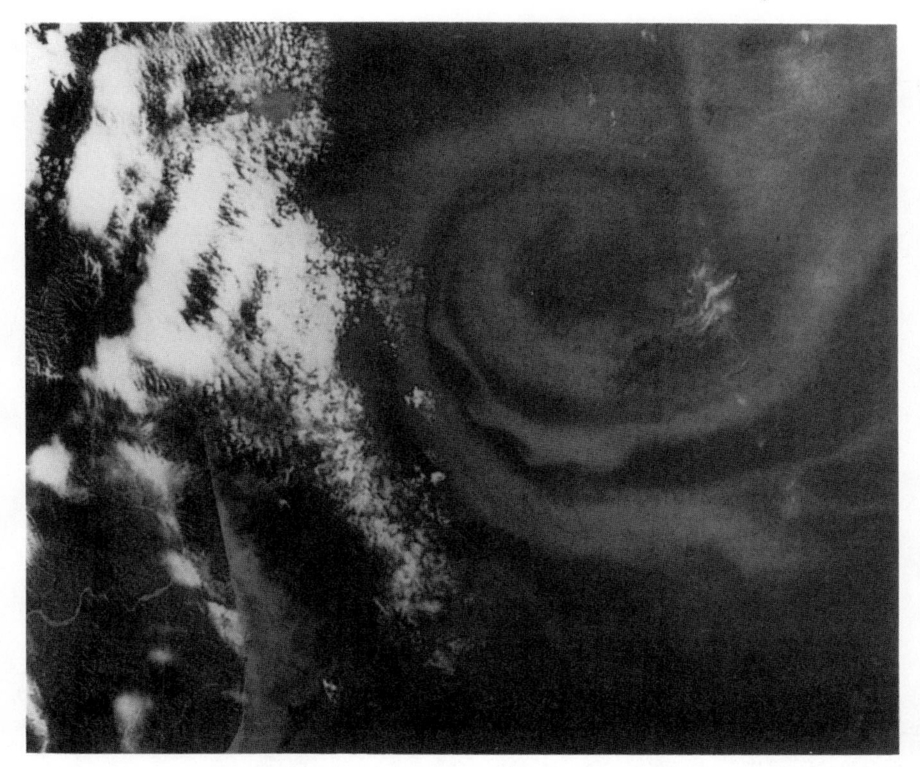

Fig. 7.3 A small oceanic eddy off New Zealand's South Island, northeast of Christchurch. The variations in water color are associated with the abundance of plankton. The brightest areas are clouds. (Challenger 9, 61A, NASA, October 30, 1985–November 6, 1985.)

their interaction with waves on the surface of the water. In the mixed layer of the ocean, heat transport by convection and turbulent mixing is so efficient that the temperature, salinity, and other properties of the seawater are almost independent of depth (Fig. 7.4).

A schematic diagram showing the processes important in the oceanic mixed layer is presented in Fig. 7.5. The depth of the mixed layer depends on the rate of buoyancy generation and the rate at which kinetic energy is supplied to the ocean surface by winds. If the surface is cooled very strongly, such as at high latitudes during fall and winter, then cold, dense water is formed near the surface at a rapid rate and buoyancy forces will drive convection, with sinking of cold water and rising of warmer water in the mixed layer. When the surface is cooled only weakly or actually heated, such as during summer, when surface solar heating rates are greatest, then the generation of mixing by buoyancy is less and the mixed layer will become thinner and warmer. Buoyancy can be generated by the effect of evaporation on surface

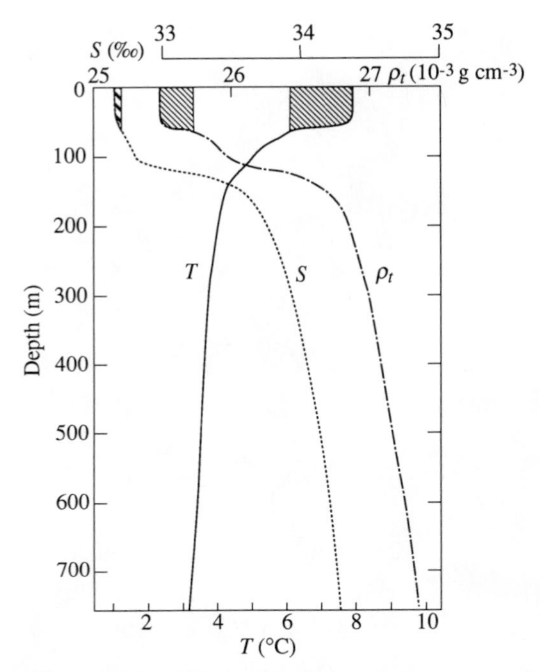

Fig. 7.4 Vertical profiles of temperature (T, °C), salinity (S, ‰), and potential density (ρ_t – 1000, kg m^{-3}) at Ocean Station P, 50°N, 145°W, on June 23, 1970 showing the mixed layer in the top 50 m. The hatched area shows the change since May 19, 1970 and indicates the springtime warming and thinning of the mixed layer. [From Denman and Miyake (1973). Reprinted with permission from the American Meteorological Society.]

salinity, even when surface temperatures are increasing with time. The density increase associated with increasingly saline surface waters can balance or overcome thermal stratification and encourage mixing. Rainfall represents an input of freshwater at the surface, which acts to decrease the density of the surface waters. Winds blowing over the ocean waves transfer kinetic energy to the water that results in turbulent water motion as well as mean ocean currents. The supply of turbulent kinetic energy to the upper ocean by winds can induce mixing even in the presence of stable density stratification. If the intensity of turbulence in the mixed layer is great enough, cool, dense water can be entrained into the mixed layer from below. This implies a downward heat transport, which cools and deepens the mixed layer.

The heat, momentum, and moisture exchanges between the atmosphere and the ocean are accomplished through contact of the atmospheric boundary layer with the mixed layer of the ocean. Storage and removal of heat from the ocean on time scales of less than a year are confined to the mixed layer over much of the ocean. The depth of the oceanic mixed layer varies from a few meters in regions where subsurface

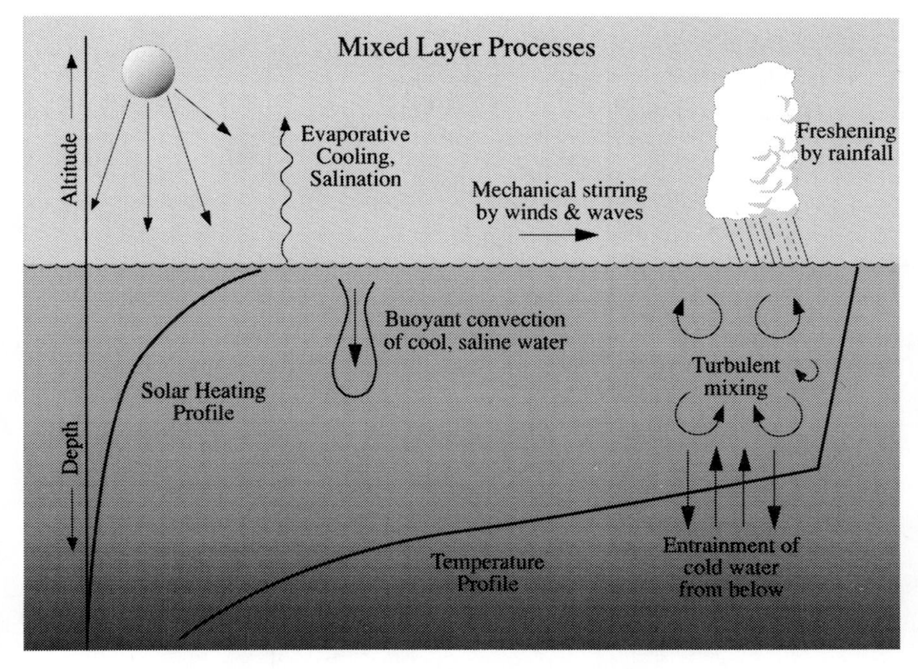

Fig. 7.5 Diagram showing important mixed-layer processes.

water upwells, as along the equator and in eastern boundary currents, to the depth of the ocean in high-latitude regions where cold, saline surface water can sink all the way to the ocean bottom. Regions where the mixed layer is deeper than 500 m constitute a small fraction of the global ocean area, however. In general, as one would expect, the mixed layer is thin where the ocean is being heated and thick where the ocean gives up its energy to the atmosphere. The global-average depth of the mixed layer is about 70 m. The mixed layer responds fairly quickly to changes in surface wind and temperature, whereas the ocean below the mixed layer does not. The thermal capacity of the mixed layer is the effective heat capacity of the ocean on time scales of years to a decade, and is about 30 times the heat capacity of the atmosphere (see Chapter 4).

The oceanic mixed layer responds strongly to the annual cycle of insolation and surface weather. Figure 7.6 shows an example from the midlatitude Pacific Ocean. The mixed layer is warmest and thinnest in late summer near the end of the period of greatest insolation and least intense stirring of the ocean by winds. After August the surface begins to cool, the storminess increases, and the mixed layer begins to deepen and cool. The mixed layer continues to deepen and cool throughout the winter, and by the end of winter may extend to a depth of several hundred meters and merge

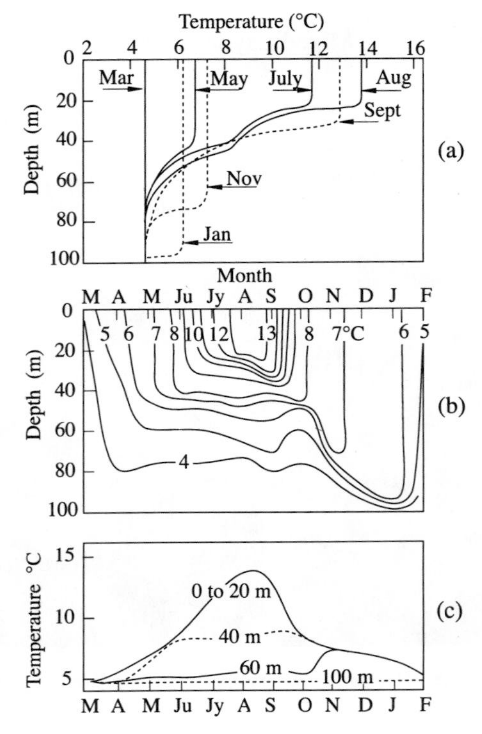

Fig. 7.6 Seasonal variation of temperature in the upper ocean at 50°N, 145°W in the eastern north Pacific. (a) Vertical profiles of temperature by months, (b) temperature contours, and (c) temperatures at various depths versus time of year. [From Pickard and Emery (1990). Reprinted with permission from Pergamon Press, Ltd., Oxford, England.]

smoothly into the permanent thermocline. During most of the rest of the year a seasonal thermocline with steep temperature gradients links the permanent thermocline with the base of the mixed layer. In spring and summer this seasonal thermocline develops and the mixed layer becomes thinner and warmer. Seasonal variations in temperature are confined primarily to the mixed layer and the seasonal thermocline, so those temperatures at depths below the deepest extent of the mixed layer experience little seasonal variation.

7.4 The Wind-Driven Circulation

The transfer of momentum from winds to ocean currents plays a critical role in driving the circulation of the ocean. This is particularly true for the currents near the

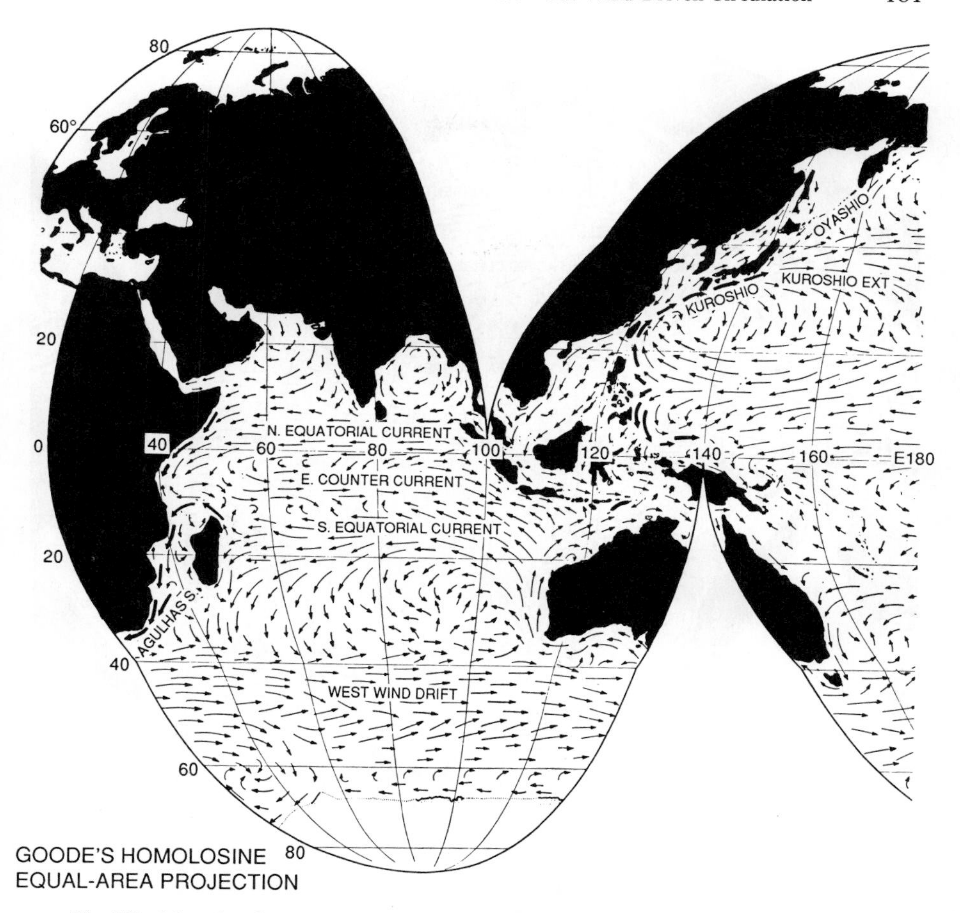

Fig. 7.7 Map of surface currents. (*Figure continues.*) [Adapted from Sverdrup *et al.* (1942).]

ocean surface. The general character of the large-scale surface ocean currents is shown in Fig. 7.7. The surface currents are arranged in coherent patterns with large circulations called *gyres* occupying the major ocean basins. In addition, many narrow but persistent currents appear in time-averaged maps.

7.4.1 Western Boundary Currents

Some of the most visible current structures are the large clockwise circulations in the northern Pacific and Atlantic oceans. Along the western boundaries of the Pacific and Atlantic Ocean basins strong poleward-flowing currents exist in a narrow zone

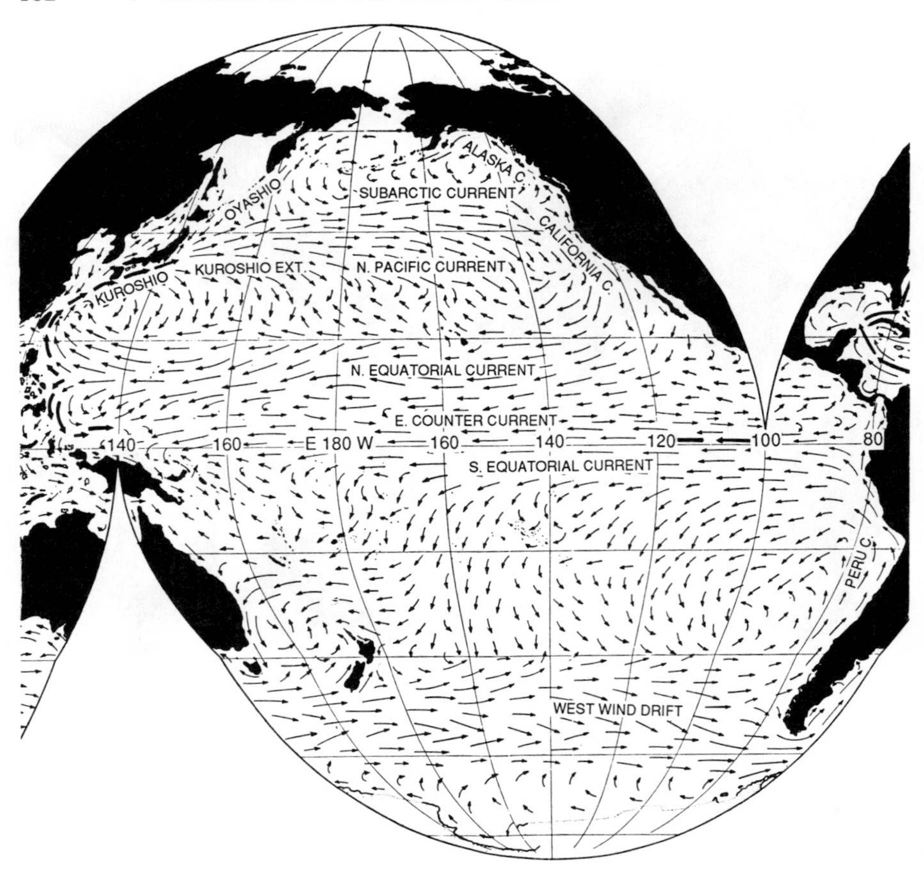

Fig. 7.7—*Continued*

very near the continents. These currents are called the Kuroshio and the Gulf Stream, respectively, and may be referred to generically as western boundary currents. Western boundary currents also occur in the Southern Hemisphere along South America (Brazil Current) and along Africa (Agulhas Stream). They are generally less sharply defined and extensive in the Southern Hemisphere, perhaps because of the different ocean geometry that allows the Antarctic circumpolar current to flow unimpeded in a continuous eastward current at about 60°S. Western boundary currents carry warm water from the tropics to middle latitudes. The speed of these currents may exceed one meter per second, which is quite fast for an ocean current. With the possible exceptions of the Antarctic circumpolar current and some zonal equatorial currents, these currents are the closest oceanographic analog to the jet streams of the atmosphere, although they flow poleward rather than eastward. The return flow of water

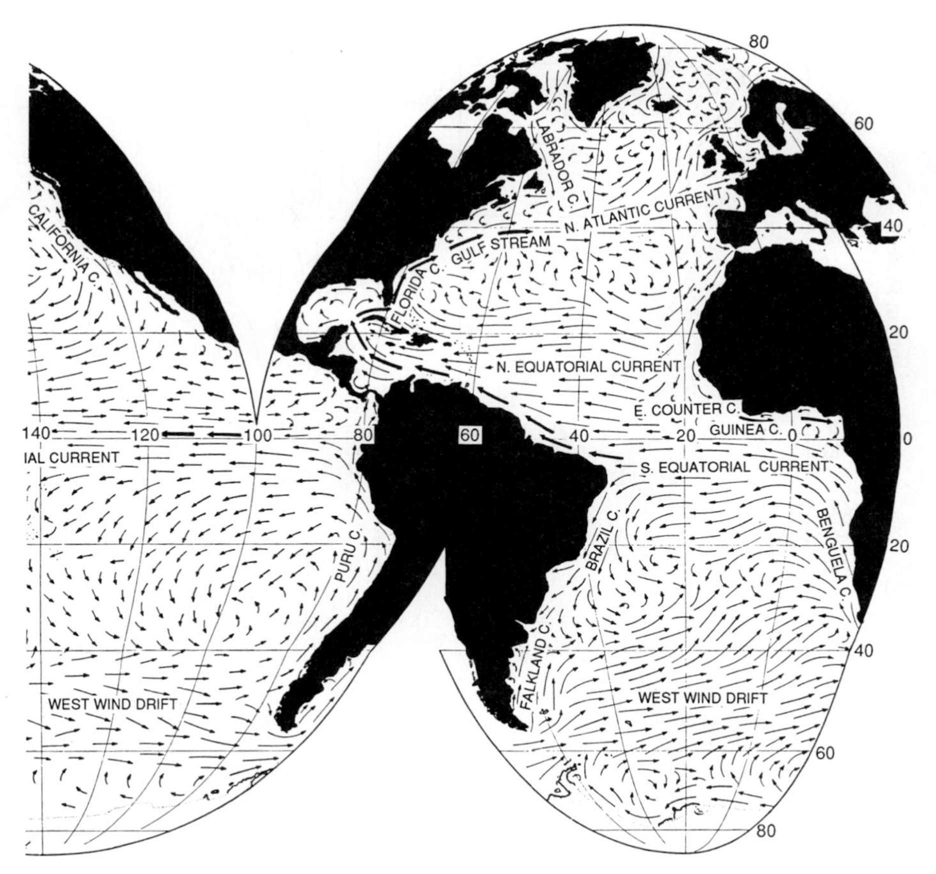

Fig. 7.7—*Continued*

from midlatitudes to the equator is much more gradual and occurs in a broad expanse across the center of each basin.

The thermal–current structure of the Gulf Stream after it has left the coast at Cape Hatteras and is flowing approximately east is shown in Fig. 7.8. The warmest water occurs near the surface coincident with the strongest current velocities, which are near 2 m s^{-1}. These strong currents are accompanied by strong subsurface temperature gradients across the stream, with warmer water to the south and east of the current and cooler water to the north and west. These temperature gradients persist when the current leaves the western margin of the ocean and flows into the interior of the ocean basin. The current is not straight or steady, but breaks down into meanders and rings and eventually loses a clear identity as the flow expands eastward across the basin (Fig. 7.9).

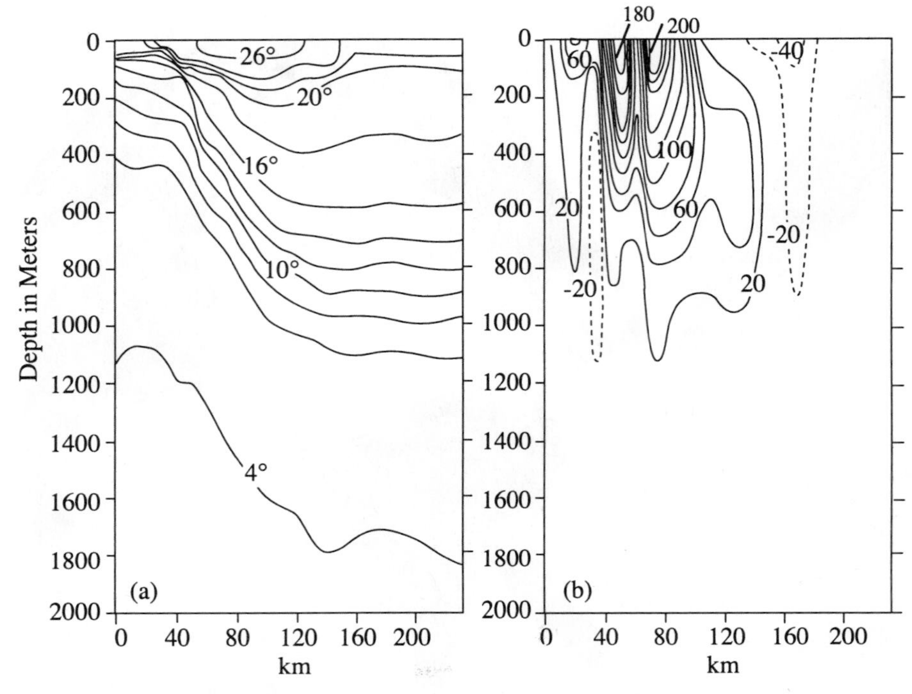

Fig. 7.8 Cross section of temperature (contour interval 2°C) and geostrophic current (contour interval 20 cm s^{-1}) across the Gulf Stream at about 38°N, 68°W. [From Stommel (1965), adapted from Worthington (1954). Reprinted with permission from Munksgaard International Publishers Ltd.]

The poleward flux of warm water in the Gulf Stream and Kuroshio currents has a profound effect on the *sea surface temperature* (SST) and the climate of the land areas bordering the oceans, especially the lands immediately downwind of the oceans. The averaged sea surface temperature distribution for DJF [Fig. 7.10(a)] shows a strong gradient in the north Atlantic Ocean aligned approximately with the mean position of the Gulf Stream. This strong temperature gradient extends northeastward from the mid-Atlantic coast of North America to the Norwegian Sea in the vicinity of Spitzbergen. It appears that some of the heat carried northward by the Gulf Stream is picked up by the Norwegian Current and carried into polar latitudes. As a result, at middle and high latitudes the eastern Atlantic is much warmer at the surface than the western Atlantic Ocean. This asymmetry in the Atlantic sea surface temperature contributes to the milder winter climates of western European land areas compared to eastern North American land areas at the same latitude. Another major contribution to this climate asymmetry is the eastward advection of temperature in the atmosphere.

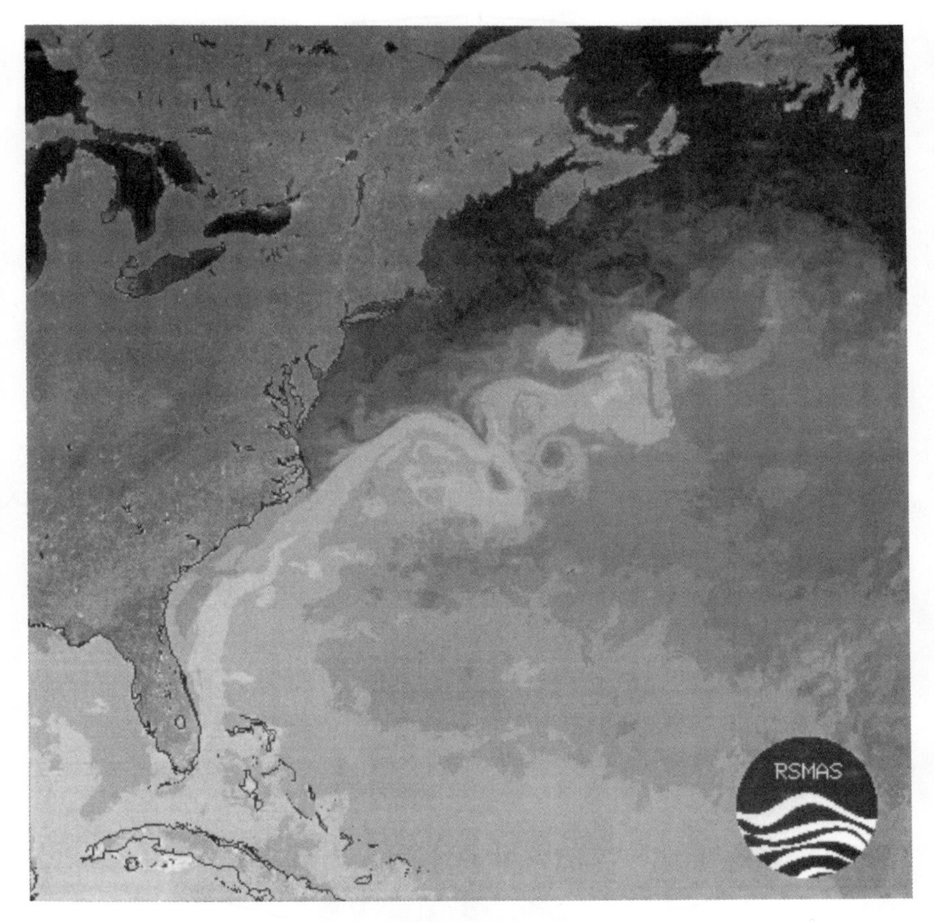

Fig. 7.9 Gray-scale image of SST in the northwestern Atlantic showing meanders and rings in the Gulf Stream. Warm water is light in color. (Courtesy of Dr. O. Brown, University of Miami.)

7.4.2 Eastern Boundary Currents

Also important for climate are eastern boundary currents, which occur in tropical and subtropical latitudes at the eastern margins of the oceans. The names given to the eastern boundary currents in these geographic areas are the California Current off North America, the Peru Current off South America, the Leuwin Current off eastern Australia, the Canary Current off northern Africa, and the Benguela Current off southern Africa. In each of these regions a wind-driven current flows along the coast toward the equator and then turns westward toward the center of the basin. These currents are associated with cold SST, which can be illustrated by

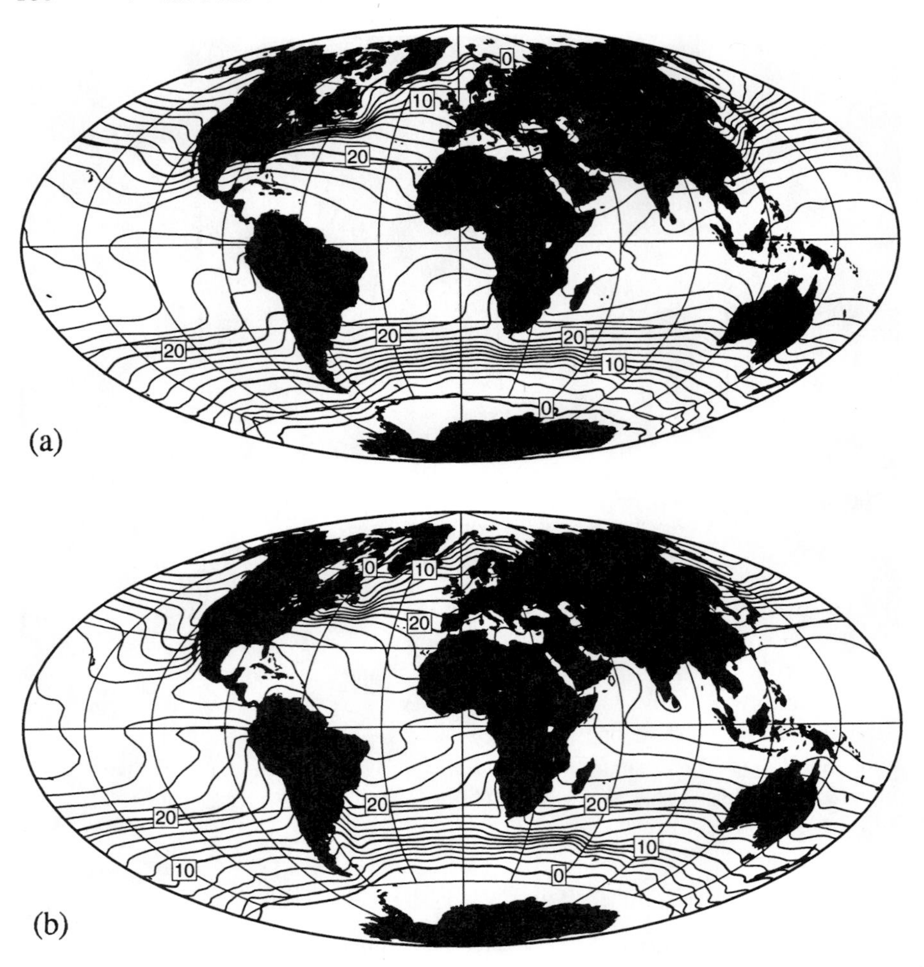

Fig. 7.10 Average (a) December–February (DJF) and (b) June–August (JJA) sea surface temperature (°C).

plotting the deviation of SST from its average at each latitude (Fig. 7.11). The SST in the subtropics to the west of the continents in the Atlantic and Pacific Oceans is much colder than the zonal average at each latitude. The coldest water occurs very near the coast and extends westward and equatorward into the oceans. The low SST near the coast is produced by upwelling of cold subsurface waters, which is driven by alongshore or offshore winds in these regions. These low-level winds are associated with the surface high pressure systems in the atmosphere above the eastern subtropical ocean areas, as shown in Fig. 6.18. The wind systems and the associated currents and cool SSTs are best developed during the summer in

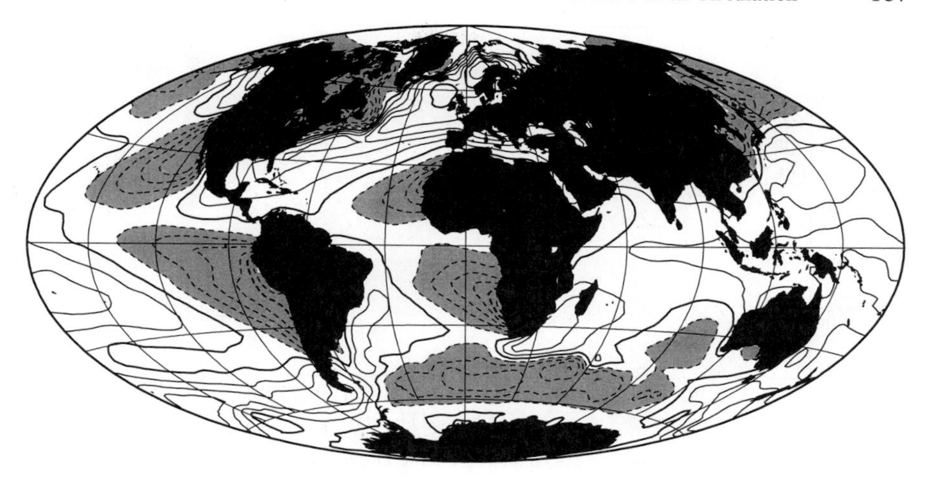

Fig. 7.11 The deviation of the July sea surface temperature from its zonal average at each latitude. Contour interval is 1°C, and values less than −1°C are shaded.

the Northern Hemisphere, and are more nearly year-round phenomena in the Southern Hemisphere. It is believed that the geometry of the coastlines in the two oceans, and in particular the northwest–southeast slope of the coastlines of South America and Africa, causes the eastern boundary currents in the Southern Hemisphere to be better developed and to extend to the equator and then westward along the equator. The cooler than average SST in eastern boundary current regions is often associated with atmospheric subsidence and persistent stratiform cloud [Fig. 3.21(b)].

7.4.3 Interannual Variability in the Equatorial Pacific: ENSO

Large east-to-west gradients in SST and subsurface thermal structure exist in the tropical Pacific and Atlantic Oceans. These gradients are associated with the upwelling of cold water at the eastern margins of the oceans and along the equator in the eastern part of the basin. The thermocline is deepest in the western tropical oceans and rises toward the surface in the eastern equatorial regions where upwelling occurs (Fig. 7.12). The density and pressure gradients associated with the east–west slope of the thermocline in the Pacific are supported by the westward wind stress applied by the tropical easterly winds. Once every several years this balance is disrupted and warm water spreads toward the east, causing SST and climatic anomalies that may persist for a year or more. Such events are called *warm anomalies* of the tropical Pacific Ocean. If warm waters appear near the coast of South America, where the waters are normally very cold, the event is known locally as an *El Niño*. The appearance of warm waters at the coast of South America is associated with a deepening of the thermocline that normally intersects the surface of the eastern Pacific. Under normal

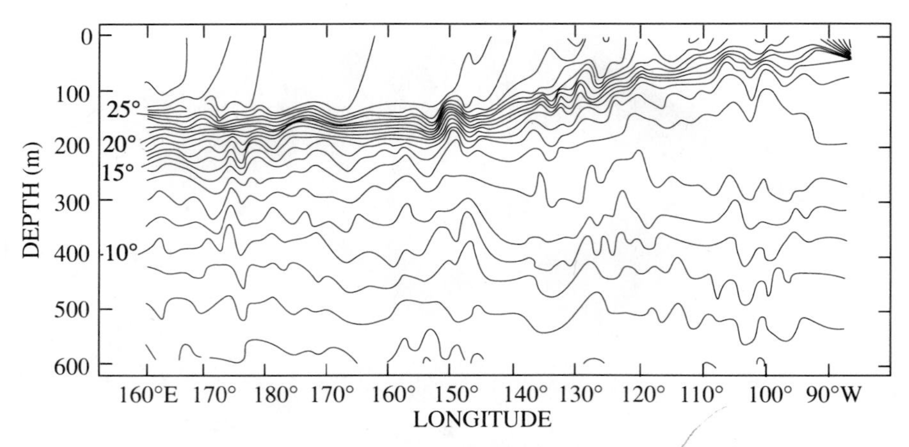

Fig. 7.12 Thermal structure of the equatorial Pacific Ocean showing the slope of the thermocline (°C). [From Colin *et al.* (1971).]

conditions upwelling of cold, nutrient-rich water from below the thermocline supports a very rich fishery. Deepening of the thermocline during an El Niño cuts off the supply of nutrients to the surface, and the fishery off the coast of equatorial South America is adversely affected. Changes in the SST are coupled with changes in the convection and large-scale atmospheric flow in the tropics. The warming of the SST near the South American coast is often associated with substantial precipitation in what is otherwise a generally arid coastal climate. Changes in SST distribution in the central and western Pacific are associated with changes in the surface pressure and large-scale wind distributions, which can have effects that extend into middle latitudes. The related oceanic and atmospheric variations that accompany warm and cold events in the equatorial Pacific are referred to jointly as the *El Niño-Southern Oscillation* (ENSO) phenomenon.

7.5 Theories for Wind-Driven Circulations

7.5.1 The Ekman Layer, Wind-Driven Transport, and Upwelling

Because of the rotation of Earth, the frictional component of the vertically integrated transport of water in the surface layer of the ocean is not in the direction of the applied wind stress, but 90 degrees to the right of it in the Northern Hemisphere and 90 degrees to the left of it in the Southern Hemisphere. This wind-driven near-surface water transport plays a critical role in determining the relatively cold surface temperatures in the eastern boundary current regions and along the equator, and it also plays an important role in driving the subtropical gyres that feed the western boundary currents.

To show the relationship between wind stress driving, currents, and transport we may consider a homogenous ocean of constant density and pressure, and assume that it is driven by a uniform wind stress with eastward component τ_x and northward component τ_y. We seek a steady solution in which frictional stresses and Coriolis accelerations are in balance.[1]

$$fv = -\nu \frac{d^2u}{dz^2} \tag{7.1}$$

$$fu = \nu \frac{d^2v}{dz^2} \tag{7.2}$$

The *Coriolis parameter* ($f = 2\Omega \sin \phi$) measures twice the local vertical component of the rotation rate (Ω) of Earth. The frictional forces have been described such that the frictional stress is proportional to the shear of the current velocity times a momentum diffusion coefficient ν. The specified wind stress thus enters as a boundary condition on the current shear at the surface, and we assume that the current goes to zero at large depths, so that the boundary conditions on (7.1) and (7.2) are

$$\left.\begin{aligned} \nu\frac{du}{dz} &= \frac{\tau_x}{\rho_o} \\ \nu\frac{dv}{dz} &= \frac{\tau_y}{\rho_o} \end{aligned}\right\} \quad \text{at } z = 0; \qquad u = v = 0 \text{ at } z \to -\infty \tag{7.3}$$

where ρ_o is the density of the seawater and assumed constant. The solution for the velocities under these conditions is

$$u_E = \frac{e^{\delta z}}{\rho_o\sqrt{f\nu}}\left\{\tau_y \cos\left(\delta z + \frac{\pi}{4}\right) + \tau_x \cos\left(\delta z - \frac{\pi}{4}\right)\right\} \tag{7.4}$$

$$v_E = \frac{e^{\delta z}}{\sqrt{f\nu}}\left\{\tau_y \cos\left(\delta z - \frac{\pi}{4}\right) - \tau_x \cos\left(\delta z + \frac{\pi}{4}\right)\right\} \tag{7.5}$$

where $\delta = \sqrt{f/2\nu} = z_E^{-1}$.

The steady solution (7.4)–(7.5) describes the Ekman spiral. The current vector has its maximum magnitude at the surface where it is directed at an angle of $\pi/4$ (45°) to the right of the wind stress vector in the Northern Hemisphere ($f > 0$). The current vector turns toward the right with increasing depth, and its magnitude decreases exponentially with depth. The magnitude of the current decreases by a factor of e^{-1} for every increase of depth equal to $z_E = \delta^{-1} = \sqrt{2\nu/f}$.[2] If we integrate the currents over the depth range in which the currents are significant, we obtain the integrated transport in the Ekman layer.

$$U_E = \int_{-\infty}^{0} u_E \, dz = \frac{\tau_y}{\rho_o f}; \qquad V_E = \int_{-\infty}^{0} v_E \, dz = -\frac{\tau_x}{\rho_o f} \tag{7.6}$$

[1] See, *e.g.*, Gill (1982).

[2] An appropriate value of $\nu = 30$ m^2 s^{-1} gives an Ekman depth of ~800 m.

The net horizontal water transport in the Ekman layer is directed at a 90° angle to the right of the applied wind stress in the Northern Hemisphere, so that if the wind stress is toward the east at the surface ($\tau_x > 0$), the Ekman layer transport is toward the south ($V_E < 0$). If a westward wind stress is applied near the equator, the Ekman layer transport will be northward in the Northern Hemisphere and southward in the Southern Hemisphere, because of the change of sign of f at the equator, so that a net divergence of surface flow will be generated. Conservation of mass requires upwelling along the equator to balance the Ekman layer transport away from the equator. The cold tongue of SST in the eastern equatorial Pacific Ocean during July [Fig. 7.10(b)] is caused largely by this wind-driven upwelling. The cold SST anomalies associated with the eastern boundary currents (Fig. 7.11) are associated with offshore Ekman transport driven by equatorward alongshore surface winds. Offshore Ekman transport near an ocean boundary requires upwelling to replace the exported water. Since water temperatures decrease with depth, upwelling is generally accompanied by cold sea-surface temperatures.

Wind stress driving can also cause vertical motions in the open ocean away from boundaries and the equator, if the wind stress has spatial gradients. If we consider the mass continuity equation for an incompressible fluid

$$\frac{\partial u}{\partial x} + \frac{\partial v}{\partial y} + \frac{\partial w}{\partial z} = 0 \tag{7.7}$$

and integrate it over the depth of the Ekman layer, we can derive a relationship between the applied wind stress and the vertical motion at the base of the Ekman layer.

$$w_E(-\infty) - w_E(0) = \int_{-\infty}^{0} \left(\frac{\partial u_E}{\partial x} + \frac{\partial v_E}{\partial y} \right) dz \tag{7.8}$$

Utilizing (7.6) in (7.8) and assuming that the vertical current vanishes at the surface yields an expression for the vertical velocity at the bottom of the Ekman layer in terms of the wind stress applied at the surface. We obtain

$$w_E = \frac{\partial}{\partial x}\left(\frac{\tau_y}{\rho_o f} \right) - \frac{\partial}{\partial y}\left(\frac{\tau_x}{\rho_o f} \right) = \vec{k} \cdot \vec{\nabla} \times \left(\frac{\vec{\tau}}{\rho_o f} \right) \tag{7.9}$$

where $\vec{\tau} = \vec{i}\,\tau_x + \vec{j}\,\tau_y$, and $\vec{i}$, $\vec{j}$, and $\vec{k}$ are unit vectors in the eastward, northward, and upward directions respectively. The vertical velocity at the base of the Ekman layer in the open ocean is thus seen to be proportional to the curl of the wind stress vector divided by the Coriolis parameter. Where lateral boundaries are present, the dependence of upwelling on the wind stress is more complex, but wind stress near boundaries can produce large upwelling even without significant wind stress curl.

7.5.2 Sverdrup Flow and Western Boundary Currents

To understand the large-scale response of the ocean to wind stress forcing it is useful to consider the balance of vorticity in the ocean. *Vorticity* is the curl of the velocity vector and is a measure of the local rotation of the fluid. For the large-scale motions of the atmosphere and the ocean it is the vertical component of absolute vorticity that is of most interest.[3]

$$\zeta_a = 2\Omega \sin \phi + \vec{k} \cdot \vec{\nabla} \times \vec{V} = f + \zeta_r \tag{7.10}$$

The absolute vorticity is the sum of planetary vorticity (f), which is associated with the rotation of Earth, and relative vorticity (ζ_r), which is associated with the fluid motion relative to the surface of Earth. For flow without friction, the absolute vorticity remains constant unless a parcel of fluid changes its shape. If a parcel of fluid maintains its shape while moving equatorward to a latitude where Earth's rotation is less, then the fluid parcel must exhibit a change in relative vorticity in order to maintain a constant absolute vorticity. Stretching of fluid parcels along the direction of the rotation vector will cause the absolute rotation rate to increase.

The famous oceanographer H. U. Sverdrup showed that in the interior of the ocean an approximate balance exists between the meridional advection of planetary vorticity and the stretching of planetary vorticity by divergent motions:

$$\beta \upsilon = f \frac{\partial w}{\partial z} \tag{7.11}$$

where $\beta = \partial f / \partial y$. If we integrate (7.11) from the bottom of the ocean to the bottom of the Ekman layer and use (7.9), we obtain

$$\beta V_I = f \vec{k} \cdot \nabla \times \left(\frac{\vec{\tau}}{\rho_o f} \right) \tag{7.12}$$

where

$$V_I = \int_{-D_o}^{-z_E} \upsilon \, dz \tag{7.13}$$

and it has been assumed that the Ekman layer is thin compared to the depth of the ocean, and that the vertical velocity is zero at the bottom of the ocean, where $z = -D_o$.

If we add the interior meridional transport velocity (7.13) to the Ekman layer meridional transport in (7.6) we obtain

$$V_I + V_E = \frac{1}{\beta} \vec{k} \cdot \nabla \times \left(\frac{\vec{\tau}}{\rho_o} \right) \tag{7.14}$$

so that the total meridional mass transport is proportional to the curl of the wind stress.

If we consider the ocean circulation between the tropics and midlatitudes, the

[3]See, *e.g.*, Pedlosky (1987).

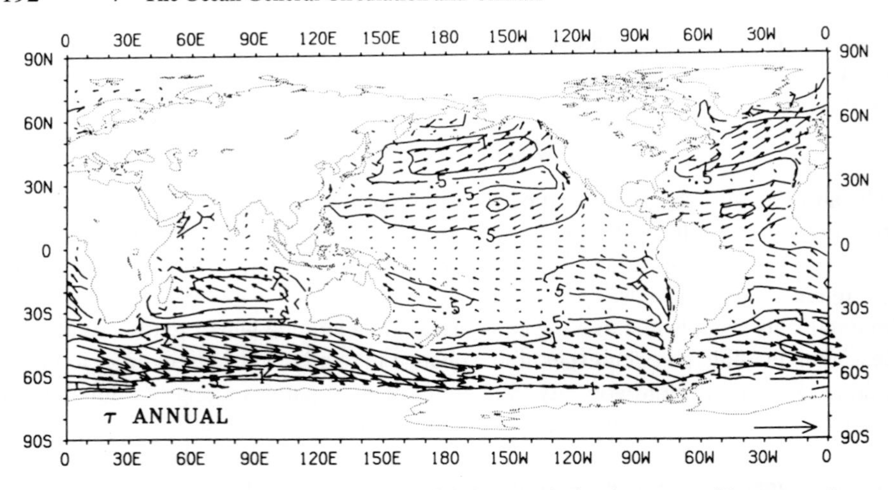

Fig. 7.13 Annual mean wind stress over the global oceans depicted as vectors. The arrow at bottom right corresponds to 5 dyn cm^{-2} and contours of magnitude of 0.5, 1, 2, and 3 dyn cm^{-2} are plotted (1 dyn cm^{-2} = 0.1 N m^{-2}). [From Trenberth *et al.* (1990). Reprinted with permission from the American Meteorological Society.]

wind stress varies from westward in the tropical easterlies to eastward in the midlatitude westerlies (Fig. 7.13). Thus a negative wind stress curl is applied to the ocean, and according to (7.14), we should expect the water transport in the ocean to be equatorward. Physically, the wind stresses are causing the water to rotate about a vertical axis in a direction that is opposite to the rotation of Earth. To maintain a steady state in the face of this application of anticyclonic rotation, water must drift toward lower latitudes. The reduction in absolute vorticity is thus expressed as a decrease in the planetary vorticity of fluid parcels, and a steady state with constant relative vorticity can be maintained.

According to (7.14), the meridional transport in the ocean will be equatorward everywhere, so long as the wind stress curl is negative. How, then, can the conservation of mass and vorticity be jointly satisfied if the wind stress curl is everywhere negative? How does the water transported equatorward return to high latitudes and close the circulation of mass and vorticity? The western boundary currents observed in the midlatitude oceans are the solution to this dilemma.

A simple model can be constructed by adding a lateral diffusion term to the vorticity equation (7.11) that produces a steady gyre circulation with the northward-flowing return current intensified along the western margin of the ocean basin,[4] much like the observed northward flow is intensified in the western boundary currents. As the water flows poleward along this western margin, the planetary component of vorticity (f) increases because of its dependence on latitude. If the absolute vorticity of the fluid parcels were to be conserved, then their relative vorticity would

[4]Stommel (1948); Munk (1950); Pedlosky (1987).

have to change to compensate for the increase in planetary vorticity. By flowing along the western margin of the ocean basin in a narrow current, the poleward return flow of the wind-driven circulation is able to collect enough vorticity in the same sense as Earth's rotation to arrive at middle latitudes with a vertical component of absolute vorticity near that of the planetary vorticity at these latitudes, so that the magnitude of the relative vorticity remains reasonably small and steady. In a simple linear model with lateral diffusion of momentum, the mechanism for collecting the necessary vorticity is through the lateral friction stresses, which generate vorticity of the proper sign only along the western boundary of the ocean basin. The warm, rapidly flowing western boundary currents of the Atlantic and Pacific Oceans are thus seen to be a response to the wind stress driving the bounded oceans of the Northern Hemisphere.

7.6 The Deep Thermohaline Circulation

The term *thermohaline circulation* is used to denote that part of the oceanic circulation that is driven by water density variations, which are in turn related to sources and sinks of heat and salt. It is traditional in oceanography to organize the discussion of the oceanic circulation into separate wind-driven and density-driven components, although the circulation of the ocean is not a simple addition of the effects of these two types of forcing. Wind driving influences the sources and sinks of heat and salt for the ocean by transporting surface water from the tropics to latitudes where cooling and evaporation can increase its density to very large values. The heat transport associated with the thermohaline circulation affects the SST gradients that help to drive atmospheric winds. Wind driving and density driving of the oceanic circulation are therefore very closely coupled and cannot be easily separated. It is generally true, however, that wind driving is the strongest influence on currents near the surface, and density driving dominates the flow at depth.

Below the thermocline there exist slow circulations driven primarily by density gradients in the deep ocean. These circulations are difficult to measure directly, since the currents associated with them are very weak, but their nature can be inferred from the distributions of trace constituents of seawater. Away from the surface, temperature and salinity of water masses change very slowly, so that the water masses and their origins can be inferred from the particular combination of temperature and salinity that characterizes them.

Most gases are soluble in water, so that the concentrations of particular gases can also be used to characterize water sources. The saturation concentration of a gas in seawater is the amount that would exist in solution at equilibrium, if seawater at a particular temperature and salinity were exposed to the gas. Saturation concentrations of gases in seawater increase as the water gets colder. For example, the saturation concentrations of oxygen and carbon dioxide in seawater at 0°C are about 1.6 and 2.2 times their values at 24°C, respectively. The concentration of oxygen in

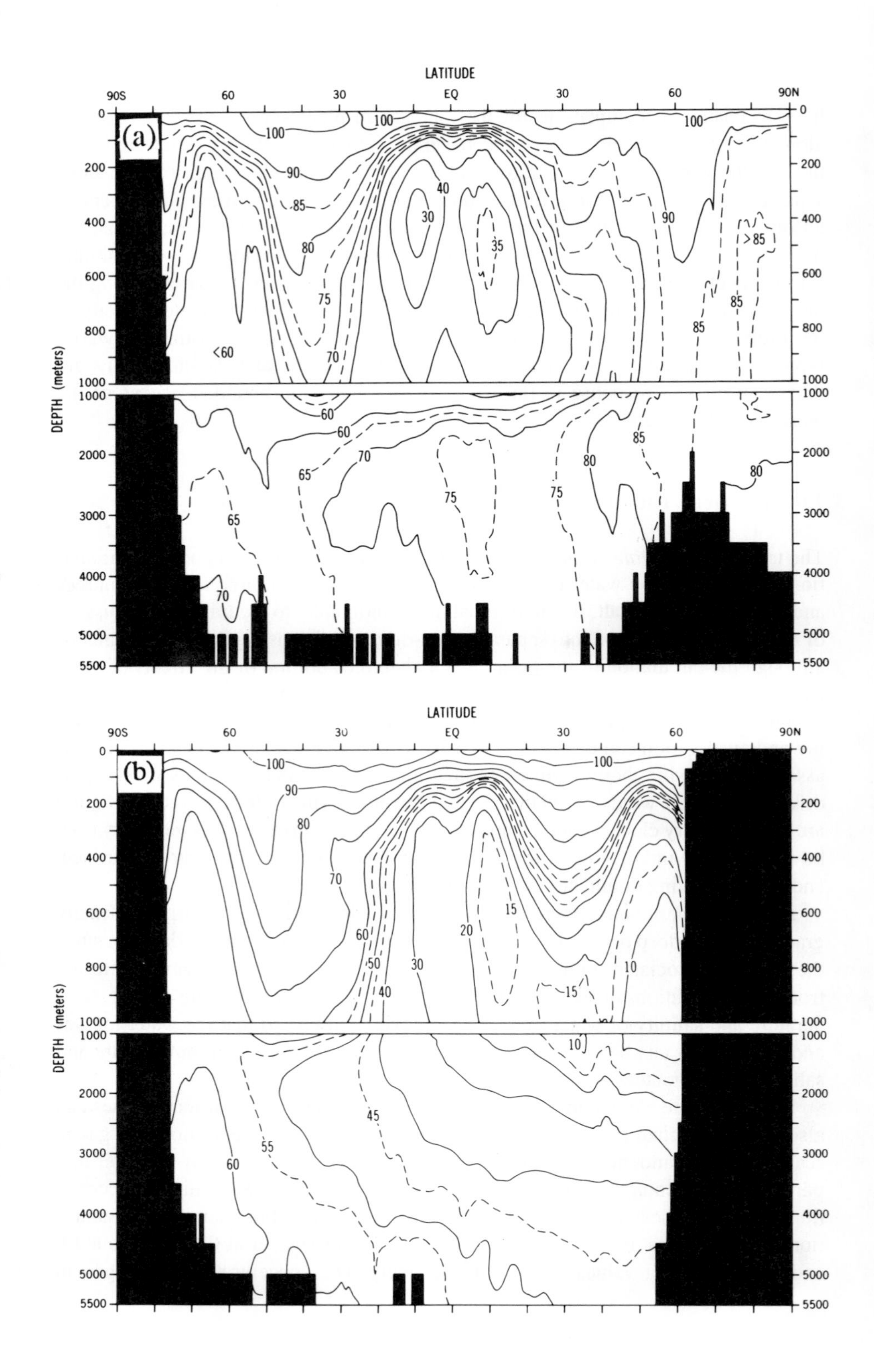

LATITUDE
90S
60
30
EQ
30
60
90N
(a)
DEPTH (meters)
100
90
85
80
75
70
<60
40
30
35
>85
65
60
(b)
20
15
10
50
45
55

surface water is always slightly greater than its saturation value, probably as a result of efficient mixing of bubbles of air into the surface water and production of oxygen in surface waters by photosynthesis. When surface water sinks into the deeper levels of the ocean its source of oxygen is cut off, and the oxygen is slowly consumed by bacteria as they feed on organic matter at depth. One may therefore use the depletion of the oxygen concentration below its saturation value as a measure of the time since the water has been at the surface.

Figure 7.14 shows the oxygen saturation versus depth and latitude in the Atlantic and Pacific Oceans. In the north Atlantic Ocean we observe that high saturation values extend to great depths, and that these high values extend toward the Southern Hemisphere at depths below ~1500 m. We infer then that significant downwelling of water occurs in the north Atlantic and that this water sinks most of the way to the bottom of the ocean and then spreads southward. The distribution of oxygen saturation in the north Pacific Ocean is very different from that in the north Atlantic. In the north Pacific we see no evidence of downwelling, and in fact the oxygen at depth is severely depleted with saturations about 10–15% at latitudes and depths where the oxygen saturation is about 85% in the north Atlantic.

From the oxygen saturation alone we can infer that water from the surface sinks relatively quickly to the deep ocean in the north Atlantic, but that this does not occur in the Pacific. We cannot infer the full circulation of the deep ocean from oxygen alone, nor can we directly infer the subsidence rate, since the rate of oxygen depletion depends on the biological activity at depth, which in turn depends on the rate at which nutrients are supplied to them by deposition from above. The inferences from temperature, salinity, oxygen, and many other tracers suggest a deep-water circulation in the Atlantic like that shown in Fig. 7.15. A large mass of deep water is formed in the northern margin of the ocean, which then flows southward to fill a large fraction of the deep Atlantic (so-called north Atlantic deep water). This water rises toward the surface again in the vicinity of 60°S. Cold, but lower salinity water is formed in midlatitudes of the Southern Hemisphere and wedges itself between the warm surface water and the north Atlantic deep water below. Bottom water is formed around Antarctica, mostly in the Weddell Sea.

The mechanisms of deep-water formation in the north and south Atlantic are believed to be somewhat different. In the north Atlantic, warm, saline water flows poleward from midlatitudes, where the Gulf Stream provides an important source of such water. This water is carried farther poleward into the Norwegian and Greenland Seas, where it is exposed to very cold atmospheric temperatures. The cooling of this saline water produces water that is dense enough to sink to great depths. The surface water in high latitudes of the Southern Hemisphere is relatively fresh, because of the excess of precipitation over evaporation in those latitudes, and there is no warm

Fig. 7.14 Oxygen saturation in percent for the (a) Atlantic and (b) Pacific Oceans. [From Levitus (1982).]

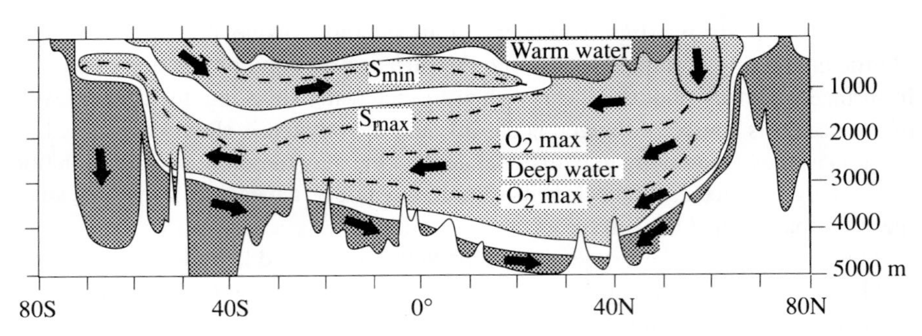

Fig. 7.15 Deep-water flow in the Atlantic Ocean inferred from temperature, salinity, and oxygen measurements. [Adapted from Dietrich *et al.* (1980). Reprinted with permission from Wiley and Sons, Inc.]

western boundary current to carry warm, saline water poleward, since the circumpolar current inhibits efficient transport of water from middle to polar latitudes in the southern oceans. Some saline water reaches the surface of the southern polar oceans by flowing southward at intermediate depths and rising at high latitudes in the south Atlantic. This water is also enhanced in nutrients, since it has spent some time at intermediate depths, where nutrients can be dissolved from falling detritus and photosynthetic organisms do not exist to consume the nutrients. In the Weddell Sea, which is the source region for much of the Antarctic bottom water, the formation of very dense water is more dependent on sea ice production. When ice is formed from seawater, salt is rejected from the crystal structure, resulting in the formation of brine, which adds salt to the water immediately under the ice and thereby increases its density. This cold, saline water is dense enough to sink to the bottom of the Atlantic.

To infer the rate of downwelling it is necessary to use tracers with known decay times such as carbon-14 (^{14}C). Carbon-14 is a radioactive isotope produced naturally in the atmosphere by cosmic rays and by the explosion of atomic bombs in the atmosphere. Since the rate of decay of ^{14}C to ^{12}C is precisely known, the ratio of these isotopes can be used to estimate how long water has been below the surface. The rate and spatial distribution of downwelling can also be inferred from transient trace gases such as chlorofluorocarbons, which are man-made and have been introduced into the atmosphere only in the last 50 years or so.

By combining the evidence available from tracers of seawater movement it has been convincingly shown that at the present time deep ocean water is formed only at high latitudes in the north and south Atlantic. Only in these two locations can water of sufficient density be formed to sink to the deep ocean. From these two locations water spreads out at depth to fill the Pacific and Indian Oceans, where the water gradually rises toward the surface. Since the north Pacific is the farthest from either of these two locations, the water at intermediate depths in the north Pacific is the "oldest" ocean water in the sense that it has been the longest time since this water was exposed to the atmosphere. The fact that the oldest water is not at the ocean bottom suggests that the deep water formed in the Atlantic slowly rises elsewhere, as would be required by the conservation of water mass. The regions of the ocean where deep water can be formed constitute a small fraction of the total surface area of the ocean. For example, 75% of the ocean has potential density greater than 27.4, but only 4% of the surface water has a density that high.[5] It is estimated that the time required to replace the water in the deep ocean through downwelling in the regions of deep water formation is on the order of 1000 years. We may call this the turnover time of the ocean. The thermal, chemical, and biological properties of the deep ocean therefore constitute a potential source of long-term memory for the climate system on time scales up to a millennium. Some chemical properties of the ocean take longer than one turnover time to change significantly, so that the potential exists for ocean memory on time scales longer than the ocean turnover time.

7.7 Transport of Energy in the Ocean

The general circulation of the ocean produces horizontal transport of energy from the tropics to the polar regions that is important for climate. It is not easy to measure this heat transport directly, however. It is difficult and expensive to obtain simultaneous current and temperature measurements from the surface to the bottom of the ocean. Such measurements require a ship or a large buoy and a cable with thermistors and current meters that extends from the surface to the ocean bottom, which is a distance of ~4 km on average. The spatial scales of the motions that are important for heat transport in the ocean are often small compared to the great expanse of an ocean basin, so that it is beyond our means to simultaneously measure current and temperature at enough spatial points and frequently enough in time to continuously monitor the product of velocity and temperature that produces most of the heat transport in the ocean. Attempts have been made to measure a series of profiles across a basin at a particular latitude, but these estimates must be assigned a rather large uncertainty.[6]

An alternative to direct measurement of currents and temperatures is to infer the heat transport of the ocean from the energy balance of Earth or of the ocean.[7] The change in the energy content of a region on Earth ($\partial E_{ao}/\partial t$), following (2.19), is the excess of the net incoming radiation at the top of the atmosphere (R_{TOA}) over the energy exported from that region by transport in the ocean and atmosphere ($\nabla \bullet \vec{F}_{ao}$).

$$\frac{\partial E_{ao}}{\partial t} = R_{TOA} - \nabla \bullet \vec{F}_{ao} \tag{7.15}$$

[5] Sarmiento and Toggweiler (1984).

[6] Bryden and Hall (1980).

[7] Vonder Haar and Oort (1973); Oort and Vonder Haar (1976).

We may assume that the divergence of the horizontal transport can be decomposed into contributions from the atmosphere ($\nabla \bullet \vec{F}_a$) and the ocean ($\nabla \bullet \vec{F}_o$), and then rearrange (7.15) to obtain an expression for the divergence of transport in the ocean.

$$\nabla \bullet \vec{F}_o = R_{\mathrm{TOA}} - \frac{\partial E_{\mathrm{ao}}}{\partial t} - \nabla \bullet \vec{F}_a \tag{7.16}$$

To estimate the effect of ocean heat transport on the energy balance, we need to know the net radiation entering at the top of the atmosphere, the rate of local energy storage, and the rate at which the atmosphere is transporting energy out of the region. The net radiative energy input at the top of the atmosphere may be estimated from measurements taken from satellites. Mapped analyses of winds, temperatures, geopotential energy, and humidity from balloon and satellite measurements are good enough to give reasonable estimates of energy transport in the atmosphere. Storage of energy in the climate system can be estimated from observations, but if we average over an integral number of annual cycles, the energy storage is generally small and can be safely ignored. In this case we may use (7.15) and (7.16) to write

$$R_{\mathrm{TOA}} = \nabla \bullet \vec{F}_{\mathrm{ao}} = \nabla \bullet \vec{F}_a + \nabla \bullet \vec{F}_o \tag{7.17}$$

By integrating the net radiation over latitude, as described in Section 2.9, we may derive the required total meridional transport and subtract from it the atmospheric transport to obtain the oceanic transport.

$$\vec{F}_o = \vec{F}_{\mathrm{ao}} - \vec{F}_a \tag{7.18}$$

Estimates of the total annual mean meridional energy transport required to balance the radiative forcing, atmospheric transport, and the oceanic transport are shown in Fig. 7.16. Such estimates imply that the maximum meridional energy transport by the oceans in the Northern Hemisphere is about the same magnitude as the atmospheric energy transport, but that it occurs at a lower latitude. The total required energy transport is nearly 6 petawatts (PW) and peaks near 45°N. The atmospheric transport has a broad maximum in middle latitudes at ~4 PW and the oceanic flux peaks near 20°N at ~3.2 PW. The total required energy transport is reasonably well measured, but the atmospheric and oceanic fluxes have uncertainties as large as 30% or ~1 PW. These large uncertainties notwithstanding, it is still interesting that the estimated ocean transport is not as sharply peaked in the Southern Hemisphere, and reaches a maximum there of only ~2 PW (1 PW = 10^{15} W). It is possible that the ocean transports are different because of the different land–sea geometry in the two hemispheres and the more developed western boundary currents in the Northern Hemisphere.

The oceanic energy flux can also be estimated from the energy balance at the surface of the ocean. Following (4.1) the energy balance at the surface can be written

$$\nabla \bullet \vec{F}_o = R_s - \mathrm{LE} - \mathrm{SH} - \frac{\partial E_s}{\partial t} \tag{7.19}$$

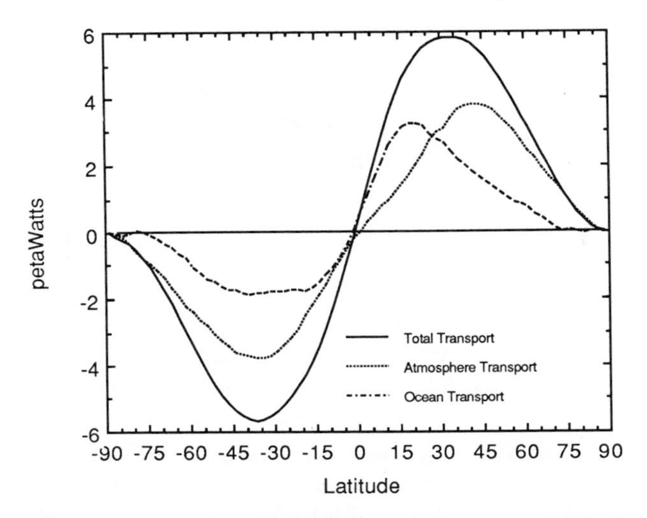

Fig. 7.16 Estimates of the annual mean meridional energy transport required by the energy balance at the top of the atmosphere, estimated from observations in the atmosphere, and the oceanic transport obtained by subtracting the atmospheric energy transport from the total transport required by the annual energy balance (7.16). Net transport inferred from Earth Radiation Budget Experiment data. [Atmospheric transport data from Peixóto and Oort (1984). Used with permission from the American Physical Society.]

The divergence of energy transport in the ocean can be estimated from the ocean surface energy balance if the net radiative heating, the evaporative cooling, the sensible cooling, and the energy storage in the ocean can be estimated. All of these terms are discussed in Chapter 4, and maps of the inferred oceanic flux divergence appear in Fig. 4.18(d). Estimates of meridional energy transport in the ocean obtained from (7.19) are in general agreement with estimates derived from (7.16), in that they show a maximum transport by the oceans at about 20°N (Fig. 7.17). The magnitudes of the estimated transports are considerably less than those of the estimates derived from the residual in the planetary energy balance, however. The surface energy balance method can also provide estimates for individual regions and oceans. The estimates indicate that the Atlantic Ocean transports energy northward across the equator, while the Indian Ocean transports energy southward.

7.8 Mechanisms of Transport in the Ocean

The meridional transport of energy in the oceans is clearly important for climate, but because the transports are not measured directly it is uncertain what types of circulations contribute most to the transport. There are three generic types of circulations that are candidates: wind-driven currents, thermohaline circulations, and midocean eddies.

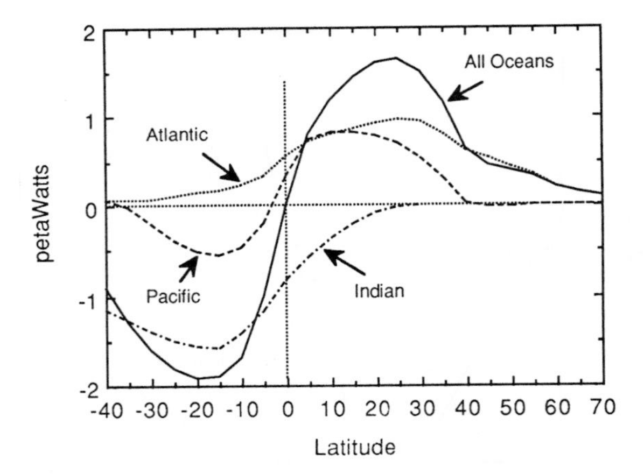

Fig. 7.17 Estimates of the annual mean northward energy transport in the global ocean, and the Atlantic, Pacific, and Indian Oceans derived from the surface energy balance of the oceans (7.18). [Adapted from Hsiung (1985). Reprinted with permission from the American Meteorological Society.]

7.8.1 Wind-Driven Currents

The warm western boundary currents such as the Gulf Stream and the Kuroshio and their associated mid-ocean drift currents seem to play an important role in meridional energy transport in the oceans. The swift, warm currents that flow poleward along the western boundaries of the Atlantic and Pacific Oceans are capable of carrying large amounts of heat poleward. The equatorward flow of relatively cold water in eastern boundary currents also contributes to the poleward energy flux. We can estimate the poleward heat flux associated with the Gulf Stream by considering the product of the mass flux of water and the temperature difference between the Gulf Stream and the average near-surface temperature at the same latitude. The mass flow is the velocity of the current times its area and the water density. From Fig. 7.8(b) we estimate that the Gulf Stream is 60 km wide and 500 m deep and has an average current of 1 m s^{-1}. Assuming a density of 10^3 kg m^{-3}, we can obtain an estimate of the mass flux in the Gulf Stream of 3.0×10^{10} kg s^{-1}.

$$\text{Density} \bullet \text{width} \bullet \text{depth} \bullet \text{speed} = 10^3 \text{ kg m}^{-3} \bullet 60 \text{ km} \bullet 500 \text{ m} \bullet 1 \text{ m s}^{-1}$$
$$= 3 \times 10^{10} \text{ kg s}^{-1}$$

This estimate of 30 Sverdrups[8] agrees with more detailed calculations of the flow through the Florida Straits.[9] If we assume that the Kuroshio has a similar mass flux, then the total poleward mass flux in Northern Hemisphere western boundary currents is 6×10^{10} kg s^{-1}. To calculate the energy flux associated with this mass transport,

[8]The Sverdrup (= 10^6 m^3 s^{-1}) is the traditional oceanographic unit of water flow.
[9]Bryden and Hall (1980).

we need the heat capacity of water (4281 J K^{-1} kg^{-1}) and the temperature difference between the poleward-flowing boundary currents and the equatorward-flowing water at the same latitude. We do not know this temperature difference precisely, and it is very dependent on whether the equatorward flow is above or below the thermocline. It is interesting to consider how big this temperature difference must be in order for the western boundary currents to produce a meridional heat transport of comparable magnitude to the maximum oceanic flux displayed in Fig. 7.16. To obtain an oceanic flux of 3.2 PW requires a temperature difference between the poleward-flowing and equatorward-flowing water of about 12°C.

$$c_w \, \rho_w \, v_w \, \text{area}_w \, \Delta T = 4218 \text{ J K}^{-1} \text{ kg}^{-1} \bullet 6 \times 10^{10} \text{kg s}^{-1} \bullet 12.6 \text{ K}$$

$$= 3.2 \text{ petaWatts}$$

From Fig. 7.8(a) we estimate the average temperature of the Gulf Stream water to be about 22°C. This is considerably warmer than the average temperature of the ocean at these latitudes, which is somewhere near 5°C. If the equatorward return flow is primarily in the interior below the thermocline, then it would be very easy to produce the observed oceanic heat flux with the western boundary currents and an associated colder interior return flow. In any case, it is reasonable to suppose that wind-driven western boundary currents play an important role in meridional energy transport in the oceans.

7.8.2 The Deep Thermohaline Circulation

The mass flow of the deep thermohaline circulation is governed by the rate at which deep water can be formed at high latitudes. In the Northern Hemisphere deep water is formed only in the Atlantic at high latitudes, and the formation rate is quite slow, since it takes several centuries to replace the deep water in the Atlantic. It is estimated that the average rate of deep-water formation in the North Atlantic is $1.5-2 \times 10^{10}$ kg s^{-1} and in the Antarctic Ocean about 1×10^{10} kg s^{-1}. The deep thermohaline circulation is critical for the climate of the far North Atlantic and for deep heat storage, and it may be an important contributor to the energy flow across 20°N.

7.8.3 Mid-ocean Eddies

The Gulf Stream and the Kuroshio spin off long-lived eddies via baroclinic and barotropic instabilities (Fig. 7.9). These are the oceanic analogs to the eddies that produce most of the atmospheric meridional energy transport in midlatitudes. The role of eddies for heat transport in the ocean is likely much less than in the atmosphere, however, because of their smaller spatial scales compared with the scale of the oceans. Moreover, the oceanic eddies are best developed well poleward of the latitude of the maximum oceanic transport. The wind-driven and thermohaline circulations are likely to provide much more important contributions to the meridional heat flux in the subtropics.

Exercises

1. Use the data in Figs. 7.1 and 7.2 to estimate how much the salinity of the surface water of the Arctic Ocean would need to increase before the surface density would equal the potential density at 1000-m depth. How does this compare with the average salinity of the ocean?
2. What depth of seawater would need to freeze in the Arctic Ocean to produce the increase in salinity of problem 1 in the top 100 m of water? Assume that all salt is rejected from sea ice and enters the 100-m layer.
3. With the negative wind stress curl characteristic of today's climate, the wind-driven meridional flow (7.14) is an equatorward drift, which we can hypothesize occurs mostly in the thermocline or above it. How would the net heat transport produced by this drift and its return flow be different if, rather than a warm western boundary jet, the return flow were a slow poleward drift near the bottom of the ocean?
4. Discuss the ways in which the extension of the warm, saline Gulf Stream into the Norwegian and Labrador Seas assists in the formation of dense water that can sink to the depths of the Atlantic Ocean.
5. Use Fig. 7.2 to estimate the initial and final density values of a kilogram of water that starts in the tropics with a temperature of 28°C and a salinity of 35‰ and flows on the surface in the Gulf Stream to the Norwegian Sea, where it arrives with a temperature of –1°C. Assume the water conserves its salinity en route and loses heat by sensible heat transfer.
6. As an alternative to problem 5, assume that the kilogram of water starts in the tropics, but is cooled by evaporation along its route rather than by sensible heat loss. Estimate the mass of water that is lost by evaporation en route, if the parcel arrives in the Norwegian Sea with a temperature of –1°C. Calculate the salinity on arrival, assuming that no horizontal mixing or precipitation occurs, and the salinity is well mixed through the top 100 m of the ocean. What is the density on arrival? When you compare the final density with that obtained in problem 5, is the effect of evaporation on the final density significant? Is it important to know the Bowen ratio for the parcel along its route?
7. Suppose that a wind stress is applied to the ocean, taking the following simple form.

$$\tau_x = \begin{cases} A \cos\left(\dfrac{\pi y}{L}\right) & -L < y < L \\ A & |y| > L \end{cases}$$

Derive an equation for the vertical velocity at the bottom of the Ekman layer assuming a constant Coriolis parameter $f = f_o = 2\Omega \sin 30°$. Derive an equation for the integrated meridional transport V_I, using (7.12) with the f and β appropriate for 30°N latitude. Determine a numeric value for the maximum w_E and V_I using the following constants: A = 2 dyn cm^{-2} = 0.2 N m^{-2}, L = 1500 km, ρ_o = 1025 kg m^{-3}. Plot τ_x, w_E, and V_I on the interval $-L < y < L$. Assuming that the ocean basin is 5000 km wide, calculate the water mass flux at 30°N associated with the interior flow. Compare this number with the estimate for the Gulf Stream mass flux given in Section 7.8.

Chapter 8 History and Evolution of Earth's Climate

8.1 Past Is Prologue[1]

The seasons come and go in an orderly cycle, and plants, animals, and humans adapt to this regular rhythm. Though occasional anomalies and extreme weather events occur, it is natural to think of climate as a constant influence to which life has adapted. In the grossest sense, this view of climate is correct. Life has existed on Earth for at least 3.5 billion years, and the climate has been hospitable enough over that great span of time for life to continue. If we look in more detail at the climates of the past by sifting through the evidence recorded by nature, we find that climate is not so invariable and passive as it may seem. Past variations in climate are of great interest to the climatologist, since they provide clues to the inner workings of the climate system that are difficult to infer in any other way. If past variations in climate can be understood thoroughly, then our chances of anticipating how climate will evolve in the future are greatly increased.

The history of Earth's climate is very complex, and most of this history was not experienced by humans, who have appeared only at the last instant of geologic time. Human ingenuity led to the definition and measurement of temperature, pressure, and other climatic variables only during the last several centuries. Climate variables that may be directly measured using modern instruments constitute what we may call the instrumental record. In addition to the instrumental record, we have historical data that stretch back farther in time, but are less quantitative. Such records include grape harvests in France, wheat harvests in ancient Egypt, and many bits of climate information contained in written accounts. A large amount of the latter type of information, going back several thousand years, is contained in the ancient Chinese literature. Most of what we know of the deeper history of Earth's climate, before the invention of writing, has come from deducing climate variations from information left behind by various natural recording systems. These recording systems include physical, biological and chemical information

[1] "We all were sea-swallowed, though some cast again,
And, by that destiny, to perform an act
Whereof what's past is prologue, what to come,
In yours and my discharge."—Antonio in William Shakespeare's *The Tempest*, Act 2, Scene 1.

contained in lake and ocean sediments, in terrestrial sediments, in ice sheets and in tree rings. Information of this type, which we may call paleoclimatic data, can be used to derive time series of climatic information for many thousands of years into the past.

8.2 The Instrumental Record

Much of the early development and use of the thermometer took place in Florence in the mid-seventeenth century; however, the usefulness of early measurements was limited by the lack of a standard scale and calibration. In 1742, Anders Celsius invented the Celsius temperature scale (called the *centigrade scale* until 1948), but regular measurements of air temperature came into fashion considerably after its acceptance. The barometer was invented by Torricelli in 1644. It came more quickly into widespread use, because of greater ease of its calibration and the perceived relationship between pressure and weather changes, which gives it a predictive capability. For a while, owning and operating a barometer was a status symbol.

Temperature records as long as 150–200 years are available for only a few locations. Manley (1974) constructed a temperature record for central England going back to 1659. Other temperature time series include those for, Berlin, Germany beginning in 1700; de Bilt, Netherlands in 1706; Germantown, Pennsylvania in 1731; Milan, Italy in 1740; and Stockholm, Sweden in 1756. Sufficient measurements to define the hemispheric or global-mean surface air temperature are available for only about the last century. Some would debate whether the quality of the network of instruments for observing temperature is adequate for climate purposes even today. Most of the variance of temperature is associated with seasonal and latitudinal variations, or weather. Any climatic trends are a small signal among much larger magnitude variations, particularly within the rather short period for which the instrumental record is available.

Because of concern over human-induced climate change, considerable effort has been devoted to developing estimates of global-mean surface air temperature based on the instrumental record and evaluating temporal trends in those estimates over the past century. The instrumental temperature record shows generally increasing temperatures since the 1880s, with a warming of about 0.5°C culminating in about 1940 (Fig. 8.1). After 1940, global-mean surface temperature appears to have declined about 0.2°C until 1975, when it began to increase again. The decade of the 1980s was the warmest in the instrumental record, up to that time. The interpretation of the apparent changes in the instrumental record of global-mean surface air temperature during the last century is a subject of some controversy. If one accepts that the instrumental record is a good measure of real global climate variations, it is

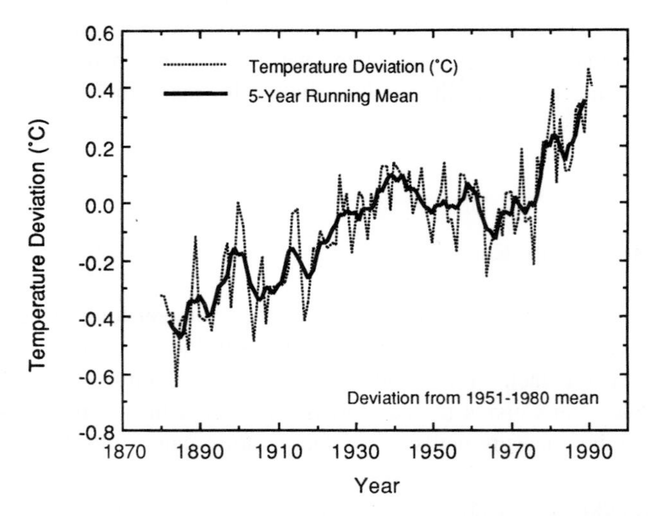

Fig. 8.1 Instrumental temperature record as analyzed by Hansen and Lebedeff (1988) with additions. Values given are departures from the long-term mean in °C.

still a difficult matter to assign causes to these variations, since so many possible causes can be enumerated and relatively few methods are available for distinguishing among them.

A portion of the warming in the instrumental record can be ascribed to urban heat island effects, which are not indicative of a global climate warming. As cities surrounding the observing sites became more densely populated and built up, the local environment was warmed by waste heat and the conversion of the land surface from forest or grassland to asphalt roads and tall buildings. About 0.1°C of the warming in the instrumental record during the twentieth century can be attributed to urban heat island effects, leaving a substantial fraction of the approximately 0.5°C apparent warming over this period to be explained by other causes. Another trend of interest in the instrumental record is the significant decrease with time of the diurnal temperature range over North America.[2] Much of this reduction in day–night temperature difference is associated with higher minimum nighttime temperatures. The reason for this trend in the diurnal temperature cycle over land is unclear, but it is consistent with increasing greenhouse gases [less IR (infrared) cooling of the surface at all times] and increasing haze (less daytime solar heating of the surface), which can each be associated with an increasingly industrialized world. The same temperature trends might be caused by an increase of cloudiness, however, which could be either natural or of human origin.

[2]Karl *et al.* (1984).

8.3 The Historical Record

The historical record consists of written or oral accounts of past events such as agricultural data (records of grape harvest dates in France, wheat yields in Egypt, cherry blossoming dates in Japan, etc.), weather diaries, diaries of a few intellectuals with an interest in weather (e.g., Aristotle, Tycho Brahe, and Johannes Kepler), ship logs, and ancient writings. Other more quantitative information might include river levels and flooding information. The flood levels for the Nile have been recorded for millennia, with especially careful records since the foundation of Islam in 622 A.D. Historical climate information is sometimes subjective, often sketchy, and limited to a few geographic areas. Often extreme events find their way into historical accounts, and these may not be representative of the climate of the time.

Although evaluation of the climatic information in ancient art and literature requires great care and skill, much useful and well-supported information can be deduced from historical accounts. Interesting examples include written and archeological evidence of cities that once flourished and then disappeared because of changes in the environment, including the climate. In some cases it is clear that these changes were in part natural and in part human-induced. An example is the city of Ephesus, an ancient Greek city in what is now Turkey. During the fourth century B.C., Ephesus was a major economic power and a center of learning, with an important port and a thriving local agriculture. An amphitheater in Ephesus had a seating capacity of 24,000. Before the city developed, the surrounding hills were covered with oak trees. As the city grew, these hills were cleared and given over to pasture and to cultivation of wheat. At the same time the climate appears to have become more arid. The combination of changing land use and changing climate ultimately led to the demise of the city. Erosion led to the filling of the harbor with silt from the surrounding hills. By the ninth century the harbor was unusable, despite considerable dredging and several moves of the port. Ephesus was left out of the economic life of the Mediterranean and fell into ruin.

Climatic information can also be contained in art. Rock paintings in the Sahara Desert dating from about 7000 years ago show pictures of the hippopotamus and grazing animals that can no longer exist there (Fig. 8.2). This suggests that the central and southern Sahara was much wetter during the warm epoch following the end of the last glaciation. Paleoclimatic evidence, such as hippopotamus bones dating from the same period and evidence of much higher lake levels, confirms the inference derived from the paintings.

8.4 Natural Recording Systems: The Paleoclimatic Record

Another very important class of information on past climates exists, for which the recording does not require a human attendant, and from which information for

Fig. 8.2 Cave painting of a hippopotamus and baby taken from the Tassili-n-Ajjer region of the Sahara Desert. Created in the Pastoral period, it may date from 5000 B.C. (Photo by Kazuyoshi Nomachi. © Photo Researchers, Inc., New York.)

thousands or millions of years into the past can be obtained. This information is catalogued in various types of "natural" recording systems. These give continuous time histories that go back a few million years and are especially good for the last 100,000 years. The sources of paleoclimatic data are outlined in Table 8.1.

The data with the most accurate time chronology come from analysis of annual tree rings (dendrochronology). The width and structure of the tree rings give some information on the climatic conditions when the tree ring was formed. By correlating tree-ring characteristics with contemporaneous instrumental data on temperature and precipitation, a transfer function can be developed to convert tree-ring characteristics into weather information. Once this transfer function is established and verified, tree-ring data can be used to estimate characteristics of the climate on an annual basis for thousands of years into the past, well before instrumental data became available. A reconstruction of precipitation in Iowa based on a 300-year-old tree correctly shows the "dust bowl" period of the 1930s and the lesser drought of the 1950s (Fig. 8.3). It also shows four decades of dryness comparable to the 1930s that occurred prior to the beginning of the instrumental record.

Table 8.1

Characteristics of Some Paleoclimatic Data Sources

Proxy data source	Variable measured	Continuity of evidence	Potential geographic coverage	Period open to study (year B.P.)	Miminum sampling interval (year)	Usual dating accuracy (year)	Climatic inference
Layered ice cores	Oxygen isotope concentration, thickness (short cores)	Continuous	Antarctica, Greenland	10,000	1–10	±1–100	Temperature, accumulation
	Oxygen isotope concentration (long cores)	Continuous	Antarctica, Greenland	100,000+	Variable	Variable	Temperature
Tree rings	Ring-width anomaly, density, isotopic composition	Continuous	Midlatitude and high-latitude continents	1,000 (common) 8,000 (rare)	1	±1	Temperature, runoff, precipitation, soil moisture
Fossil pollen	Pollen-type concentration (varved core)	Continuous	Midlatitude continents	12,000	1–10	±10	Temperature, precipitation, soil moisture
	Pollen-type concentration (normal core)	Continuous	50° S to 70° N	12,000 (common) 200,000 (rare)	200	±5%	Temperature, precipitation, soil moisture
Mountain glaciers	Terminal positions	Episodic	45° S to 70° N	40,000	—	±5%	Extent of mountain glaciers
Ice sheets	Terminal positions	Episodic	Midlatitude to high latitudes	25,000 (common) 1,000,000 (rare)	—	Variable	Area of ice sheets
Ancient soils	Soil type	Episodic	Lower and midlatitudes	1,000,000	200	±5%	Temperature, precipitation, drainage
Closed-basin lakes	Lake level	Episodic	Midlatitudes	50,000	1–100 (variable)	±5% ±1%	Evaporation, runoff, precipitation, temperature
Lake sediments	Varve thickness	Continuous	Midlatitudes	5,000	1	±5%	Temperature, precipitation
Ocean sediments (common deep-sea cores, 2-5 cm/1000 years)	Ash and sand accumulation rates	Continuous	Global ocean (outside red clay areas)	200,000	500+		Wind direction
	Fossil plankton composition	Continuous	Global ocean (outside red clay areas)	200,000	500+	±5%	Sea-surface temperature, surface salinity, sea ice extent
	Isotopic composition of planktonic fossils; benthic fossils; mineralogic composition	Continuous	Global ocean (above $CaCO_3$ compensation level)	200,000	500+	±5%	Surface temperature, global ice volume; bottom temperature and bottom water flux; bottom water chemistry
(rare cores, >10 cm/1000 years)	As above	Continuous	Along continental margins	10,000+	20	±5%	As above
(cores, <2 cm/1000 years)	As above	Continuous	Global ocean	1,000,000+	1000+	±5%	As above
Marine shorelines	Coastal features, reef growth	Episodic	Stable coasts, oceanic islands	400,000	—	±5%	Sea level, ice volume

(From National Research Council, 1975.)

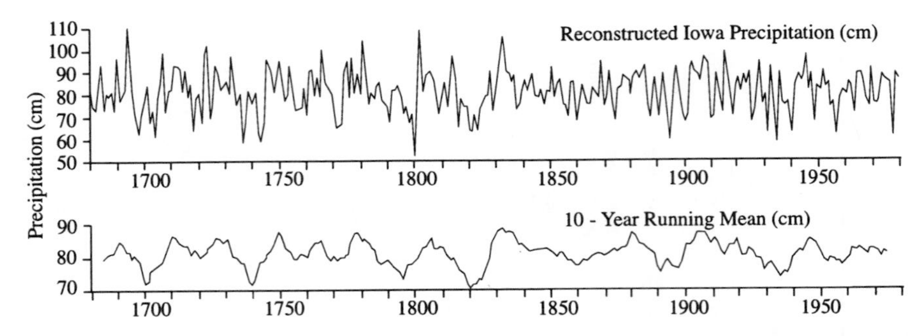

Fig. 8.3 Time history of precipitation in Iowa derived from tree-ring analysis. [From Duvick and Blasing (1981).]

The paleoclimatic data source yielding the longest continuous record is the ocean sediment core. Ocean sediments are laid down over time, so that a core drilled into the sea bottom contains a time history of the environment at the time the layers in the core were formed. The time resolution attainable with sediment cores is determined by the sedimentation rate and the degree of stirring of the most recent sediment by bottom-dwelling animal life such as worms, a process called *bioturbation*. Ocean cores measure the history of the ocean ecology as laid down in the sediment. The sediment includes organic matter and the shells of tiny sea creatures (foraminifera, coccoliths, etc.). Each sea creature has a particular niche in the ecology of the ocean. The relative abundance of some species is related to sea surface temperature (SST), so that relative abundance can be used to estimate SST in the past. In addition, the relative abundance of oxygen isotopes in deep-sea sediment cores provides an indication of the global mass of water tied up in terrestrial ice. Because lighter isotopes (^{16}O) are more readily evaporated, they are more likely to be removed from the ocean and incorporated in ice sheets. Ice ages are therefore marked by a relatively rich mixture of heavy isotopes (^{18}O) in ocean waters. These higher ^{18}O ratios are imprinted in the shells of sea creatures and find their way into sediments that collect on the ocean bottom. The oxygen isotope abundance in ocean sediment can thus be used to estimate the global volume of ocean water that is tied up in continental ice sheets and mountain glaciers.

Similar stratigraphic evidence can be collected in ice cores. Aerosols trapped in ice give evidence of past atmospheric dust loading and aerosol chemistry. Air bubbles in ice provide information on the gaseous composition of the atmosphere at the time the bubbles were formed. In the upper parts of ice sheets, annual layers can be identified. Deeper into the ice, and thus farther back in time, the ice becomes compacted and stretched out as the ice flows away from the accumulation regions toward the regions where the ice sheet melts or breaks off into icebergs at a marine boundary. Thus the ability to resolve rapid changes decreases with age in the core.

8.5 A Brief Survey of Earth's Climate History

8.5.1 Early Earth

Earth is believed to have formed nearly 5 billion years ago by the accretion of solid materials and gases that formed the solar nebula. The modern atmospheres of Earth, Mars, and Venus have less abundance of noble gases (e.g., argon, neon) than is typically present in stellar nebula, so it is assumed that any primordial atmosphere collected from the gases in the solar nebula was removed early in Solar System history. The primordial atmosphere could have been removed during the collisions of large planetesimals with Earth or been swept away by the more intense solar wind of the early sun. The modern atmosphere is thus a secondary one, which has resulted from the release of gases that were mechanically or chemically trapped inside the solid Earth during its formation and were released slowly over time. The process of releasing gases from the interior of a planet may be called *outgassing,* and it continues on Earth today, most obviously in the form of volcanic eruptions. The gases released are primarily water vapor, carbon dioxide, and nitrogen.

We can hypothesize that immediately after the removal of their primordial atmospheres, the inner planets each consisted of an essentially bare ball of rock, although some gases may have continued to be collected from space during and after the sun's T-Tauri phase. Assuming that the release of heat from within the planets was negligible, their surface temperatures would have equilibrated at the emission temperature of a sphere with the albedo of bare rock and with a solar constant appropriate to the distance of the respective planet from the sun. Thus Venus, because of its greater proximity to the sun, started out at a higher temperature than Earth. Being farther from the sun, Mars would have started at a lower temperature than Earth. With time, the temperatures of the three planets would gradually increase, because of the greenhouse effect of the water vapor and carbon dioxide that would steadily build up in their atmospheres through outgassing.

On Venus the surface temperature stayed well above the condensation point for water during its evolution, so that all of the water stayed in the atmosphere. Eventually much of the hydrogen in the atmosphere escaped to space, and a thick atmosphere of primarily carbon dioxide remains. This thick atmosphere gives a very high surface temperature of about 700 K. The process that led to these conditions on Venus is often termed the *runaway greenhouse effect,* but it is also possible that oceans did not form because Venus simply received less water during its formation.

On Mars, it has been speculated that the temperature mostly stayed below the freezing point for water so that the greenhouse effect may never have significantly influenced the surface temperatures there. Because of its relatively low temperature, the partial pressure of water reaches the freezing point before substantial amounts of water vapor can accumulate in the atmosphere. The greenhouse effect on Mars may therefore never have really gotten established, and most of the outgassed water could have frozen on the surface. A considerable amount of frozen carbon dioxide may also exist on Mars. An alternative to this view of Mars is suggested by some surface terrain with erosional features that appear to have been formed by a running fluid, presumably water, very early in Mars' history.

The distance of Earth from the sun is such that outgassed water vapor condensed into oceans and the surface temperature remained near the triple point of water, where liquid water, water vapor, and ice can exist simultaneously. The collection of outgassed water vapor in the oceans stabilized the climate and provided the environment that is appropriate for the development of life as we know it. Once the condensation point was reached on Earth, any additional water vapor that was outgassed went into the oceans and the infrared opacity of the atmosphere stopped increasing. The carbon dioxide was dissolved into the ocean and eventually reached an equilibrium with carbonate rocks. At this point the atmosphere was composed mostly of molecular nitrogen. Within about a billion years after the formation of Earth, life developed, leading to green plant photosynthesis, which produces molecular oxygen in the atmosphere. With the development of an oxygen-rich atmosphere came the stratospheric ozone layer, which by protecting the surface from harmful ultraviolet-B radiation, allowed life to emerge from the water and occupy the land surface.

Thus Earth stayed cool enough to avoid the runaway greenhouse effect that occurred on Venus, largely because of its favorable distance from the sun, which allowed the oceans to form. Life developed quickly in the conditions provided by early Earth, and temperatures favorable for the life forms we are most familiar with ($0 < T < 42°C$) have been maintained since. Theories for the life cycles of stars suggest that, early in Earth history, the solar constant would have been as much as 30% less than its current value. This suggests that some compensating changes in atmospheric radiative properties may have occurred over time to keep the surface temperature in a favorable range for life to progress.

8.5.2 The Last Billion Years

The continents have probably been drifting as part of lithospheric plates for the last several billion years. Geologists have attempted to reconstruct the positions of the continents for the past 500 million years or so, but little is known about the positions of the continents before that. Little is known about climate back farther than about a billion years, except for fossil evidence that life existed then and that liquid water was present on the surface (Fig. 8.4).

The climate was cold enough for large ice sheets to form during at least three periods over the last billion years. There is some evidence for ice sheets during the late Precambrian 600 million years ago, more extensive evidence for glaciation during the late Paleozoic 300 million years ago, and fairly detailed documentation for the late Cenozoic glaciation of the last several million years or so, which we are still experiencing. The late Paleozoic glaciations appear to have occurred in southern Africa, South America, and Australia, at a time when all these continents were bunched

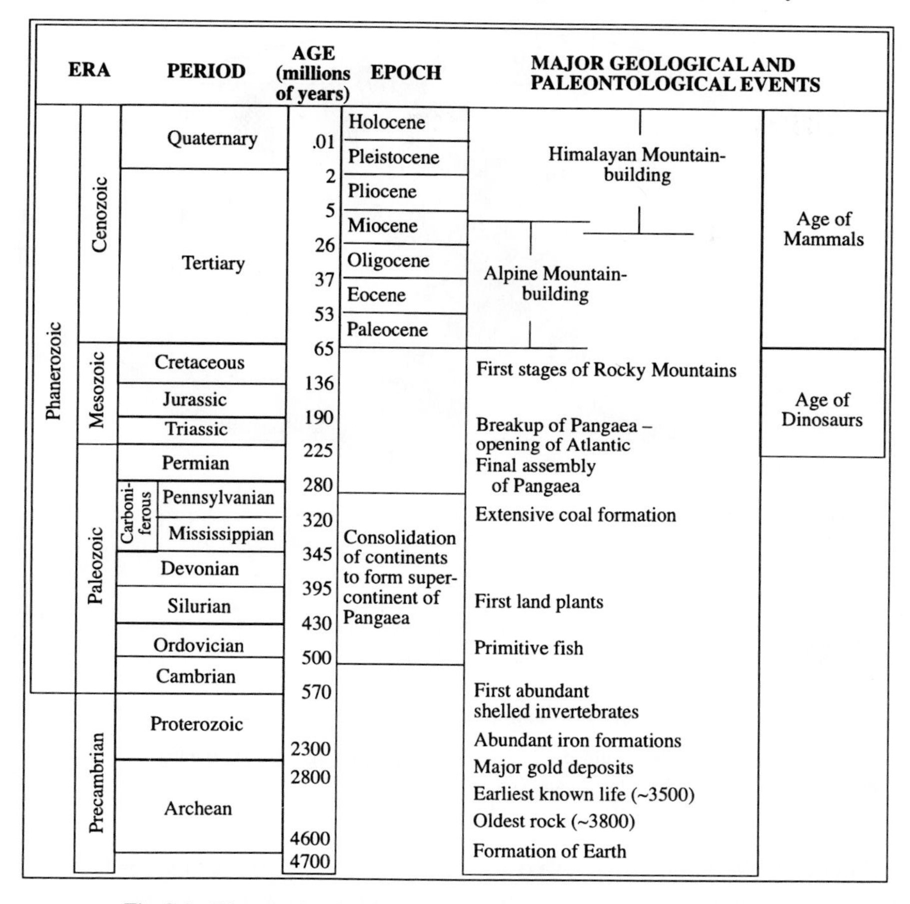

Fig. 8.4 Diagram showing the major geologic ages. [From Crowley (1983).]

together into a single, large, land mass in high southern latitudes (Fig. 8.5). The glaciated areas were attached to Antarctica and were not far from the South Pole. About 300 million years ago, the continents moved apart and began to drift slowly toward their present positions.

The most studied of the nonglacial climates that occurred between these major glacial ages is the most recent one, the middle Cretaceous, spanning the period from about 120 to 90 million years ago. The continents were in a configuration very different from that seen today. North America and Europe were close together, as were South America and Africa, so that the Atlantic Ocean did not yet exist. Africa had not yet joined Europe, India was a large island in the midlatitudes of the Southern Hemisphere, and a shallow tropical sea, called the Tethys Sea, extended from Central America to Indochina. Australia was in middle to high latitudes and still attached

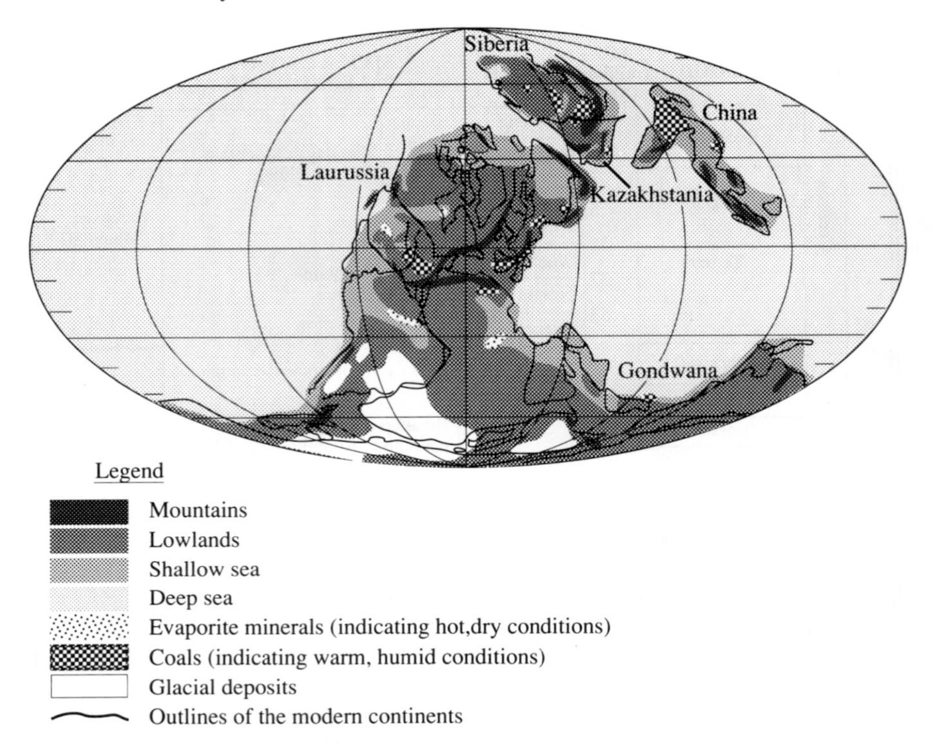

Fig. 8.5 Position of the continents 310–300 million years ago. [From Bambach *et al.* (1981). Reprinted with permission from the Sigma Xi Scientific Research Society.]

to Antarctica. Very little land ice existed during this period, so that the sea level was about 100 m higher than now. This greater mass of ocean water flooded about 20% of the continental areas that are now above water, including large portions of western Europe, northern Africa, and North America.

Abundant fossil evidence exists to suggest that the middle Cretaceous was substantially warmer than the present. Plant habitats moved up to 15 degrees of latitude poleward of their current positions, as did the ranges of many animals. Sedimentary evidence also suggests that coal and other deposits indicating warm, moist climate formed during this period in lands that were then above the Arctic Circle, and dinosaurs ranged there also. Isotopic evidence taken from bottom-dwelling organisms suggests that the deep ocean temperature was between 15 and 20°C. Such warm temperatures in the deep ocean are inconsistent with the presence of much marine ice in high latitudes.

Unusually large amounts of carbon in the form of coal and oil deposits were laid down during the Cretaceous. Many of the major oil deposits were formed during this period. This high rate of coal and oil formation indicates that the cycling of carbon was very different than it is now. A change in the cycling of carbon might logically be associated with the apparently different oceanic circulation, deep ocean temperatures, and the great expanse of relatively warm, shallow seas. The warmth of this period might have been associated with enhanced levels of carbon dioxide in the atmosphere, and climate models suggest that the altered continental positions combined with atmospheric CO_2 concentration 4 times higher than modern values come close to explaining the apparent warmth in polar latitudes during the Cretaceous.[3] Direct paleoclimatic evidence of the atmospheric CO_2 abundance during the Cretaceous is inconclusive, though studies of ancient soils suggest that CO_2 was 4–6 times higher during the Mesozoic than at present.[4] Increased CO_2 levels may have been associated in part with increased volcanism during this period, for which considerable suggestive evidence also exists.

The Cretaceous ended with a bang about 65 million years ago. About 75% of the total number of living species became extinct, some of them very abruptly. The dinosaurs were among several groups of species that disappeared entirely, and the ascent of mammals as the dominant group of large vertebrates followed and persists today. The so-called *K-T boundary,* marking the end of the Cretaceous (K) and the beginning of the Tertiary (T) periods, is identified in sedimentary records by a well-defined layer of clay deposits. This layer contains anomalously large amounts of iridium. The iridium content of Earth's crust is lower than the amount in the stuff from which the solar system was formed, because much of Earth's iridium was carried to the core with molten iron during Earth's formative stages. Comets and meteors retain their cosmic abundance of iridium, and the clay layer with its high iridium content could have come from a comet or meteor with a diameter of about 10 km.[5] This iridium anomaly at the K-T boundary can be found over a large portion of the globe and is a rare event in Earth history. The evidence is strong that a large meteorite or comet collided with Earth about 65 million years ago, and evidence of a likely impact site has been found near the Yucatan Peninsula.

The impact of such a large bolide with Earth would be a spectacular event with very serious environmental consequences. The shock would heat the atmosphere in the vicinity of the impact, which would produce large amounts of nitrogen oxides. In the stratosphere nitrogen oxides would lead to a loss of ozone. The nitrogen oxides would likely leave the atmosphere as highly acidic rains. The impact of the projectile and its penetration of Earth's crust would cast fine particles high into the atmosphere and even into low orbits, where they might persist for several weeks or months. This pall of dust could block out the sunlight needed by photosynthetic organisms and would lead to a cooling of the surface after the heat of the impact had dissipated. All of these effects—shock heating, fires, acid rain, ozone loss, dark and cold—would be stressful to many species and may have been the cause of the rapid species extinctions at the end of the Cretaceous.

[3]Barron and Washington (1985).
[4]See, e.g., Cerling (1991).
[5]Alvarez (1987).

8.5.3 The Last 50 Million Years

Changes over the last 50 million years seem to be related mostly to the movement of the continents away from Antarctica. During this period, South America and Australia moved northward from the edge of Antarctica to their present positions as a result of sea floor spreading and movement of the continental plates. As the Drake Passage between South America and Antarctica was opened, the southern ocean circulation became circumpolar, and the temperature of the surface and deep waters gradually cooled by more than 10°C. Glaciers developed over this period on Antarctica, and an east Antarctic ice sheet formed about 14 million years ago.

Pollen from ocean cores shows that cool temperate forests existed on the Antarctic continent until 20 million years ago. Ice volume over Antarctica increased to reach its present value about 5 million years ago, and the current Northern Hemisphere polar ice sheets first appeared about 3 million years ago. The overall trend in the last 50–100 million years was for the climate to cool from the warm climate of the middle Cretaceous to our present Quaternary ice age.

8.5.4 The Last 2 Million Years

The last 1.8 million years constitute the Quaternary period, the most recent geologic age and the one during which *Homo sapiens* developed. This period is characterized by the presence of a large amount of land ice, which varied from the amount we have today to much larger amounts during periods of glacier advance. The advance and retreat of this land ice can be inferred from glacial deposits and from the ratios of oxygen isotopes in ocean sediment cores. The deviation of the ratio of the heavier isotope to the lighter isotope from a standard ratio is denoted by $\delta^{18}O$,[6] which is normally given in parts per thousand. Figure 8.6 shows a time series of $\delta^{18}O$ in ocean cores for the last 2.5 million years. The last 700,000 years were marked by wide swings that indicate a large shift in the amount of land ice present. These swings had a characteristic interval of about 100,000 years between succeeding periods of maximum glaciation. Glaciers advanced and retreated in both hemispheres simultaneously. Prior to about 700,000 years ago the swings were more frequent and less extreme. The dominant period of the glacial–interglacial swings in the early part of the record is about 41,000 years. The spectrum of $\delta^{18}O$ variance for the record in Fig. 8.6 is shown in Fig. 8.7. Predominant contributions to the variability are made by variations with periods around 41, 100, and 480 thousand years, with an additional accumulation of variance around 23 thousand

---

[6] $$\delta^{18}O = \frac{\left({}^{18}O/{}^{16}O\right)_{\text{sample}} - \left({}^{18}O/{}^{16}O\right)_{\text{standard}}}{\left({}^{18}O/{}^{16}O\right)_{\text{standard}}} \times 1000.$$

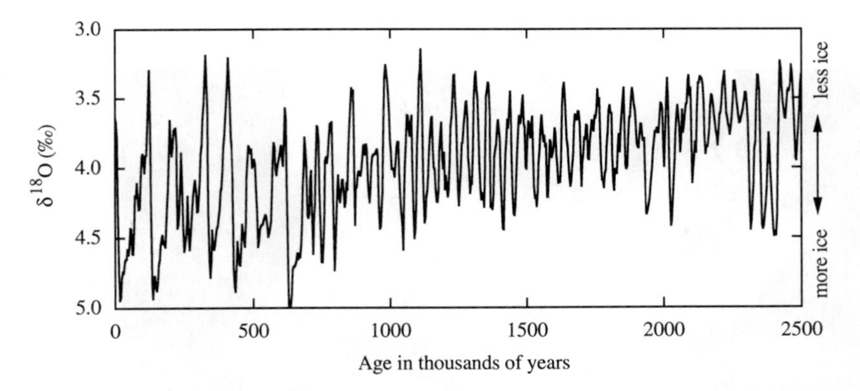

Fig. 8.6 History of δ^{18}O over the last 2.5 million years derived from several ice cores. [Plot made from data provided by M. E. Raymo and previously published in Raymo *et al.* (1990). Reprinted with permission from Elsevier Scientific Publishers.]

years. We will see later that some parameters of Earth's orbit vary approximately with these periods.

8.5.5 The Last 150,000 Years

About 125,000 years ago the land ice on Earth reached a minimum amount, comparable to today's interglacial conditions. In the intervening period the land ice gradually increased, while undergoing some minor oscillations, until the ice volume

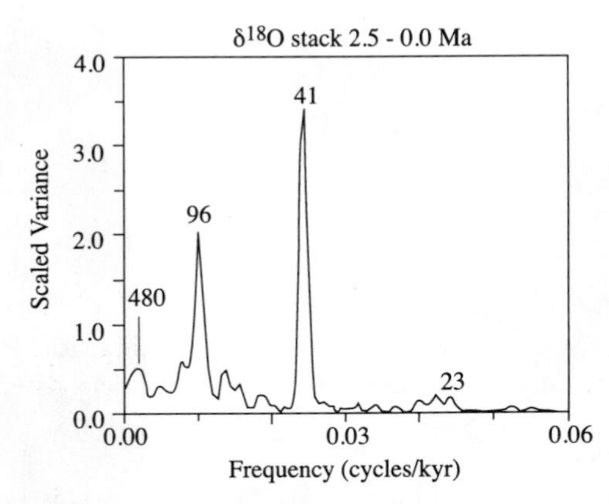

Fig. 8.7 Frequency spectrum of the variance of the record in Fig. 8.6 for the period from 2.5 million years ago to the present. Periods corresponding to peaks are indicated in thousands of years. [From Raymo *et al.* (1990). Reprinted with permission from Elsevier Scientific Publishers.]

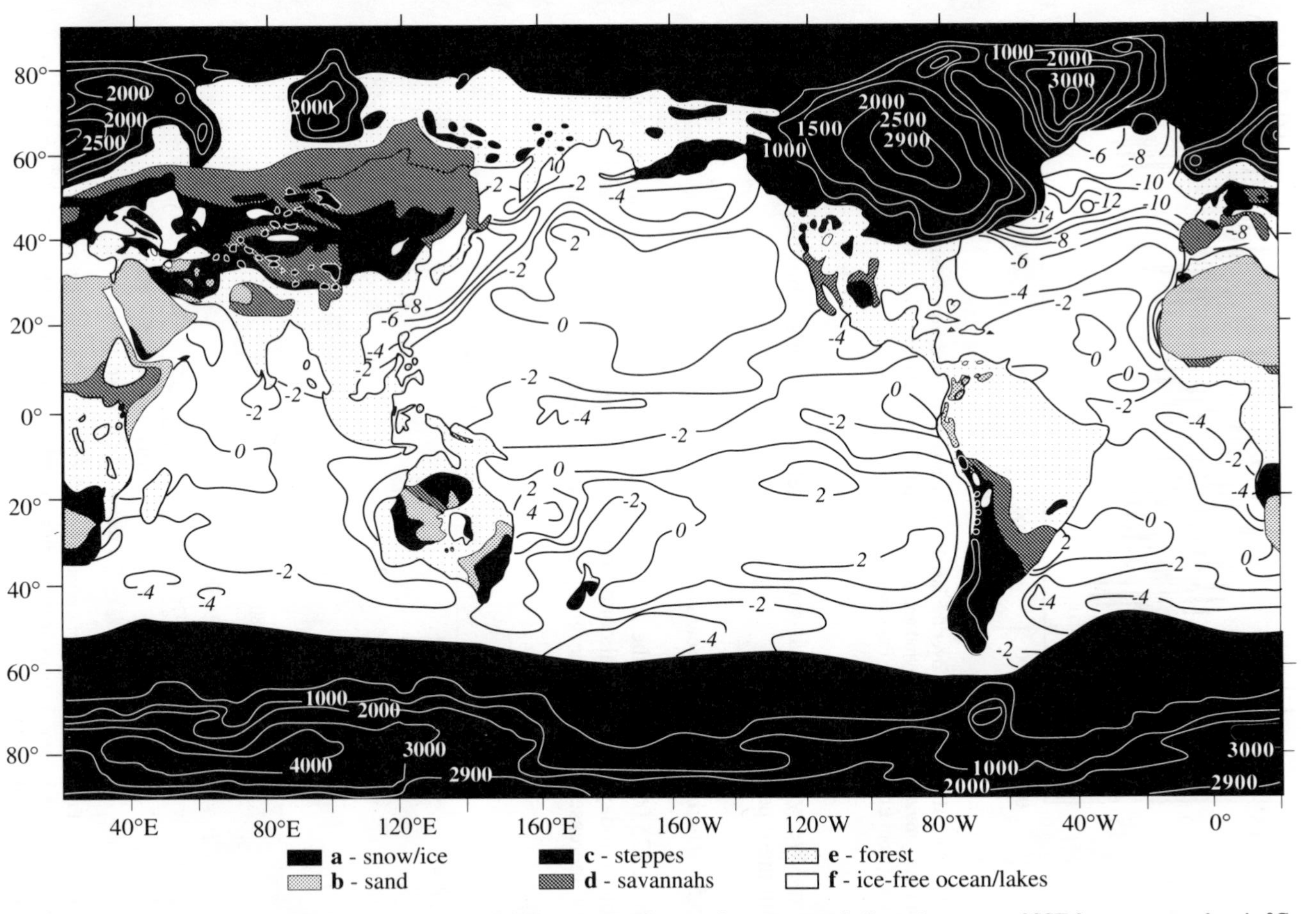

Fig. 8.8 Earth's surface conditions during August 18,000 years ago. Contours in ocean areas indicate departures of SST from present values in °C. Contours in snow and ice regions indicate depth of ice in meters. [From CLIMAP Project Members (1976), © by the AAAS.]

reached a maximum about 20,000 years ago. The most recent and one of the more abrupt swings in global ice volume occurred in the last 20,000 years, between the last glacial maximum and the current interglacial period. The extent of ice cover at the last glacial maximum can be inferred from a variety of paleoclimatic evidence, including the physical evidence left behind by the huge ice sheets. A synthesis of this information indicates a giant ice sheet covering much of northern North America, and another large ice sheet in northern Eurasia, which were 3–4 km thick (Fig. 8.8). The weight of these ice sheets was so great that the supporting crust was depressed by nearly a kilometer. The plastic deformation of the crust under these massive ice sheets may have been one reason for the rapid removal of ice at the end of major glacial maxima during the last 700,000 years. As the crust yields under the ice, the altitude of the ice is lowered, thus bringing the ice surface to higher ambient-air temperature. Also, glacial lakes may form in the depression made in the crust by the ice sheet, and seawater may flow into the depression and assist in the melting of glacier ice. The crust near the locations of the maximum thickness of the major ice sheets is still rebounding from the removal of the ice sheets. Because so much water was tied up in these large continental ice sheets, sea level was about 120 m lower 20,000 years ago than at present. The $\delta^{18}O$ records and modeling studies suggest that this was about the maximum amount of water that could be moved to continental ice sheets, given the continental positions and climate regime of the late Quaternary.

The high-latitude oceans also experienced major changes during the last glacial maximum. The Gulf Stream turned more sharply eastward at about 45°N and warm currents did not extend northward into the Greenland and Norwegian Seas as at present. It is likely that during winter the sea ice extended equatorward of 50°N, so that a vast area of icecovered surface extended from about 45°N to the pole in the Northern Hemisphere, except in the Pacific Ocean and in those parts of Asia that were too dry to sustain substantial surface icecover. A simultaneous glacier advance was experienced in the Southern Hemisphere, where sea ice around Antarctica was greatly expanded, and mountain glaciers covered parts of Australia, Africa, and South America. During this time tropical continental regions appear to have been drier than at present, but snowlines on tropical mountains dropped about 1000 m. Midlatitude continental regions near the ice sheets were also drier for the most part, and wind-blown dust deposits in these areas indicate dry windy conditions at the equatorward margins of the great ice sheets.

Careful analysis of ice cores from the Greenland and Antarctic ice sheets can reveal much about changes in the climate system over the last major glacial-interglacial cycle. From air bubbles trapped in ice, the composition of the atmosphere as a function of depth in a core can be inferred. Figure 8.9 shows a 160,000-year record of atmospheric CO_2 concentration as determined from a 2200-m-deep ice core from Vostok, Antarctica (78°S, 107°E). The history of CO_2 concentration closely tracks the temperature. The last glacial maximum (about 20,000 years ago) was marked by CO_2 concentrations of about 190 ppmv, compared to the modern preindustrial level

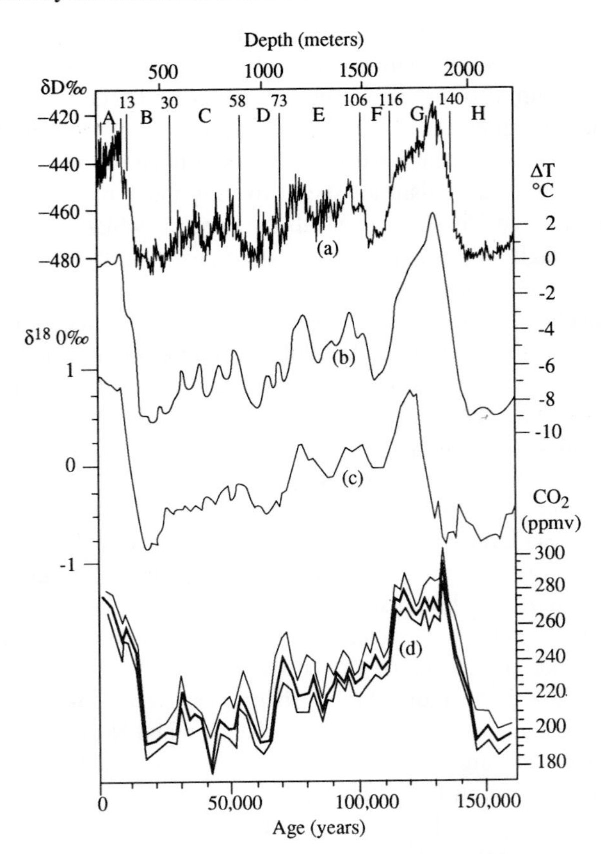

Fig. 8.9 160,000-year time history of (a) deuterium anomaly (δD) of the Vostok, Antarctica ice core [From Lorius *et al.* (1985), © Macmillan Magazines Limited]; (b) smoothed air temperature inferred from the Vostok (δD) record [From Jouzel *et al.* (1987), © Macmillan Magazines Limited]; (c) marine $\delta^{18}O$ record which is indicative of global ice volume [From Martinson *et al.* (1987)]; (d) CO_2 concentration from the air bubbles in the Vostok ice core with uncertainty limits indicated by the envelope of light lines [From Barnola *et al.* (1987), © Macmillan Magazines Limited].

of about 280 ppmv and the 1990 value of about 350 ppmv. Air bubbles trapped in ice have also shown that methane (CH_4), another radiatively active gas, has also undergone significant variations. During the last glaciation, methane concentrations were only 350 ppbv, compared to the preindustrial level of 650 ppbv and the 1990 value of about 1650 ppbv. According to measurements taken in ice samples, methane in the atmosphere remained near 650 ppbv for thousands of years before an obviously anthropogenic increase started about 1800 A.D. These large changes of atmospheric carbon dioxide and methane between glacial and nonglacial times indicate major changes in the cycling of carbon between ocean, atmosphere, and land, which accompanied the growth and decay of ice sheets.

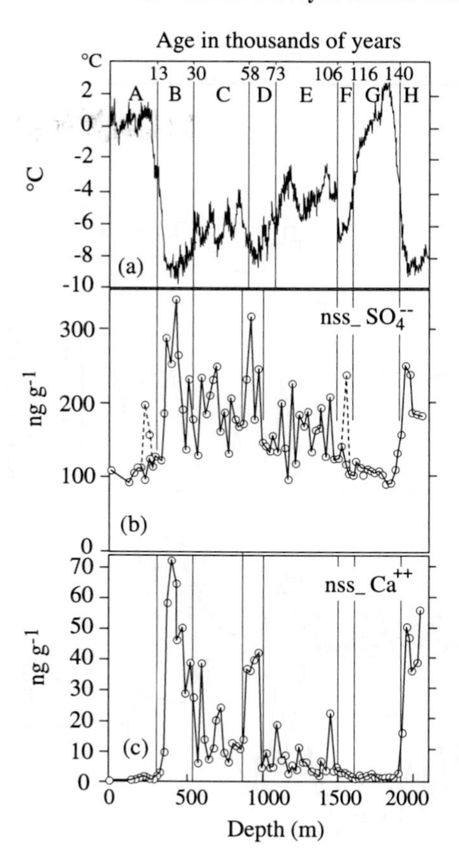

Fig. 8.10 Temperature, non sea-salt sulfate, and non sea-salt calcium in the Vostok ice core. [From Legrand *et al.* (1988), © Macmillan Magazines Limited.]

Ice cores also contain information on past variations in the amount and chemical composition of aerosols deposited in high latitudes. Figure 8.10 shows time series of temperature, non-sea-salt sulfate (nss-sulfate), and non-sea-salt calcium in the Vostok ice core. The nss-sulfate and terrestrial calcium are much higher during glacial than interglacial ages. After correcting for the effects of changing accumulation rates, these changes indicate that nss-sulfate concentration in the atmosphere was 20–40% higher during the glacial maxima. The increased atmospheric composition of sulfate is thought to be related to increased production of sulfur-bearing gases by life in the ocean, most probably dimethyl sulfide gas. Changes in nss-calcium are believed to be related to atmospheric dust, principally in the form of calcium carbonate ($CaCO_3$).

Variations in atmospheric carbon dioxide of the magnitude observed must be related to changes in the gross biological productivity of the oceans through photosynthesis.

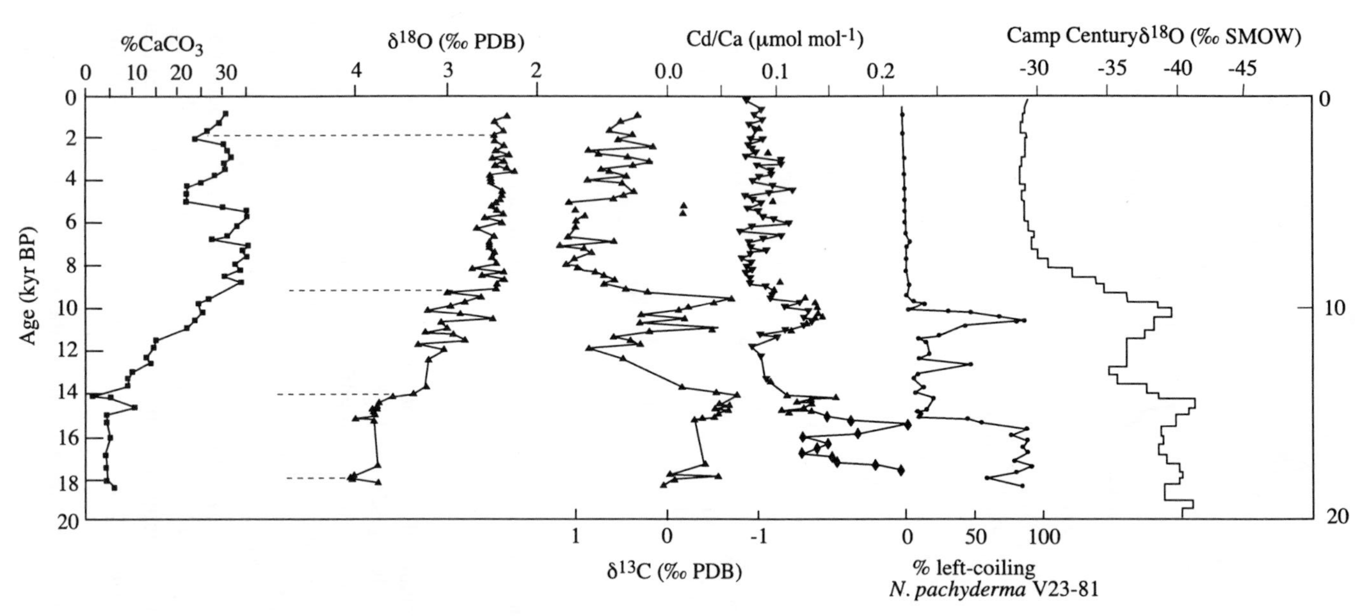

Fig. 8.11 Time histories taken from an ocean sediment core raised from 4450-m depth at 34°N, 58°W. Shown are the calcium carbonate content, ^{18}O isotope abundance, ^{13}C isotope abundance, and the cadmium/calcium ratio. Also shown are the *N. pachyderma* abundance from a north Atlantic sediment core and ^{18}O isotope fraction from a Greenland ice core (*N. pachyderma* are planktonic foraminifera that prefer cold water). Note the brief cool episode about 11,000 years ago during a period when the global ice volume was declining. [From Boyle and Keigwin (1987), © Macmillan Magazines Limited.]

An increase in productivity during the ice age would reduce the partial pressure of CO_2 in the water and thereby reduce the CO_2 content of the atmosphere, which is constrained by the ocean concentration on these time scales. If it is assumed that the nss-sulfate is produced biogenically through dimethyl sulfide (DMS) emissions by phytoplankton, then the sulfate increase is consistent with an increase of gross ocean productivity during the ice age, assuming that the rate at which organisms in the ocean produce DMS rises and falls with gross productivity.

As discussed in Chapter 7, at the present time most of the deep water in the global ocean is formed in the north Atlantic Ocean. Because it is formed by cooling water that has spent some time near the surface where nutrients are efficiently consumed, north Atlantic deep water (NADW) is generally deficient in nutrients. In contrast, the source of deep water formed around Antarctica is water that has spent considerable time at intermediate depths collecting nutrients and then rises to the surface in the southern ocean. Antarctic deep water is therefore comparatively nutrient rich. Because most deep water is currently formed from low-nutrient surface water in the north Atlantic and gradually gathers nutrients as it flows away from the formation region, a substantial gradient in deep water nutrient content exists between the Atlantic and the Pacific oceans.

Evidence from ocean sediment cores suggests that the contribution of the north Atlantic to deep-water formation was greatly diminished during the glacial maximum of about 20,000 years ago. Associated with the low nutrient content of NADW are relatively high levels of $\delta^{13}C$, a measure of the ratio of ^{13}C to ^{12}C, and low levels of the cadmium to calcium ratio, Cd/Ca. A time history of past levels of $\delta^{13}C$ and Cd/Ca in deep waters can be obtained from analysis of sediment cores. The shells of bottom-dwelling organisms contained in sediment cores retain a signature of the deep-ocean properties at the time the shells were formed. Records of $\delta^{13}C$ and Cd/Ca from ocean sediments indicate that the rate of deep-water formation in the north Atlantic Ocean was greatly reduced during the last glacial maximum, and nutrients in the deep ocean were more uniformly mixed (Fig. 8.11).

The formation of deep water in the north Atlantic and the associated heat and water transports that were suppressed during the last glacial maximum began again about 14,000 years ago at a time when the $\delta^{18}O$ in ocean cores indicates that the land ice volume began to decrease rapidly. The proxy data indicate that deep-water formation in the north Atlantic ceased again for a period between about 13,000 and 11,000 years ago. During the same interval fossil data indicate that the climate in Europe returned nearly to glacial conditions after having warmed considerably during the period from 15,000 to 13,000 years ago. This return to cold conditions about 12,000 years ago is documented in many ways. The cold event is known as the *Younger Dryas,* because pollen data indicate that forests that had recently developed in Europe during the aborted warming following the ice age were suddenly replaced again by arctic shrubs, herbs and grasses, including the herbaceous plant *Dryas octopetella*. This cold event was confined primarily to the lands bordering the North

Atlantic, and ended abruptly about 11,000 years ago. The event is also evident in time series of $\delta^{18}O$ in ice cores from Greenland (Fig. 8.11). $\delta^{18}O$ in ice cores is a proxy for the air temperature in the vicinity of the glacier. When the temperature near the ice sheet is low, it is more likely that the heavier isotope of oxygen would have condensed out before reaching the ice sheet, because of its lower saturation vapor pressure, so lower $\delta^{18}O$ in glacial ice is associated with colder temperatures. Greenland ice cores indicate a local cooling of about 6°C during the Younger Dryas cold event.

It has been hypothesized that the brief shutdown of the deep thermohaline circulation of the North Atlantic during the Younger Dryas event was associated with the melting of the North American Laurentide ice sheet. The ice sheet melted most rapidly from its southern flank at first and the majority of this melted water flowed down the Mississippi River draining into the Gulf of Mexico. Several great meltwater lakes formed in the depression of Earth's surface left by the retreating ice sheet, and one of these occurred in what is now southern Manitoba. This paleo-lake has been named Lake Agassiz after the geologist, Louis Agassiz, who in 1837 was an early and ardent proponent of the idea that Earth had undergone an ice age. As the ice sheet retreated farther north, a channel was opened that allowed the meltwater lake to drain eastward down the St. Lawrence River to the North Atlantic Ocean.[7] This diversion is dated at about 12,000 years ago from land evidence and by $\delta^{18}O$ records in sediment cores from the Gulf of Mexico, which went from isotopically light to heavy as the low $\delta^{18}O$ meltwater was suddenly diverted away. The supply of freshwater to the north Atlantic via the St. Lawrence reduced the salinity of the surface ocean waters. Since high salinities are critical to attaining the densities required to form deep water, the supply of freshwater was sufficient to cut off the thermohaline circulation of the north Atlantic. With the thermohaline circulation went its associated heat transport, resulting in large local cooling of the climate. The thermohaline circulation restarted about 11,000 years ago when the meltwater was again directed down the Mississippi, and has continued to operate as the climate warmed to today's interglacial conditions.

8.5.6 The Last 10,000 Years

The global climate warmed following the last glacial maximum. Melting of the great ice sheets took about 7000 years from about 14,000 to about 7000 years ago, with considerable variation in the dates of final melting for different ice sheets. The early Holocene from about 10,000–5500 years ago was a time when Northern Hemisphere summer climates were warmer than today's. This period was also characterized by remarkable changes in the hydrology of the monsoonal climates of Africa and Asia. Evidence from closed-basin lakes indicates that the period from about 9000–6000 years ago was the wettest in northern Africa in the last 25,000 years. It

[7]Broecker *et al.* (1988); Broecker and Denton (1990).

was during this interval that the hippopotamus, the crocodile, and a variety of hooved animals occupied regions in the Sahara and in Saudi Arabia that are now some of the most arid on Earth. Sediment cores from the Arabian Sea indicate that monsoonal winds were stronger during the early Holocene than at present, and sediments from the Indus and Ganges Rivers indicate that precipitation over India was greater then than now. All of this evidence is suggestive of stronger summer monsoons during the early Holocene, which has been related to changes in Earth's orbital parameters, as will be discussed in Chapter 11.

Minor advances of mountain glaciers culminated about 5300, 2800, and 150–600 years ago. The last of these falls within the Little Ice Age, circa 1250–1850 A.D. (Fig. 8.12). Between these cold periods there were relatively warm periods, the best known being around 900–1200 A.D. and during the present century. During the former period, the Norsemen populated Iceland, Greenland, and North America. The Norse Greenland colonies were abandoned at the onset of the Little Ice Age. Fields in coastal Greenland that had grown crops became perennially frozen. This seemingly drastic change, compared to what has occurred elsewhere, is related to the fact that climate changes appear to have larger amplitudes at high latitudes.

8.6 Uses of Paleoclimatic Data

Paleoclimatic data are very useful in developing a scientific understanding of climate change that will be essential to anticipating the nature of future climate changes associated with human activities and natural processes. The description of past climates derived from paleoclimatic data is useful for three reasons: it gives us perspective on the range of climate changes that are possible and likely, it provides clues about how the climate system works, and it provides a data set for testing theories and models of how climate changes.

Paleoclimatic data give us a perspective on what constitutes a "significant" global climate change. During a full-glacial episode like the one 20,000 years ago, the global-mean surface temperature was about 5°C colder than now. Therefore, 1°C in global-mean surface temperature represents about 20% of the temperature change between glacial and interglacial conditions. The current climate is very near the warmest that has been observed in the last million years, so we do not have very good analog information about what constitutes a really warm world climate. The warm period about 9000–6000 years ago was nearly the warmest in the last million years and was probably not as much as 1°C warmer than today. To get climates that were substantially warmer than today one must return to the early Pliocene, about 5 million years ago, or to the Cretaceous period, about 65 million years ago, when continental positions and the configuration of the world ocean were very different from today. The dominant time scales for natural climate changes during the current geologic epoch, the Holocene, seem to be in the tens of thousands to hundreds of thousands of years. Although climate changes on a variety of time scales, the largest

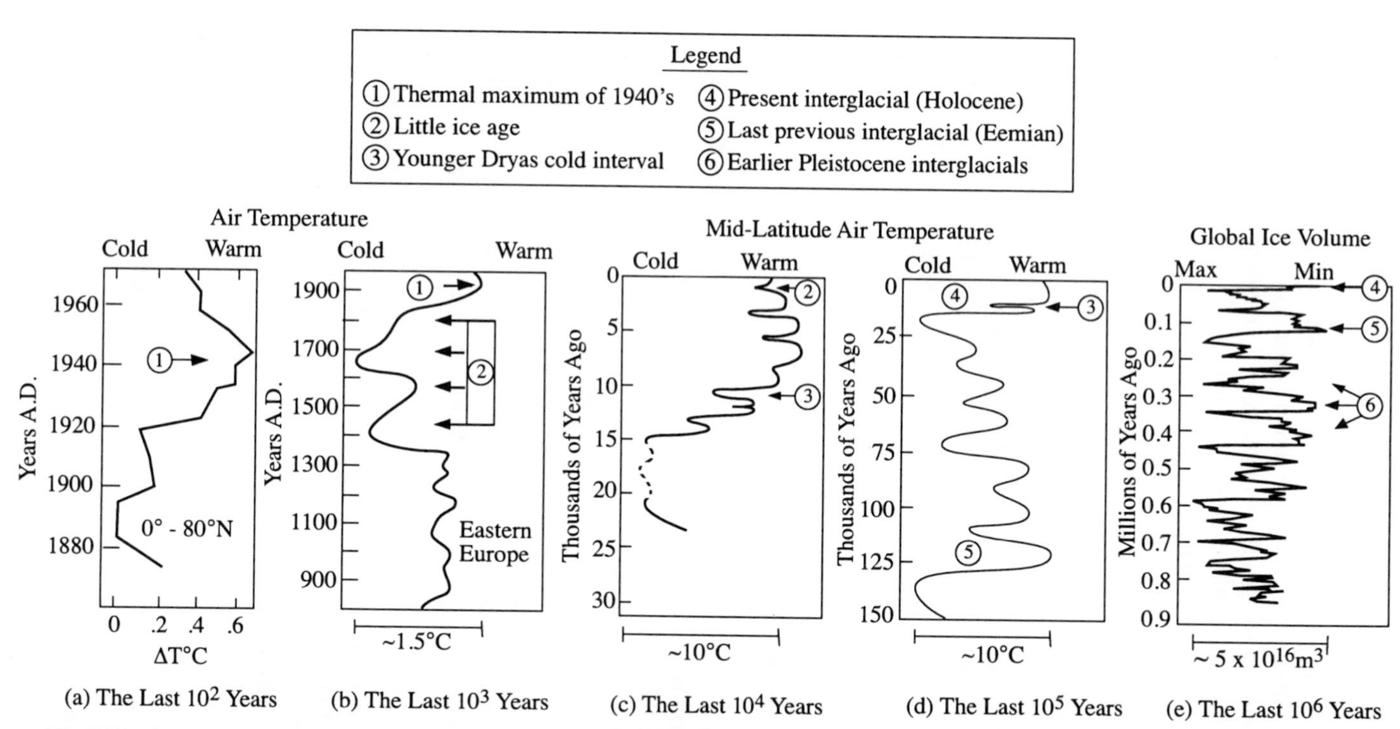

Fig. 8.12 General trends in global climate for a variety of time scales. (a) Changes in the 5-year average surface temperatures averaged from instrumental records over the region 0–80°N [from Mitchell (1963)]. (b) Winter severity index for eastern Europe during the last 1000 years [from Lamb (1969)]. (c) Midlatitude Northern Hemisphere air temperature trends during the last 15,000 years based on changes in tree lines [from LaMarche (1974), © by the AAAS], marginal fluctuations in alpine and continental glaciers [from Denton and Karlén (1973)], and shifts in vegetation patterns recorded in pollen spectra [from van der Hammen *et al.* (1971), © Yale University Press]. (d) Northern Hemisphere air temperature trends during the last 100,000 years based on midlatitude sea surface temperature and pollen records and on worldwide sea-level records. (e) Fluctuations in global ice volume during the last million years as recorded in changes in isotopic composition of fossil plankton in deep-sea core V28-238 [from Shackleton and Opdyke (1973)].

variations appear to occur on longer time scales. Nonetheless, the record also contains evidence of rapid changes of substantial magnitude on time scales of tens to hundreds of years, such as those associated with the Younger Dryas event and the Little Ice Age.

Paleoclimatic data can provide clues about how the climate system works and how the observed variations were produced. The thick ice sheets over the continents during glacial maxima suggest that continental ice sheets may play an active role in climate change, through their albedo, topography, and effects on ocean circulation and biogeochemical cycles. Temperature changes during an ice age seem to be largest at high latitudes and rather small near the equator. Why is this and what does it tell us about the mechanisms of climate change? The tropical land regions appear to have been drier on average during glacial maxima than during warm periods. What are the reasons for these hydrological and vegetation changes and do they play an active role in climate variability? Air trapped in bubbles in glacier ice shows that the carbon dioxide content of the atmosphere was lower (~190 ppmv) during the last ice age than 200 years ago before the industrial revolution (~280 ppmv). This suggests that global ocean productivity was enhanced during the ice age. Does this information indicate a role for biology in determining climate variability?

Paleoclimatic data provide evidence for testing theories and models of global climate change. Time series of global ice volume can be compared to time series of insolation variations associated with cycles in Earth's orbital parameters. Can we explain the relationship between Earth's orbital parameters and global ice volume? If we put inferred distributions of SST, surface ice, vegetation, and atmospheric composition into our best climate models, do they produce a consistent climate in balance with these conditions? How important are the changes in atmospheric composition for explaining glacial–interglacial cycles of climate change? In Chapter 9 we examine the question of how sensitive the climate is to ice sheets and atmospheric composition, and in Chapter 11 we examine the role of orbital parameter variations and other natural climate forcings in determining climate variability.

Exercises

1. Discuss the difficulties with interpreting a curve like Fig. 8.1 or extrapolating it into the future.
2. One of the earliest examples of primitive humans, *Australopithecus,* lived in Africa between about 4 and 1 million years ago. Hominids began using stone tools about 2 million years ago. Place *Australopithecus* and tool use in their proper place in Fig. 8.4. For what fraction of Earth history have hominids been around?
3. How would the differences of atmospheric CO_2 and CH_4 during the last glacial maximum and the present preindustrial era have contributed to the differences in the climates between then and now?

4. Using a current physical map of Earth and the description in Section 8.5.2 as a guide, draw a physical map of Earth for the middle Cretaceous era and label the Tethys Sea.
5. Estimate the mass of the North American and Greenland ice sheets from the data presented in Fig. 8.8.
6. Estimate the rate of freshwater mass production (kg s^{-1}) by the melting of the North American ice sheet, assuming that its melting took 500 years. How does this compare with the rate of deep-water formation in the North Atlantic, which is estimated to be $1.5–2 \times 10^{10}$ kg s^{-1}?
7. Estimate by how much the freshwater flux calculated in problem 6 would decrease the salinity of the surface waters of the far north Atlantic if the flow of the 35‰ salinity water from the south is 2×10^{10} kg s^{-1}?
8. What fraction of the North American ice sheet at its maximum would need to melt in 100 years in order that the flow of freshwater over this period would decrease the salinity by 2‰? How would such a burst of freshwater flux affect the formation of deep water? What would be the response of the climate in the north Atlantic region? How would the rate of melting respond to the climate changes? Can you imagine how the interactions between the melting rate, deep water formation, and climate variations might cause an oscillation during the decline of the ice sheet?

Chapter 9 Climate Sensitivity and Feedback Mechanisms

9.1 Fools' Experiments[1]

Understanding and forecasting of climate change presents a great challenge to geoscientists. The surface climate of Earth is determined by a complex set of interactions among the atmosphere, the ocean, and the land, and these interactions involve physical, chemical, and biological processes. Many of these interactions are only dimly understood, and important interactions and processes have probably yet to be discovered. Because it is apparent that humans have the capability to alter climate, it is necessary that we attempt to understand how the climate system works and make quantitative estimates of how our past, present, and future actions will change it. To make progress on a problem with the intimidating complexity of climate change, the proper response of a scientist is to begin by considering simpler questions and then add complexity as understanding is gained. The lessons drawn from these simple models must be taken seriously, but with the full realization that they may not be a faithful representation of nature.

In the climate modeling problem, complexity is determined by the number and kind of processes or interactions among system components that are included. Once a starting point is chosen, one must decide in what order to bring additional processes or interactions into consideration, and with how much detail it is necessary to represent these processes. These decisions can be made on the basis of how important the processes are for the maintenance of climate, or on the basis of their importance for determining the magnitude of a climate change response to a climate forcing of a given measure. The relationship between the measure of forcing and the magnitude of the climate change response defines what we will call the *climate sensitivity*. A process that changes the sensitivity of the climate response is called a *feedback mechanism*. The strength of a feedback can also be quantified, and will be termed a *positive feedback* if the process increases the magnitude of the response, and *negative* if the feedback reduces the magnitude of the response. To construct a model of climate change, it is logical to order the feedback processes according to importance, and then include the most important feedbacks first. If positive and negative feedbacks of similar strength are possible, then it is important to include both

[1] "I love fools' experiments. I am always making them."—Charles R. Darwin

simultaneously, or otherwise a very inaccurate estimate of the overall system sensitivity can be obtained.

9.2 Objective Measures of Climate Sensitivity and Feedback

The concept of sensitivity can be used to judge which climate processes should be considered first and which can be neglected until later. Sensitivity can also be used as an objective measure of the performance of a model. Quantitatively, sensitivity is the amount by which an objective measure of climate changes when one of the assumed independent variables controlling the climate is varied. For example, we might ask how much the global mean surface temperature, T_s, changes if we vary the solar constant, S_0; however, the global mean temperature is a function of more than the solar constant. Suppose it depends on a number of variables y_j, which are all parametrically related to S_0, $y_j = y_j(S_0)$. Secondary variables would be water vapor, other greenhouse gases, cloudiness, ice cover, land vegetation, and many others. Then the chain rule yields,

$$\frac{dT_s}{dS_0} = \frac{\partial T_s}{\partial S_0} + \sum_{j=1}^{N} \frac{\partial T_s}{\partial y_j} \frac{dy_j}{dS_0} \tag{9.1}$$

where partial and total derivatives are denoted by ∂ and d, respectively, and N is the number of important subsidiary variables. The total change of surface temperature with respect to solar constant is the sum of many contributions each made up of the partial derivative of surface temperature with respect to a secondary variable that controls surface temperature, times the total change of that secondary variable with respect to the solar constant. As an example, if we increase the solar constant, the specific humidity of air will increase because of the dependence of saturation vapor pressure on temperature. The increase of atmospheric water vapor will in turn lead to an increase in surface temperature, because of the greenhouse effect of the added water vapor.

We may simplify the concept of climate sensitivity somewhat by supposing that some climate forcing, dQ, with units of W m^{-2} is applied to the climate system, and ask how it is related to a measure of climate change such as the change in global-mean surface temperature, dT_s. We define the ratio of the climate response to the climate forcing as one measure of climate sensitivity.

$$\frac{dT_s}{dQ} = \lambda_R \tag{9.2}$$

As an example, we can consider the global-mean energy balance at the top of the atmosphere.

$$R_{\text{TOA}} = \frac{S_0}{4}\left(1 - \alpha_p\right) - F^{\uparrow}(\infty) = 0 \tag{9.3}$$

If we imagine that some forcing dQ of the energy balance is applied, so that the climate is driven to some new equilibrium state with a new global-mean surface temperature, then we can use the chain rule to write

$$\frac{dR_{\text{TOA}}}{dQ} = \frac{\partial R_{\text{TOA}}}{\partial Q} + \frac{\partial R_{\text{TOA}}}{\partial T_s}\frac{dT_s}{dQ} = 0 \tag{9.4}$$

from which, using (9.3) we obtain

$$\frac{dT_s}{dQ} = -\left(\frac{\partial R_{\text{TOA}}}{\partial T_s}\right)^{-1} = \left(\frac{S_0}{4}\frac{\partial \alpha_p}{\partial T_s} + \frac{\partial F^{\uparrow}(\infty)}{\partial T_s}\right)^{-1} = \lambda_R \tag{9.5}$$

In this simple model, the climate sensitivity is controlled by two basic feedback processes. One feedback is measured by the dependence of the emitted terrestrial radiation flux on surface temperature, and the other is measured by the dependence of the albedo on surface temperature. If the sensitivity parameter λ_R is known, then we can estimate the global-mean surface temperature change ΔT_s expected from a given climate forcing ΔQ.

$$\Delta T_s = \lambda_R \, \Delta Q \tag{9.6}$$

9.3 Basic Radiative Feedback Processes

9.3.1 Stefan–Boltzmann Feedback

We may use the simple global-mean model of Chapter 2 to evaluate the strength of the negative feedback associated with the temperature dependence of thermal emission. This may be the most important negative feedback controlling the surface temperature of Earth. The emission temperature of a planet is defined by the balance described by (9.3) with the outgoing terrestrial energy emission determined by the Stefan–Boltzmann law, $F^{\uparrow}(\infty) = \sigma T_e^4$. If we ignore the dependence of albedo on temperature for the moment and assume that the surface temperature and the emission temperature are linearly related, then the sensitivity parameter for this model, which cools like a blackbody, is given by

$$\left.\lambda_R\right)_{\text{BB}} = \left(\frac{\partial\left(\sigma T_e^4\right)}{\partial T_s}\right)^{-1} = \left(4\sigma T_e^3\right)^{-1} = 0.26 \text{ K}\left(\text{W m}^{-2}\right)^{-1} \tag{9.7}$$

The estimate of sensitivity obtained here indicates that about a quarter of a degree change in temperature will result from a one Watt per square meter forcing of the energy balance. A 1 W m^{-2} forcing of the energy balance can be produced by about a 5.7 W m^{-2} change in the solar constant, if the albedo is 0.3. If only the Stefan–Boltzmann feedback is considered, it will take about a 22 W m^{-2} or 1.6% change in solar constant to produce a 1°C change in Earth's surface temperature. With such a stable climate it would be difficult to produce the climate changes observed in the past, so we conclude that some strong positive feedback processes must operate.

From radiative transfer model calculations, we can estimate that if we double the carbon dioxide concentration from 300 to 600 ppmv, and leave the temperature and humidity distributions as they are, then the outgoing longwave radiation would decrease by about 4 W m^{-2}, which would result in an increase of the net radiation at the top of the atmosphere by the same amount. This is equivalent to a solar constant increase of 1.67%, assuming a fixed albedo of 0.3. We can use (9.7) in (9.6) to estimate that about a 1°C global-mean surface temperature increase would result from such a climate forcing.

9.3.2 Water Vapor Feedback

One of the most powerful positive feedbacks is associated with the temperature dependence of the saturation vapor pressure of water. As the temperature increases, the amount of water vapor in saturated air increases. Since water vapor is the principal greenhouse gas, increasing water vapor content will increase the greenhouse effect of the atmosphere and raise the surface temperature even further. The dependence of atmospheric water vapor on temperature thus constitutes a positive feedback.

Because much of Earth's surface is wet, the humidity of air near the surface tends to remain close to the saturation vapor pressure of air in contact with liquid water. The temperature dependence of the saturation vapor pressure can be obtained from the Clausius–Clapeyron relationship,

$$\frac{de_s}{dT} = \frac{L}{T(\alpha_v - \alpha_l)} \tag{9.8}$$

where e_s is the saturation vapor pressure above a liquid surface, L is the latent heat of vaporization, T is the temperature, α_v is the specific volume of the vapor phase, and α_l is the specific volume of the liquid phase. From (9.8) we can show that the change in saturation specific humidity divided by the actual value of specific humidity is related to the change in temperature divided by the actual temperature.

$$\frac{dq^*}{q^*} = \frac{de_s}{e_s} = \left(\frac{L}{R_v T}\right)\frac{dT}{T} \tag{9.9}$$

For terrestrial conditions $(L/R_v T) \approx 20$, so that a 1% change in temperature, which is about 3°C, is associated with about a 20% change in saturation specific humidity.

It is observed that the relative humidity of the atmosphere, which is the ratio of the actual to the saturation humidity, tends to remain constant, even when the air temperature goes through large seasonal variations in middle to high latitudes. This is not too surprising, since we know that the relative humidity cannot exceed 100% and vapor is extracted very efficiently from the ocean when the relative humidity of

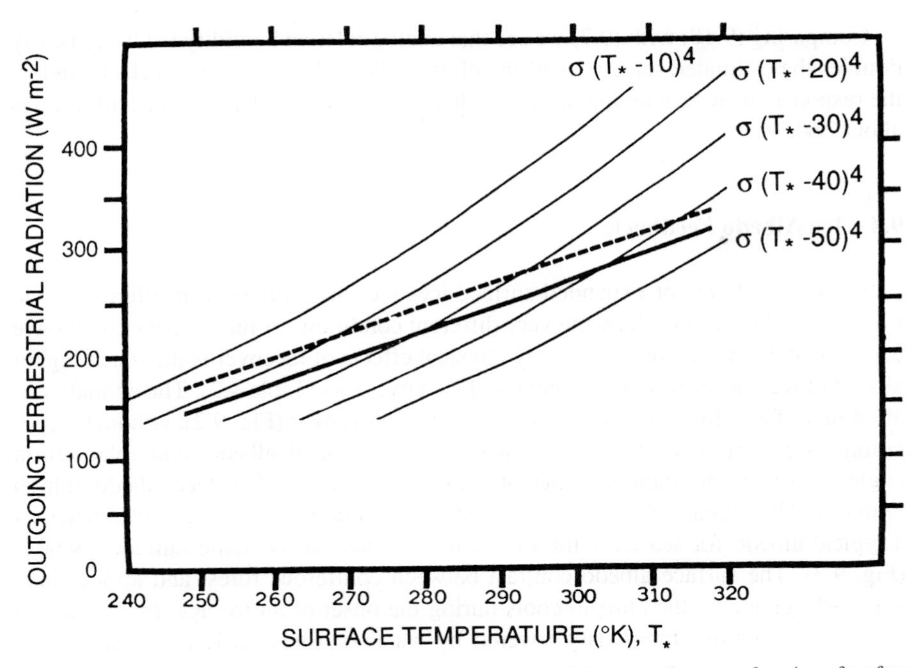

Fig. 9.1 Graph of outgoing terrestrial emission at the top of the atmosphere as a function of surface temperature calculated from a radiative–convective equilibrium model in which the relative humidity is kept fixed so that the specific humidity increases with temperature. For comparison, Stefan–Boltzmann emission curves are shown for temperatures equal to the surface temperature T_*, minus fixed amounts from 10 to 50°C. The dashed line is for clear skies and the solid line is for average cloudiness. [Adapted from Manabe and Wetherald (1967). Reprinted with permission from the American Meteorological Society.]

air is low. As an estimate of the effect of water vapor feedback on climate sensitivity, we may utilize a one-dimensional radiative–convective equilibrium model (see Section 3.10) in which we assume that the relative humidity remains fixed at observed values while the temperature changes. In this case we find that the terrestrial radiation emitted from the planet increases much less rapidly with temperature than would be indicated by the Stefan–Boltzmann relationship. Moreover, the terrestrial emission increases linearly with surface temperature, rather than as the fourth power of temperature (Fig. 9.1). From radiative–convective equilibrium calculations with fixed relative humidity (FRH) we can obtain an estimate of the climate sensitivity parameter with only the Stefan–Boltzmann and relative humidity feedbacks included.

$$\lambda_R)_{\mathrm{FRH}} = \left(\left(\frac{dF^{\uparrow}(\infty)}{dT_s} \right)_{\mathrm{FRH}} \right)^{-1} \approx 0.5 \ \mathrm{K} \left(\mathrm{W\ m^{-2}} \right)^{-1} \tag{9.10}$$

Comparing (9.10) with (9.7) we see that adding relative humidity feedback nearly doubles the estimated sensitivity of the climate. With these two feedbacks included, the response of surface temperature to a doubling of carbon dioxide concentration is about 2°C.

9.4 Ice Albedo Feedback

The physical effects of expanded surface icecover have often been offered as one possible explanation for how the very different conditions of the ice ages could have been maintained. One of the primary physical effects of icecover is the much higher albedo of ice and snow than all other surface coverings (Table 4.2). The annual variation of surface albedo is controlled largely by snowcover (Fig. 9.2), although vegetation cover can also influence the seasonal variation of albedo, and solar zenith angle is also an important influence on seasonal variations of surface albedo in high latitudes. The albedo of ocean surfaces at high latitudes is typically 10%, whereas a typical albedo for sea ice with a covering of snow at the same latitudes is 60% (Fig. 9.3). The surface albedo contrast between coniferous forest and an ice sheet is equally great. As the climate cools during the onset of an ice age, the ice would expand into regions previously covered by water or forest in latitudes where the

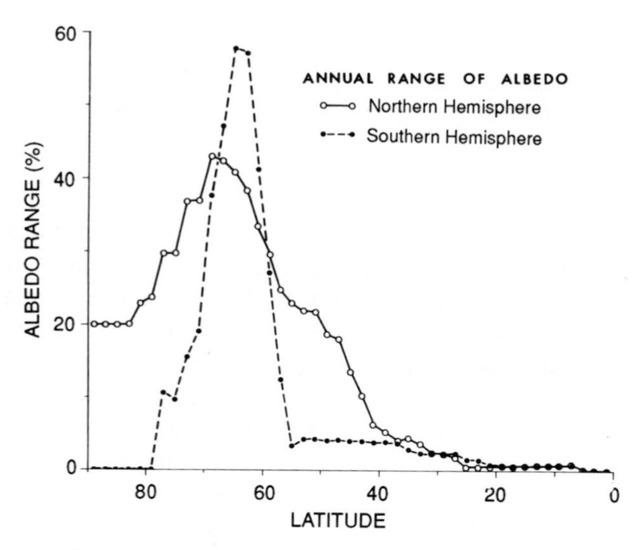

Fig. 9.2 Graph of the annual range of surface albedo in the Northern and Southern Hemispheres. The largest annual ranges occur in the latitude range of Antarctic sea ice in the Southern Hemisphere and sea ice and snowcover in the Northern Hemisphere. [From Kukla and Robinson (1980). Reprinted with permission from the American Meteorological Society.]

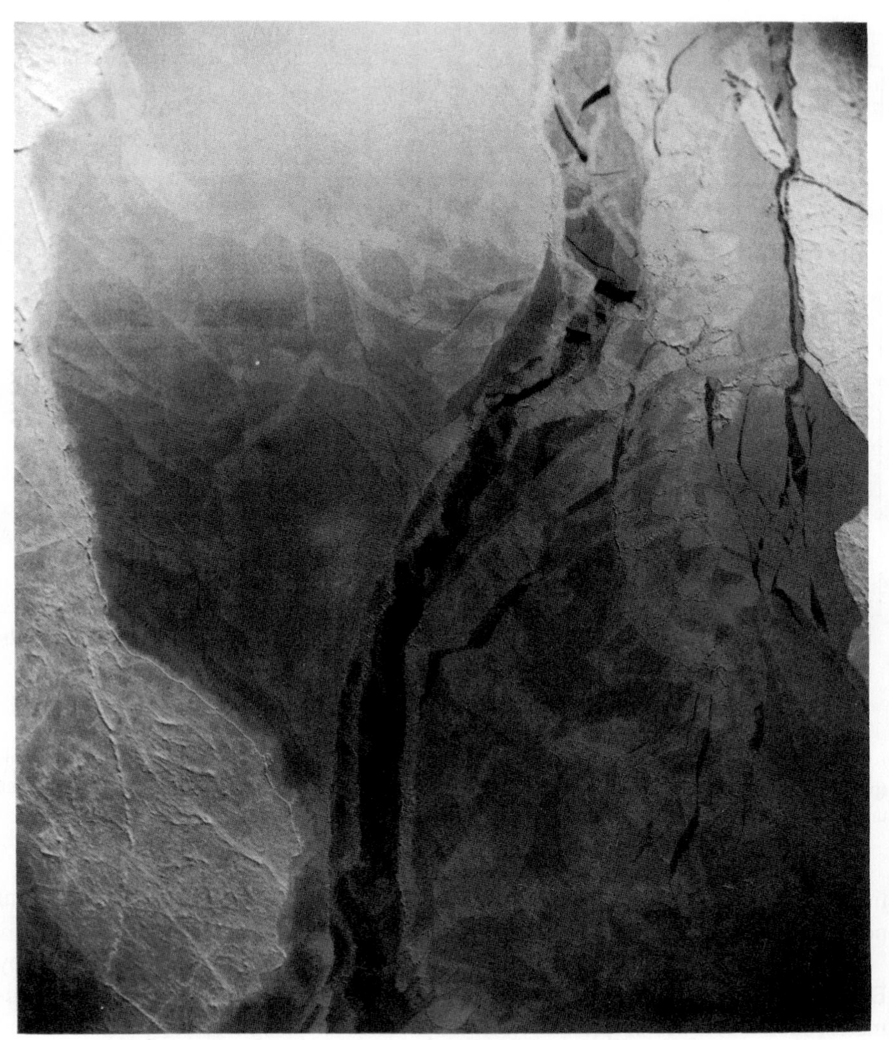

Fig. 9.3 Aerial photograph of sea ice in the Arctic, showing the relationship of albedo to ice thickness and snowcover. A large lead has refrozen with a thin layer of ice that has reopened and fingered. Fingering occurs when thin layers of ice are pushed together and slide over the top of each other. The darkest areas are open water. Image is about 15 km across. [Photo: U.S. Navy (NAVOCEANO), 1972.]

insolation is substantial, particularly in summer (Fig. 9.4). The increased surface ice-cover would raise the surface albedo and reduce the amount of solar energy absorbed by the planet. This reduction of solar energy absorption would cause further cooling and thus further ice expansion that might lead ultimately to an ice age. The associa-

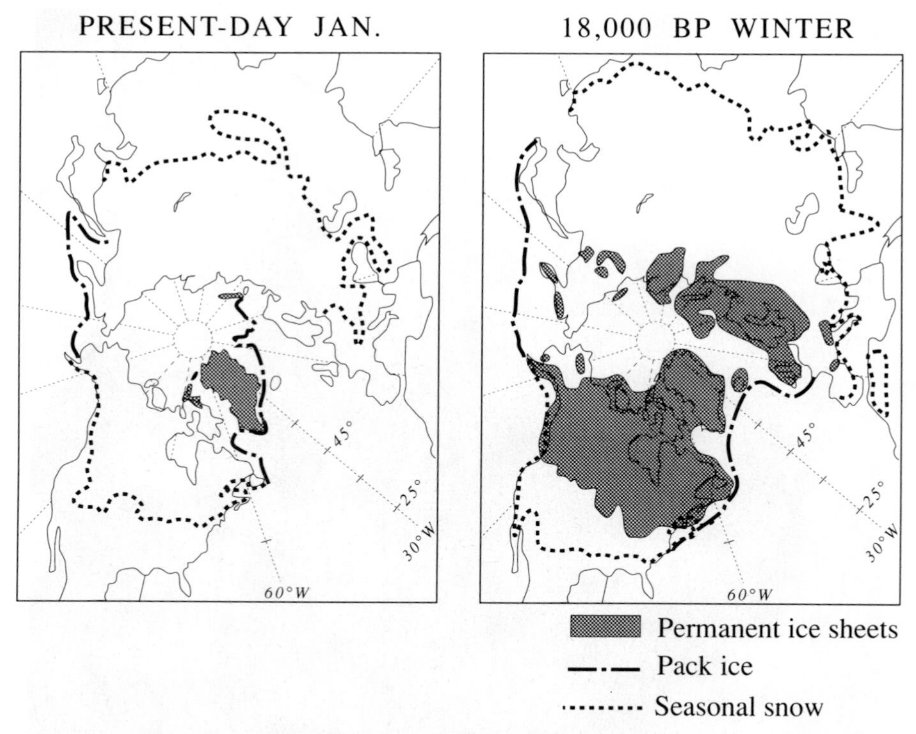

Fig. 9.4 Extent of permanent ice sheets (shaded), seasonal snowcover (dotted line), and pack ice (perennial sea ice, dash–dot line) at present and estimated for the last glacial maximum. [From Robinson, unpublished, as published in Kukla (1979).]

tion of surface glaciation with colder temperatures can constitute a very powerful positive feedback, since it modulates the direct energy input from the sun.

The albedo feedback effect was incorporated in simple climate models by Budyko (1969) and Sellers (1969). Both of their models produced very sensitive climates, such that it was surprisingly easy to produce an ice age, or even an icecovered Earth, with a relatively modest climate forcing. Their simple models for the ice caps assumed that everything about the climate could be characterized by the surface temperature, and that the only independent variable was latitude. The determination of the surface temperature was based on the conservation of energy for the system, so that these models are termed *energy-balance climate models*. The steady-state energy balance has three terms: the absorbed solar energy, the emitted terrestrial energy, and the divergence of the horizontal energy transport by the atmosphere and oceans. All of these terms are allowed to depend on latitude, since the growth of polar ice caps and their effect on global climate is the central question that these models are designed to address.

It is convenient to express the three terms in the energy balance as functions of the sine of latitude, $x = \sin\phi$, and the surface temperature T_s.

$$Q_{ABS}(x,T_s) - F_\infty^{\uparrow}(x,T_s) = \Delta F_{ao}(x,T_s) \tag{9.11}$$

Q_{ABS} is the absorbed solar radiation, $F_\infty^{\uparrow}$ is the outgoing longwave radiation, and ΔF_{ao} is the removal of energy by horizontal energy transport, which would usually be expressed as the divergence of the meridional energy transport by the atmosphere and ocean. The observed climatological zonal average values of these terms are shown in Fig. 2.12. The absorbed solar radiation is written as the product of the solar constant, S_0, a function that describes the distribution of insolation with latitude, $s(x)$, and the absorptivity for solar radiation, $a_p(x,T_s) = 1 - \alpha_p(x,T_s)$, which is one minus the planetary albedo, α_p.

$$Q_{ABS}(x,T_s) = \frac{S_0}{4}\, s(x)\, a_p(x,T_s) \tag{9.12}$$

The distribution of annual mean insolation is shown in Fig. 2.7. The distribution function $s(x)$ is defined by dividing the annual-mean insolation at each latitude by the global-average insolation. Its global area average is unity.

$$\frac{1}{2}\int_{-1}^{1} s(x)\, dx = 1 \tag{9.13}$$

The annual-mean insolation distribution function for current conditions can be approximated fairly accurately as

$$s(x) = 1.0 - 0.477\, P_2(x) \tag{9.14}$$

where

$$P_2(x) = \frac{1}{2}\left(3x^2 - 1\right) \tag{9.15}$$

is the Legendre polynomial of second order in x.

The emitted terrestrial flux is specified as a linear function of surface temperature.

$$F_\infty^{\uparrow}(x,T_s) = A + B\,T_s \tag{9.16}$$

The coefficients A and B can be obtained from Fig. 9.1, or from other theoretical or empirical estimates.

The transport term can be specified in two different ways that facilitate an easy solution. Budyko assumed a linear form for the transport term, such that at every latitude the transport relaxes the temperature back toward its global-mean value.

$$\left.\Delta F_{ao}\right)_{Budyko} = \gamma\left(T_s - \tilde{T}_s\right) \tag{9.17}$$

where

$$\tilde{T}_s = \frac{1}{2}\int_{-1}^{1} T_s\, dx \tag{9.18}$$

is the global-mean temperature.

Sellers (1969) and North (1975) used a diffusive approximation to meridional transport.

$$\left.\Delta F_{\text{ao}}\right)_{\text{Sellers}} = \nabla \bullet K_H \nabla T_s = \frac{\partial}{\partial x}\left(1 - x^2\right) K_H \frac{\partial T_s}{\partial x} \tag{9.19}$$

The linear relaxation coefficient, γ, and diffusivity, K_H, are chosen to make the model simulation as close to the observed climate as possible by producing the observed meridional energy transport when the temperature gradient is approximately as observed.

Albedo feedback is introduced by assuming that ice forms when the temperature falls below a critical value and that this is associated with an albedo increase. It is observed that the annual mean temperature at which surface icecover persists throughout the year is about −10°C currently. We therefore assume that an abrupt transition in the albedo takes place at this temperature.

$$\alpha_p = \begin{cases} \alpha_{\text{ice-free}}, & T_s > -10^\circ\text{C} \\ \alpha_{\text{ice}}, & T_s < -10^\circ\text{C} \end{cases} \tag{9.20}$$

Typical values chosen for planetary albedos with and without surface icecover are 0.62 and 0.3, respectively.

If we substitute (9.12), (9.16), and (9.17) into (9.11), we obtain the equation considered by Budyko.

$$A + BT_s \; + \; \gamma\left(T_s - \tilde{T}_s\right) = \frac{S_0}{4}\, s(x)\; a_p(x, x_i) \tag{9.21}$$

Here we have used the fact that because of (9.20), the absorptivity is only a function of sine of latitude x and the position of the iceline, x_i, the point where the temperature equals −10°C. We define a new variable I, which is the ratio of the terrestrial emission to the global-average insolation.

$$I = \frac{\left(A + BT_s\right)}{\left(1/4\right)S_0} \tag{9.22}$$

Substituting this definition into (9.21) yields a simple equation for I.

$$I + \delta\left(I - \tilde{I}\right) = s(x)\; a_p(x, x_i) \tag{9.23}$$

It is interesting that the linear parameters describing the efficiency of meridional energy transport and longwave radiative cooling occur only in the ratio $\delta = (\gamma/B)$. If

δ is large, then meridional transport is relatively efficient compared to longwave cooling, and equator-to-pole gradients in I will be small.

The global area average of (9.23) is

$$\tilde{I} = \frac{1}{2} \int_{-1}^{1} s(x)\, a_p(x, x_i)\, dx \tag{9.24}$$

which indicates that the global average value of the terrestrial emission divided by the insolation is equal to the global average of the product of the absorptivity and the distribution function for insolation. Thus if the absorptivity is high where the insolation is also high, the global-mean value $\tilde{I}$ will also be high. Since I is linearly related to T_s, the global-mean temperature also increases with the average product of insolation and absorptivity. Because the annual mean insolation is least at the poles and greatest at the equator, the albedo increase associated with icecover will have a greater effect on global-mean temperature as the icecover extends farther toward the equator.

It is interesting to solve for the latitude of the ice boundary as a function of solar constant for particular albedo specifications and values of δ. The solution is obtained by specifying the position of the iceline and then solving for I at the iceline latitude using (9.23). Because the albedo specification is discontinuous at the iceline and no diffusion is present in the model, I and T_s are discontinuous and the solution for x_i as a function of S_0 is not unique. A unique solution can be obtained by adding a small diffusive transport and taking the limit of vanishing diffusion coefficient. This procedure suggests using the average of the albedo on both sides of the ice edge when solving for I at the ice edge.[2] The general structure of such solutions is shown in Fig. 9.5. Because of the strong nonlinearity introduced by the albedo specification, the model may have between one and three solutions for a particular value of the solar constant. For sufficiently large solar constants, an ice-free planet is a stable solution to the model. For sufficiently small solar constants, an icecovered planet is a stable solution. For an intermediate range of solar constants both of these stable solutions are possible. The diagram has been scaled such that for a normalized solar constant of 1.0 the boundary of the polar ice cap is at 72°N, which is approximately the current position of the ice edge. If the solar constant is raised or lowered from this value the ice cap will recede or expand. The rate at which the ice cap expands or contracts as the solar constant is changed is a measure of the sensitivity of the climate. This rate is equivalent to the slope of the solution curve in Fig. 9.5. As the solar constant is decreased and the ice cap expands, the model becomes more sensitive until the line is vertical at some critical latitude. The latitude where the slope becomes infinite is labeled the global stability

[2]Held and Suarez (1974).

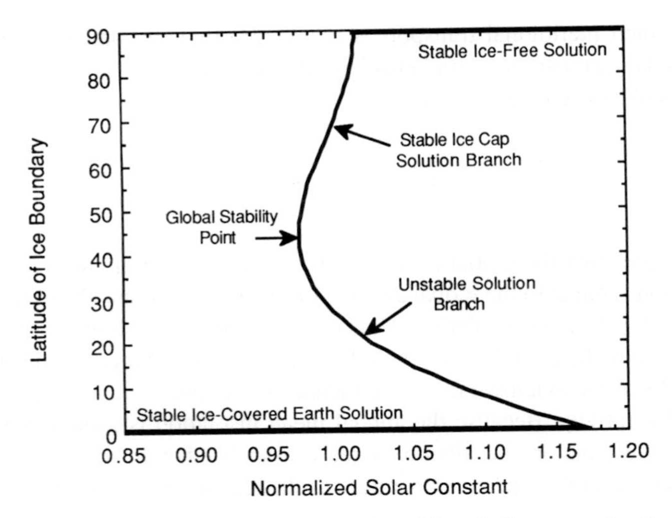

Fig. 9.5 Diagram showing the latitude of the boundary of the polar ice cap as a function of solar constant in the solution of the energy-balance climate model. The solar constant has been normalized such that the boundary of the ice cap occurs at 72° for a normalized solar constant of unity. The solution shown is for $\delta = 1.9$, with an ice-free albedo of 0.3 and an ice albedo of 0.62.

point in Fig. 9.5. If the solar constant is reduced below this value the ice abruptly expands to cover the entire planet in a kind of runaway ice albedo feedback. In this model, ice caps that extend equatorward of the global stability point are unstable and grow to cover the planet.

The parameters chosen by Budyko (1969) and Sellers (1969) make the ice cap very sensitive to small changes in solar constant. For example, Budyko calculated that a reduction in solar constant of 1.6% would lead to an icecovered Earth. This result is attractive in that it indicates that a relatively small change in climate forcing could produce the glacial–interglacial transitions that have occurred during Earth history. On the other hand, the global stability point is reached when the ice cap has expanded to about 45 degrees latitude. The inference from this simple model is thus that during glacial maxima, when the average latitude of perennial icecover was near midlatitudes, the planet is very close to making an irreversible transition to an ice-covered condition. It is very difficult to see how Earth could have avoided this transition during its history, if ice albedo feedback is as strong as indicated by Budyko's calculation and no additional feedbacks are present. Energy balance climate models are extremely simple, of course, and neglect many important processes, but it is still interesting to see how their sensitivity depends on the parameters needed for their solution.

We may first examine how the sensitivity of the model depends on the parameter δ, which measures the ratio of the meridional transport coefficient to the longwave

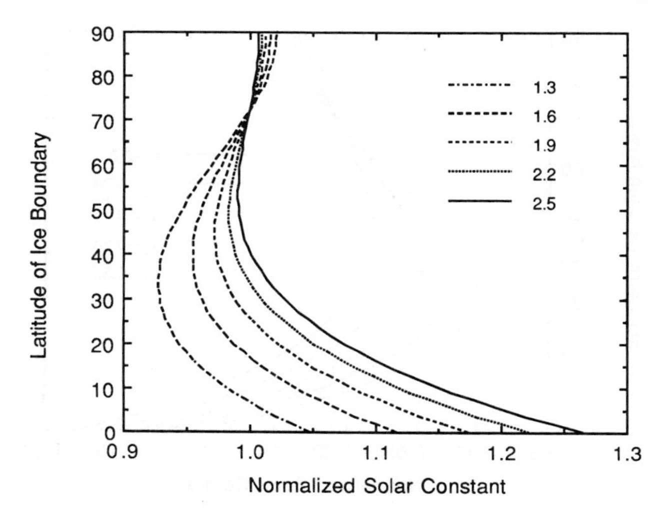

Fig. 9.6 Energy-balance climate model solution curves as in Fig. 9.5 for five values of the parameter δ, the ratio of the meridional transport coefficient to the longwave cooling coefficient. For larger δ the meridional temperature gradient is weaker and small changes in global-mean temperature are associated with larger displacements of the ice edge. Larger values of δ thus result in an ice cap whose area is more sensitive to solar constant.

radiative cooling coefficient. Figure 9.6 shows a set of solution curves for selected values of δ. For larger values of δ the boundary of a small polar ice cap is more sensitive to solar constant changes. This is related to the fact that large values of δ produce weak meridional temperature gradients, so that a small change in global-mean temperature is translated into a large displacement of the iceline and thereby a large change in absorbed solar radiation. Budyko used a value of B = 1.45 W m^{-2} K^{-1}, whereas more recent estimates would place the most appropriate value closer to 2.2 W m^{-2} K^{-1}. The resulting value of δ in Budyko's calculation was 2.6. This high value of δ contributed to the great sensitivity of his model.

The planetary albedo contrast associated with surface icecover is also a critical parameter controlling the sensitivity of energy balance climate models. The increase of planetary albedo with latitude is an observed fact, but only a portion of this meridional increase of planetary albedo is associated with surface ice. Significant contributions are also made by the increase of cloud cover with latitude and the increase of the albedo with solar zenith angle, which also increases with latitude (Fig. 2.8). As a result the ice-free albedo increases with latitude, so that in the latitudes where ice is likely to form, the contrast of planetary albedo between ice-free and icecovered conditions is not as great as one might expect. Much of the surface albedo increase is screened by clouds, which are very good reflectors of solar radiation, particularly at high solar zenith angles. One may take this into account in the energy balance

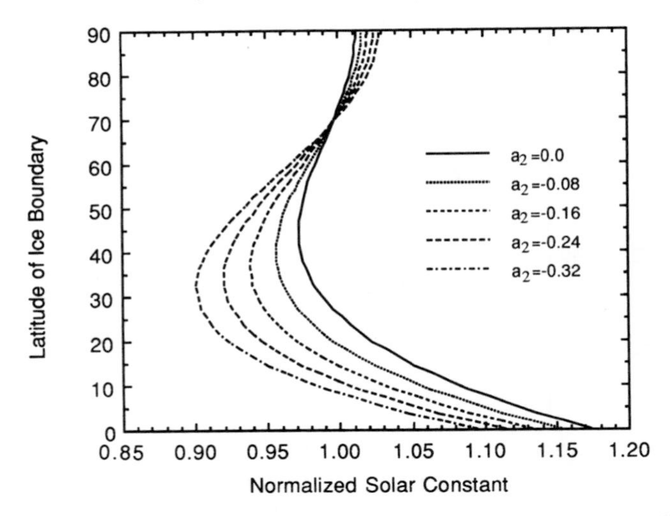

Fig. 9.7 Energy balance climate model solution curves as in Fig. 9.5 for five values of the parameter a_2, which measures the amount by which the ice-free planetary absorptivity decreases toward the poles. Larger negative values of a_2 weaken ice albedo feedback in high latitudes, since the albedo is then already fairly high even in the absence of icecover ($\delta = 1.9$).

climate model by assuming that the ice-free absorptivity for solar radiation decreases with latitude.

$$a_p(x, x_i) = \begin{cases} a_0 + a_2 P_2(x), & T_s > -10°\text{C}; \quad |x| < |x_i| \\ b_0, & T_s < -10°\text{C}; \quad |x| > |x_i| \end{cases} \tag{9.25}$$

Figure 9.7 shows solutions of the energy balance climate model for $\delta = 1.9$, $b_0 = 0.38$, and $a_0 = 0.7$. The values of a_2 range from zero to –0.32, with the latter value chosen to make the ice-free and the icecovered planetary albedos equal at the pole. Budyko used $a_2 = 0.0$, whereas a realistic value of a_2 is about –0.175.[3] With a realistic meridional decrease of ice-free planetary absorptivity for solar radiation, the energy-balance climate model indicates a much more stable climate. The sensitivity of the current iceline position to small changes in solar constant is reduced when the ice-free albedo increases with latitude, and a much larger reduction in solar constant is required to produce a transition to an icecovered planet. With the more realistic ice-free albedo specification, the transition to an icecovered Earth does not occur until the ice edge has reached a lower latitude. The energy balance climate model seems incapable of producing a climate that is relatively sensitive for modest changes of ice cap size that characterize the observed glacial–interglacial transitions, and yet is also very resistant to an irreversible transition to an icecovered Earth. A realistic

[3]Hartmann and Short (1979).

appraisal of ice albedo feedback is that, while it is an important feedback process, it is not strong enough by itself to explain the ease with which the climate system appears to change from glacial to interglacial conditions.

9.5 Dynamical Feedbacks and Meridional Energy Transport

In the energy-balance climate models, the efficiency of meridional energy transport and the resulting meridional temperature gradient play an important role. It is interesting to consider how meridional energy transport in the atmosphere and ocean might respond to a developing ice sheet or to some other forcing of the meridional temperature gradient. If we limit ourselves to midlatitudes, the biggest contribution to meridional energy transport comes from eddies in the atmosphere (Chapter 6). The dependence of the eddy fluxes of heat on the mean temperature gradients can be estimated from simple dimensional arguments. We first suppose, as was done by Sellers (1969) and North (1975), that the meridional energy transport depends on the meridional gradient of potential temperature, Θ.

$$[v'\Theta'] = -K_H \frac{\partial[\Theta]}{\partial y} \tag{9.26}$$

where square brackets indicate a zonal average and y is the northward displacement in a local Cartesian coordinate system. The transport coefficient K_H must have the units of $\mathrm{m^2\ s^{-1}}$, so we can think of K_H as the product of a characteristic wind speed V and distance L_e.

$$K_H \propto V \bullet L_e \tag{9.27}$$

The eddy heat fluxes are produced by disturbances that result from baroclinic instability of the zonal-average state of the atmosphere, so we use the theory for baroclinic instability to select the characteristic wind speed and length scales. The length scale is chosen to be the Rossby radius of deformation, which is approximately the scale of disturbances that grow most rapidly in response to the baroclinic instability of the zonal-mean flow.[4]

$$L_e \approx L_R = \frac{N d_e}{f} \tag{9.28}$$

where d_e is the depth scale for the disturbance, f is the Coriolis parameter, and N is the buoyancy frequency, which is related to the vertical gradient of the potential temperature.

$$N^2 = \frac{g}{[\Theta]} \frac{\partial[\Theta]}{\partial z} \tag{9.29}$$

[4]See, e.g., Holton (1992).

The appropriate wind speed scale is the difference in zonal-mean wind speed experienced by the eddy across its vertical extent, d_e.

$$V = d_e \left| \frac{\partial [u]}{\partial z} \right| \tag{9.30}$$

The vertical shear of the zonal-mean wind is related to the meridional gradient of potential temperature by the thermal wind approximation, which is accurate away from the equator.

$$\frac{\partial [u]}{\partial z} \approx -\frac{g}{f[\Theta]} \frac{\partial [\Theta]}{\partial y} \tag{9.31}$$

Combining (9.26) through (9.31), we obtain an expression for the meridional flux of dry static energy by transient eddies as a function of the mean state variables.

$$[v' \Theta'] \approx -\left(\frac{g}{[\Theta]} \right)^{3/2} \frac{d_e^{\,2}}{f^2} \left(\frac{\partial [\Theta]}{\partial z} \right)^{1/2} \left| \frac{\partial [\Theta]}{\partial y} \right| \frac{\partial [\Theta]}{\partial y} \tag{9.32}$$

The conservation of potential temperature can be used to scale the characteristic vertical velocity and characteristic vertical length scale to the horizontal velocity and length scales.

$$W = V \left| \frac{\partial [\Theta]}{\partial y} \right| \left| \frac{\partial [\Theta]}{\partial z} \right|^{-1} \tag{9.33}$$

Using this scaling of the vertical velocity we can obtain an expression for the vertical heat flux analogous to (9.32).

$$[w' \Theta'] \approx \left(\frac{g}{[\Theta]} \right)^{3/2} \frac{d_e^{\,2}}{f^2} \left| \frac{\partial [\Theta]}{\partial y} \right|^3 \left(\frac{\partial [\Theta]}{\partial z} \right)^{-1/2} \tag{9.34}$$

According to these scaling arguments, the magnitude of the meridional eddy heat flux is proportional to the square of the meridional temperature gradient and to the square root of the static stability (9.32). The depth scale d_e is as yet unspecified. Strongly unstable disturbances occupy the entire troposphere, so that it is reasonable to use the scale height H as the depth scale. For more weakly unstable flows the depth scale itself is proportional to the meridional temperature gradient, so that the total heat flux is even more highly sensitive to the meridional temperature gradient.[5] Because of the quadratic or higher dependence of the eddy heat flux on the meridional temperature gradient, eddy fluxes act to suppress large deviations of the meridional temperature gradient, so that, to the extent that the meridional temperature gradient is controlled by eddy fluxes, we expect the meridional gradient of temperature to be rather insensitive to altered thermal driving of the equator-to-pole temperature

[5] Held (1978).

difference. Altered meridional thermal drive will result in a large change in baroclinic eddy activity and eddy heat transport, which will suppress the response of the temperature gradient to the altered forcing. The strong sensitivity of the eddy activity and heat flux to meridional thermal driving has been verified by calculations with numerical models of the atmosphere.

We expect that the oceanic heat flux is also sensitive to the meridional temperature gradient, since we expect that both the thermal driving and the wind stress driving of oceanic circulations would increase with the temperature gradient. We do not have any simple quantitative theory for how oceanic fluxes depend on the meridional temperature gradients that corresponds to the simple dimensional arguments given above.

Dynamical models of the free atmosphere suggest that baroclinic eddies will act to maintain a nearly constant meridional temperature gradient by changing their meridional energy flux by a large amount when a small change in meridional temperature gradient is imposed. It is very interesting that paleoclimatic data show that the surface temperature has a very different behavior than expected from these dynamical considerations. During ice ages the tropical surface temperatures changed only a few degrees from today's values, while the polar temperatures cooled by 10°C or more. This "polar amplification" of climate change is an important feature that needs to be fully understood. It points to the importance of surface and boundary-layer processes in determining the surface temperature and its latitudinal gradient.

9.6 Longwave and Evaporation Feedbacks in the Surface Energy Balance

One interesting inference from analysis of ocean core data is that the average tropical sea surface temperatures do not appear to have changed by more than 1-2°C between glacial and interglacial conditions (Fig. 8.8). Why were the tropical SSTs so stable, when polar temperatures changed by tens of degrees? Is this related to ice albedo feedback, or are other processes equally important for producing the polar amplification? The sensitivity of tropical SSTs can be approached by considering the surface energy balance. It is unclear how the oceanic energy flux divergence and the solar heating of the tropical oceans depend on mean tropical temperature, but we can make good quantitative estimates of the temperature dependence of the net longwave heating of the surface and the evaporative cooling of the surface.

The net longwave heating of the surface is equal to the downward longwave from the atmosphere minus the longwave emission from the surface.

$$F_{\text{net}}^{\downarrow} = F^{\downarrow}(0) - \sigma T_s^4 \tag{9.35}$$

Here we have assumed that the surface absorbs and emits like a blackbody.

We can calculate the dependence of net radiation on surface temperature under clear-sky conditions using a radiative transfer model with some assumptions. We assume that the temperature decreases linearly with height away from the surface with

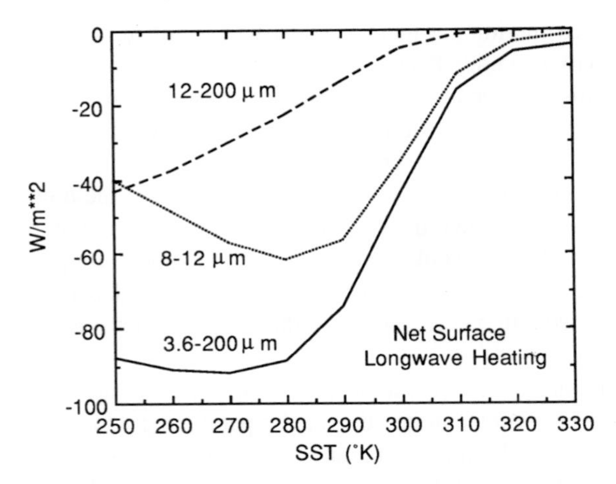

Fig. 9.8 Net longwave energy flux at the surface as a function of surface temperature, calculated using a fixed lapse rate and relative humidity distribution. Curves are shown for the total terrestrial flux and the flux within the wavelength intervals indicated. [From Hartmann and Michelsen (1993). Reprinted with permission from the American Meteorological Society.]

a lapse rate of 6.5 K km^{-1} until the height where the temperature reaches a minimum of 200 K, and that the atmosphere is isothermal with a temperature of 200 K above that level. We assume that the relative humidity is distributed linearly in pressure according to

$$RH = RH_o \left(\frac{p / p_s - 0.02}{1 - 0.02} \right) \tag{9.36}$$

where $RH_o = 77\%$ gives a reasonable approximation to the observed global-mean relative humidity.[6] The net longwave flux at the ground computed with these specifications of lapse rate and relative humidity and with current values of carbon dioxide and ozone content is shown in Fig. 9.8. The general character of this plot is not very sensitive to the details of how the relative humidity and temperature profiles are specified. For temperatures less than ~280 K the net longwave loss from the surface increases with temperature as a result of the increase of surface blackbody emission with temperature. At warmer temperatures the net longwave loss from the surface decreases with increasing temperature because the downward longwave flux from the atmosphere increases more rapidly than the surface longwave emission. This is because the water vapor content of the atmosphere is increasing, and water vapor is a very efficient emitter of terrestrial radiation. For wavelengths in the range 8–12 μm, called the water vapor window, the net longwave loss increases up to about 285 K, after which point it decreases rapidly with temperature. This is because of the

[6]Manabe and Wetherald (1967).

so-called continuum absorption by water vapor that becomes increasingly effective at high vapor pressure. The nature of this absorption is not fully understood, but it is well measured and might be associated with water vapor molecules sticking together at high vapor pressures, and thus forming a complex molecule that can efficiently absorb and emit window radiation. This closes the water vapor window at high temperatures and makes the cloud-free atmosphere virtually opaque to thermal infrared radiation. According to these calculations, upward and downward longwave radiation at the surface would become almost equal for surface temperatures in excess of ~320 K. The net surface longwave loss decreases most rapidly with temperature in the vicinity of 300 K, which is approximately the mean temperature of the tropical oceans. This constitutes a positive feedback in the surface energy budget that is most effective at tropical temperatures, since as the temperature increases the net longwave cooling of the surface decreases. If no other feedback processes were operating, the tropical SST might be quite sensitive and variable.

The approximate energy balance of the tropical ocean surface is about 200 W m^{-2} solar heating, balanced by 120 W m^{-2} evaporative cooling, 50 W m^{-2} net longwave loss, 10 W m^{-2} sensible cooling, and 20 W m^{-2} energy export by ocean currents. The evaporative cooling of tropical ocean waters is greater than the longwave cooling and is also expected to be sensitive to surface temperature. We can estimate the sensitivity of evaporative cooling to surface temperature if we are willing to make several critical assumptions. We begin with the aerodynamic formula for evaporative cooling approximated using (4.32) and (4.35).

$$\mathrm{LE} = \rho_a \, L C_{\mathrm{DE}} \, U \, q_s^* \left((1 - \mathrm{RH}) + \mathrm{RH}\left(\frac{L}{R_v T_s^{\,2}} \right)(T_s - T_a) \right) \tag{9.37}$$

If we assume again that the RH distribution and the temperature profile remain constant as we increase the surface temperature, then the evaporative cooling depends primarily on the temperature dependence of the saturation specific humidity at the surface. We can also assume that the characteristic wind speed and the exchange coefficient are independent of the surface temperature. Making these assumptions and neglecting the weaker linear and quadratic temperature dependencies compared to the exponential temperature dependence of saturation humidity, one can show from (9.37) and (4.35) that the logarithmic derivative of evaporative cooling with respect to surface temperature is only weakly dependent on temperature itself.

$$\frac{1}{\mathrm{LE}} \frac{\partial \mathrm{LE}}{\partial T_s} \approx \frac{L}{R_v \, T_s^2} \approx 0.06 \ \mathrm{K}^{-1} \tag{9.38}$$

This is a result of the exponential dependence of saturation humidity. The derivatives of exponential functions are themselves exponential functions. The numerical estimate in (9.38) is obtained for $T_s = 300$ K, which is a typical tropical oceanic surface temperature. If we multiply (9.38) by a typical evaporative cooling value of

LE = 120 W m^{-2}, then we obtain an estimate of the sensitivity that is independent of specific choices of the parameters in the aerodynamic formula (9.37).

$$\frac{\partial \mathrm{LE}}{\partial T_s} \approx 7 \ \mathrm{W\,m^{-2}\,K^{-1}} \tag{9.39}$$

For fixed relative humidity the rate at which evaporative cooling increases with temperature is large compared to the rate at which the longwave cooling of the surface decreases. Figure 9.8 can be used to calculate the dependence of net longwave cooling on temperature for cloud-free conditions. The maximum rate of change of net longwave heating of the ground occurs at about T_s = 305 K and has a numeric value of

$$\frac{\partial F^{\downarrow}_{\mathrm{net}}(z=0)}{\partial T_s} \approx 3 \ \mathrm{W\,m^{-2}\,K^{-1}} \tag{9.40}$$

The magnitude of the rate of increase of evaporative cooling with temperature is more than double the rate of decrease of the longwave cooling of the surface. The ocean surface gradually loses its ability to cool by longwave radiative emission as the temperature increases above 300 K, but the evaporative cooling increases more rapidly with temperature than the longwave cooling declines with temperature. The sensitivity parameter for tropical sea surface conditions (TSS) obtained by including only these two feedbacks is rather small.

$$\left.\lambda_R\right)_{\mathrm{TSS}} = \left\{ \frac{\partial \mathrm{LE}}{\partial T_s} - \frac{\partial F^{\downarrow}_{\mathrm{net}}(z=0)}{\partial T_s} \right\}^{-1} \approx 0.3 \ \mathrm{K}\left(\mathrm{W\,m^{-2}}\right)^{-1} \tag{9.41}$$

If surface longwave heating and evaporation vary with temperature as these simple estimates suggest, tropical SST will be relatively stable. This stability, which is rooted in the rapid change of saturation vapor pressure at tropical surface temperatures, may explain why tropical SST has changed relatively little in the past. It may also contribute to the apparent polar amplification of surface temperature changes by providing a tropical damping of temperature changes.

Evaporative cooling of the tropical ocean surface is very sensitive to the thermodynamic properties of the atmospheric boundary layer, which are determined through a complex interaction among boundary-layer turbulence, mesoscale convection, and large-scale flow. One may illustrate the potential importance of relative humidity variations by differentiating (9.37) with respect to relative humidity for fixed temperature.

$$\left.\frac{\partial \mathrm{LE}}{\partial \mathrm{RH}}\right)_{T_s} \approx \frac{-\mathrm{LE}}{(1-\mathrm{RH})} \approx -8 \ \mathrm{W\,m^{-2}\,\%^{-1}} \tag{9.42}$$

For tropical conditions, the effect on the evaporation rate of a 1 K increase in temperature at fixed relative humidity is approximately equal and opposite to the effect of a 1% increase in relative humidity at fixed temperature. The processes that control relative humidity in the tropical boundary layer are extremely important to the sensitivity of climate.

9.7 Cloud Feedback

According to the numbers presented in Table 3.3, clouds double the albedo of Earth from 15 to 30% and reduce the longwave emission by about 30 W m^{-2}. Because of the partial cancellation of these two effects, the effect of clouds on the global net radiative energy flux into the planet is a reduction of about 20 W m^{-2}. The effect of an individual cloud on the local energy balance depends sensitively on its height and optical thickness, the insolation, and the characteristics of the underlying surface. For example, a low cloud over the ocean will reduce the net radiation substantially, because it will increase the albedo without much affecting the longwave flux at the top of the atmosphere. On the other hand, a high, thin cloud can greatly change the longwave fluxes without much affecting the solar absorption, and will therefore increase the net radiation and lead to surface warming.

If the amount or type of cloud is sensitive to the state of the climate, then it is possible that clouds provide a very strong feedback to the climate system. Since the fractional area coverage of cloud, A_c, is about 50%, and the net effect of this cloud on the energy balance is about –20 W m^{-2}, we can make the crude estimate that

$$\frac{\partial R_{\mathrm{TOA}}}{\partial A_c} \approx \frac{\Delta R_{\mathrm{TOA}}}{A_c} \approx \frac{-20 \text{ W m}^{-2}}{0.5} = -40 \text{ W m}^{-2} \tag{9.43}$$

Thus, a 10% change in cloudiness has the same magnitude of effect on the energy balance as a doubling of carbon dioxide concentration. Increasing the cloudiness by 10% would offset the effect of CO_2 doubling. Decreasing the cloudiness by 10% would double the effect of CO_2 doubling. If the cloud type changed in such a way as to decrease the shortwave cloud forcing by 5% and increase the longwave cloud forcing by 5%, this would also have the same effect as doubling the CO_2 concentration. This could be achieved by decreasing the area coverage and raising the mean cloud top height. Therefore, both the cloud amount and the distribution of cloud types are extremely important for climate.

Unfortunately, our understanding of how global cloud amount and type are determined is currently insufficient, and estimates of the magnitude and sign of cloud feedback are basically unknown. Cloud feedback is one of the major uncertainties in climate sensitivity studies.

9.8 Biogeochemical Feedbacks

The climate of Earth has been shaped by life, and it is certain that biology plays a role in the sensitivity of climate. There are many ways that plants and animals can influence climate sensitivity. Perhaps the strongest and most direct means by which organisms affect climate is through their control of the composition of the atmosphere. The concentration of carbon dioxide in the atmosphere is determined by a complex set of processes, but a key process is the uptake of carbon dioxide by plants in the surface ocean and on land. It is estimated that up to half of the global cooling during the last glacial maximum was contributed by the reduction in atmospheric carbon dioxide concentration. This reduction in atmospheric carbon dioxide must have been produced by an alteration in the biology and chemistry of the oceans, since on these time scales atmospheric carbon dioxide is controlled by the partial pressure of CO_2 in surface waters.

The primary source of condensation nuclei for cloud droplets over the oceans is the production of sulfurous gases such as dimethyl sulfide by tiny organisms in the surface waters. In the atmosphere these sulfur-bearing gases are converted into sulfuric acid particles, which form the nuclei around which cloud droplets form by condensation of water vapor.[7] If more of these particles are present, more cloud droplets will form. Because the droplets also tend to be smaller when more condensation nuclei are available, the cloud droplets tend to remain in the atmosphere longer before coagulating and falling out as rain. Also, if the cloud water is spread over more droplets, the albedo of the cloud will be higher. We expect that larger production of sulfur gases that are the precursors of cloud condensation nuclei will lead to higher oceanic cloud albedos and thereby to a cooling of Earth. If the rate at which the organisms release sulfur gases is dependent on temperature, we have the elements of a feedback mechanism. The magnitude of this possible feedback mechanism has not been quantified, and the sign of the feedback is also unknown.

Some have argued that it is useful to think of Earth as a single complex entity involving the biosphere, atmosphere, ocean, and land, which can be called *Gaia*, the Greek word for "Mother Earth."[8] It is further hypothesized that the totality of these elements constitutes a feedback system, which acts to optimize the conditions for life to exist here. Active control to maintain relatively constant conditions is described by the term *homeostasis*.

Watson and Lovelock (1983) have offered Daisyworld, a simple heuristic model for how life might maintain a very stable climate. They imagine a cloudless planet with an atmosphere that is completely transparent to radiation. The only plants are two species of daisies: one dark (say, black) the other light (say, white) in color. One species reflects less solar radiation than bare ground; the other reflects more. The daisy population is governed by a differential equation for each species.

$$\frac{dA_w}{dt} = A_w \left(\beta x - \chi \right) \tag{9.44}$$

$$\frac{dA_b}{dt} = A_b \left(\beta x - \chi \right) \tag{9.45}$$

[7]Charlson *et al.* (1987).
[8]Lovelock (1979).

where: A_w, A_b, and A_g are the fractional areas covered by white daisies, black daisies, and bare ground, respectively; β is the growth rate of daisies per unit of time and area; χ is the death rate per unit time and area; and x is the fractional area of fertile ground uncolonized by either species.

The growth rate is assumed to be a parabolic function of the local temperature T_i, which has a maximum value of one at 22.5°C and goes to zero at 5 and 40°C.

$$\beta_i = 1.0 - 0.003265(295.5\text{ K} - T_i)^2 \tag{9.46}$$

The planet is assumed to be in global energy balance, such that

$$\sigma T_e^4 = \frac{S_0}{4}\left(1 - \alpha_p\right) \tag{9.47}$$

The planetary albedo α_p is the area-weighted average of the albedos of bare ground, white daisies, and black daisies.

$$\alpha_p = A_g\,\alpha_g + A_w\,\alpha_w + A_b\,\alpha_b \tag{9.48}$$

It is necessary to decide how the local temperatures T_i are to be determined. One extreme is to have all local temperatures equal to the emission temperature T_e. This corresponds to perfectly efficient horizontal transport of heat. The other extreme is for each surface type to be at its own radiative equilibrium temperature. This would correspond to no transport of heat between regions with different surface coverings, so that the local temperature is that necessary to balance the local absorption of solar radiation.

$$T_i^4 = \frac{S_0}{4\sigma}\left(1 - \alpha_i\right) \tag{9.49}$$

Combining (9.49) with (9.47) yields an expression that satisfies both the local and global energy balance.

$$T_i^4 = \frac{S_0}{4\sigma}\left(\alpha_p - \alpha_i\right) + T_e^4 \tag{9.50}$$

We can generalize (9.50) by writing

$$T_i^4 = \eta\left(\alpha_p - \alpha_i\right) + T_e^4 \tag{9.51}$$

where $0 < \eta < (S_0/4\sigma)$ represents the allowable range between the two extremes in which horizontal transport is perfectly efficient, $\eta = 0$, and horizontal transport is zero $\eta = (S_0/4\sigma)$.

Steady-state solutions for Daisyworld can be obtained for specific values of the parameters, and some of these are shown in Fig. 9.9. Because of the strong feedbacks within the system, the global-mean emission temperature is remarkably stable over a wide range of solar constants within which daisies can exist. These stable regions are bounded by values of solar constant where daisies appear or disappear almost discontinuously and temperature takes a corresponding jump. The stable region exists because changes in temperature cause changes in daisy population that produce a strong negative feedback on the temperature. For example, as the insolation increases, white

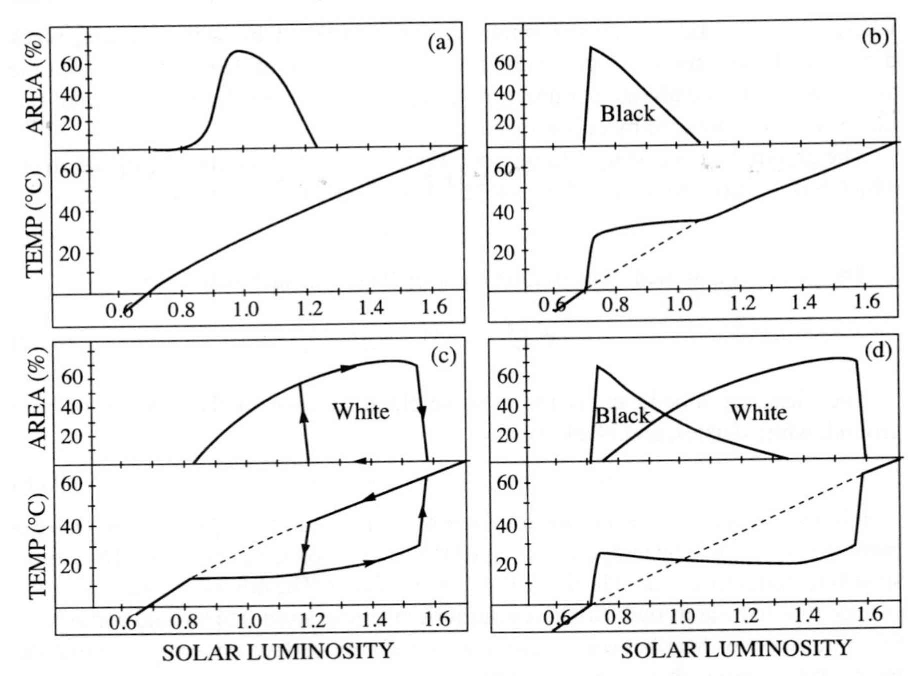

Fig. 9.9 Steady-state responses of Daisyworld. Areas of daisies and emission temperature are plotted as functions of normalized solar constant. The solutions were obtained with a solar constant of $S_0 = 1368\ \mathrm{W\ m^{-2}}$, a bare ground albedo of $\alpha_g = 0.5$, a daisy death rate of $\gamma = 0.3$, and a heat exchange parameter of $\eta = 2.0 \times 10^9\ \mathrm{K}^4 = 0.12(S_0/4\sigma)$. (a) For a population of 'neutral' daisies with $\alpha_i = 0.5$; (b) for a population of black daisies with $\alpha_b = 0.25$; (c) for a population of white daisies with $\alpha_w = 0.25$ (note that this shows the evolution for both increasing and decreasing solar constant and that the system exhibits hysteresis); (d) for a population of both black and white daisies. [Adapted from Watson and Lovelock (1983). Reprinted with permission from Munksgaard International Publishers Ltd.]

daisies are favored because they are cooler than the emission temperature. The increase in white daisy population increases the planetary albedo, which reduces the response of the emission temperature to the solar constant increase.

Daisyworld is certainly not a reasonable representation of Earth, but it serves to illustrate how biology, simply by doing what is natural for it, can influence the global climate and its stability. Biological mechanisms do operate on Earth that could play the role of the daisies on Daisyworld, although it is likely that their effect would be more subtle and muted. For example, the role of biology in controlling CO_2 might act like the white daisies, since, once photosynthetic life developed, CO_2 was drawn out of the atmosphere. Models suggest that the development of land plants led to a reduction in atmospheric CO_2 in the Paleozoic that helped to offset warming associated with increasing solar constant.[9] In the Daisyworld model biological processes

[9]Berner (1993).

are hypothesized to be strongly stabilizing, but CO_2 changes on glacial–interglacial time scales during the Quaternary have done just the opposite by contributing significantly to the sensitivity of climate. Clouds might be the earthly counterpart of the white daisies, with their biological connection coming through the production of cloud condensation nuclei by life in the sea. If the release of DMS is proportional to the temperature, then the DMS–cloud albedo connection could constitute a strong negative feedback. The sign and magnitude of the climate feedback associated with biological production of cloud condensation nuclei are both uncertain; however, since ocean biology is the primary source of cloud condensation nuclei over the ocean, it seems likely that the cloud contribution to albedo would be less without life.

Exercises

1. Derive (9.5) from (9.3).
2. Derive (9.9) from (9.8).
3. Assume that transient eddies are the primary heat transport mechanism determining vertical and horizontal potential temperature gradients in midlatitudes, and that the eddy transports respond to the mean gradients as indicated by (9.32) and (9.34). If the radiative forcing of the meridional gradient produces a larger equator-to-pole heating contrast, how will the vertical gradient of potential temperature (static stability) in midlatitudes respond? Will the change in static stability be relatively large or small compared to the change in meridional temperature gradient? Assume that potential temperature is relaxed linearly toward a radiative equilibrium value and that the relaxation rate of vertical and horizontal gradients is the same.
4. Use the Budyko energy-balance climate model (9.23) to solve for the solar constant as a function of iceline latitude. Use (9.14) and the parameters $B = 1.45\ \mathrm{W\ m^{-2}\ K^{-1}}$, $a_0 = 0.68$, $b_0 = 0.38$, and $a_2 = 0.0$. Plot the curves for two values of γ; $\gamma = 3.8\ \mathrm{W\ m^{-2}\ K^{-1}}$ and $\gamma = 2.8\ \mathrm{W\ m^{-2}\ K^{-1}}$.
5. According to the energy-balance climate models with ice albedo feedback, how does the sensitivity of climate relate to the meridional temperature gradient in equilibrium? As the meridional temperature gradient increases, does the climate become more or less sensitive, assuming an ice cap is present?
6. Discuss how the sensitivity of climate might be different if Earth's surface were 70% land and 30% ocean.
7. How do you think the behavior of the Daisyworld model might be different if the planet's sphericity were taken into account? Would it be more or less stable?
8. Program a time-dependent version of Daisyworld for the computer, but add an equation for the population of rabbits whose birth rate is proportional to the area covered by daisies. How does the approach to a steady state depend on the initial conditions and the parameters of the model? Add foxes.

Chapter 10 Global Climate Models

10.1 Mathematical Modeling

We can better understand and predict climate and its variations by incorporating the principles of physics, chemistry, and biology into a mathematical model of climate. A climate model can range in complexity from the simple energy-balance models described in Chapter 2, whose solution can be worked out on the back of a small envelope, to very complex models that require the biggest and fastest computers and the most sophisticated numerical techniques. This chapter will discuss the most complex global climate models, since these are the models that can produce the most realistic simulations of climate and are the tool on which we place our hope of predicting future climates in sufficient detail to be useful for planning purposes. For brevity we will refer to this type of model as a "GCM," which we can take to be an acronym for *general-circulation model.*[1] A hierarchy of progressively less complicated climate models exists. These less detailed models are useful also, but we will not discuss them here.

10.2 Historical Development of Climate Models

The modern GCM has its roots in the computer models that were first developed to predict weather patterns a few days in advance. L. F. Richardson[2] was the first to promote the idea that future weather could be predicted by numerically integrating the equations of fluid motion using the present weather as the initial condition. He attempted a hand calculation of a weather forecast while serving as an ambulance driver in France during World War I. The forecast was spectacularly inaccurate, because his initial conditions contained a large spurious wind convergence. The first successful numerical forecasts used a set of equations that are greatly simplified compared to Richardson's and for which the solution is less sensitive to the initial conditions.

[1]The term *general circulation model* refers to models that explicitly calculate the evolution of flow patterns and thereby replicate from first principles the complete statistical description of the large-scale motions of the atmosphere or ocean. Since in the field of climate modeling GCM now denotes the most elaborate type of climate model for which the circulation is only one of many key components, it seems appropriate to modify the source of the acronym to be *global climate model*. On the other hand, people often speak of atmospheric GCMs and oceanic GCMs, in which cases the original meaning may be more correct.

[2]Richardson (1922). This is the same man who formulated the Richardson number discussed in Chapter 4.

Numerical weather prediction was proposed as one potential use for the electronic computer developed by John von Neumann in the late 1940s. The first successful numerical weather forecast with an electronic computer was conducted at the Institute for Advanced Study in Princeton, New Jersey by a group led by Jule Charney.[3] This model had only a single atmospheric layer and described the region over the continental United States. It included only the fluid dynamical evolution of an initial velocity distribution and none of the physical forcings that drive the climate. The first numerical experiment that included radiation and dissipation was conducted using a simple model with only two levels in the vertical.[4] Later more detailed simulations of the atmospheric general circulation included a more accurate formulation of the equations of motion, more spatial resolution in the horizontal and vertical, and more realistic specifications of the physical processes that drive the atmospheric circulation, such as radiation, latent heat release, and frictional dissipation.[5] With these and succeeding models it has become possible to simulate the general circulation of the atmosphere with reasonable fidelity. The quality of the simulations obtained with atmospheric models has benefited greatly from intensive experimentation associated with providing practical weather forecasts, beginning with the first routine numerical weather forecasts that were conducted by the U.S. Joint Numerical Weather Prediction Unit beginning in 1955. Associated with the practical effort to provide weather forecasts is a substantial effort to collect routine weather observations at the surface and at upper levels of the atmosphere. These observations are capable of describing the atmospheric state in sufficient detail to justify newly initialized forecasts about every 6 hours, which are able to provide some useful information about the weather up to a week in the future.

Numerical models of oceanic-general circulation have been developed by applying the same basic techniques used for atmospheric models. Development of oceanic general-circulation models has lagged behind atmospheric models both because the observational base is less in the ocean and because the computational problems are greater for oceanic simulations. No longstanding effort has been undertaken to observe and predict the state of the global ocean on an operational basis, so oceanographers have not had a base in numerical weather forecasting on which to build an oceanic climate modeling enterprise until very recently. An observed climatology of the ocean that is sufficiently detailed and accurate for climate purposes is as yet unavailable, particularly for the deep ocean. The spatial scale of the significant motion systems is small compared to the width of an ocean basin, and observations of currents, temperature, and salinity at depths below the surface must be acquired from ships and are expensive. Therefore the number of observations taken in a year is typically less than is necessary to define the state of the ocean. Moreover, the speed of many important ocean currents is small and therefore difficult to observe with great precision. Also, a very long integration time is necessary for an oceanic model to

[3]Charney *et al.* (1950).
[4]Phillips (1956).
[5]Smagorinsky (1963); Manabe *et al.* (1965).

achieve a stable climatology. While the atmosphere will spin up to climatology from rest in a few weeks or months, the deep circulation in the oceans requires centuries to spin up from rest or to respond to changed forcings. This long integration time and the high spatial resolution required for realistic simulations of ocean circulation mean that numerical experimentation with ocean models can require substantial and sophisticated computer resources. On the other hand, the number and complexity of the physical processes required for ocean simulation are much less than in the atmosphere.

Early simulations of the atmosphere were obtained by fixing the sea surface temperature. Similarly, ocean general-circulation models assumed fixed wind stress, air temperature, air humidity, precipitation, and radiative forcing. Together these determined the flux of momentum, heat, and freshwater at the surface, which are the key driving forces of the ocean circulation. In a fully general model of the climate the exchanges of heat, moisture, and momentum between the ocean and the atmosphere must be internally determined so that the state of the coupled climate system can evolve in a consistent manner. Initial experiments with fully coupled atmosphere–ocean models often produced climates that were very different from reality, even though each component produced a reasonable climatology when run independently with realistic fixed boundary conditions. The greater freedom of fully coupled atmosphere–ocean climate models to determine their own climate has revealed many deficiencies in atmospheric and oceanic general circulation models, and these deficiencies are being addressed through current research efforts. Improved climate simulation and prediction will also require land surface processes to be treated with greater accuracy and detail.

The component parts of a global climate model are shown schematically in Fig. 10.1. All of the components and the processes represented within them are criti-

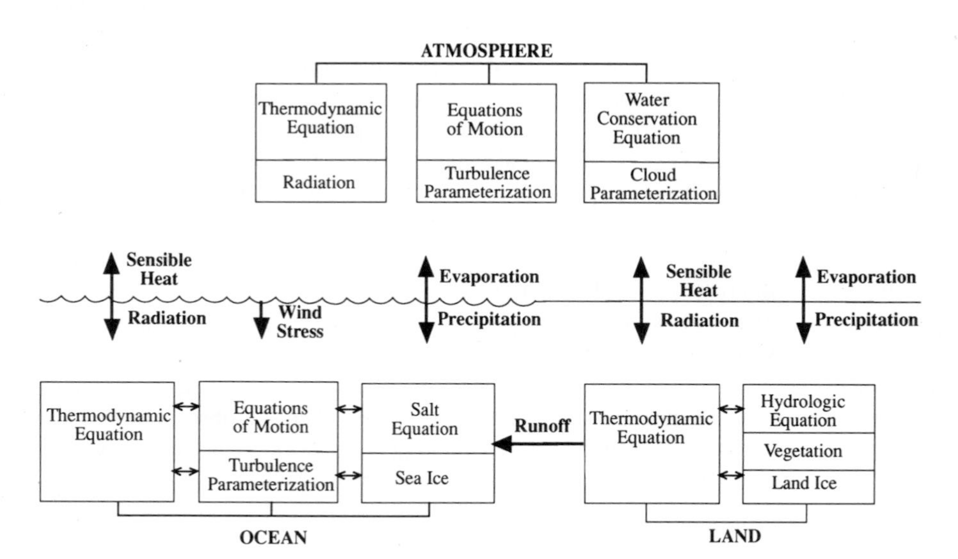

Fig. 10.1 Schematic diagram showing the components of a global climate model.

cal to the understanding and simulation of climate. Not explicitly shown are the biogeochemical feedbacks, which determine the atmospheric composition and the nature of the land vegetation. These processes have not been included in most climate model simulations completed to date, and we will not discuss them here.

10.3 The Atmospheric Component

The atmospheric component of a global climate model is basically the same as a model constructed for the purpose of weather forecasting. Improvements of numerical weather prediction systems have been fostered by efforts to predict weather more than a week in advance. For these longer predictions the similarity of the model's "climatology" with reality has become very important. The errors in the climatology of the model make a large contribution to the errors in these medium-range forecasts. Efforts to make longer weather forecasts and to accurately simulate climate are thus synergetic and together are leading to improvements in atmospheric models.

The basic framework of a numerical model of the atmosphere or ocean is a spatial gridwork on which the equations of physics are represented. Sometimes so-called spectral methods are used wherein the fields of velocity, temperature, and pressure are represented with continuous functions in order to efficiently solve the equations of motion. These methods allow very accurate calculation of the derivatives needed for following the flow of heat and momentum, but the fields must be interpolated to grid points in order to calculate radiative heating and the effects of sub-grid-scale processes such as convection. The globe is thus always divided into geographic regions, whose sides often correspond to lines of latitude and longitude. The size and number of the grid boxes are limited by the amount of computer power available and the time period over which the model must be integrated. Numerical weather prediction models are required to simulate at most a few months of weather at a time, and current models are run with about 1-degree latitude–longitude resolution. This means that the weather is specified by the values of the meteorological variables at about 6×10^4 geographic locations. Since climate models may be required to simulate decades or centuries, the horizontal resolution is often about 5 degrees (Fig. 10.2), so that only about 3×10^3 geographic locations are represented. Since the time step must also be decreased when the grid points are moved closer together, high spatial resolution simulations require large amounts of computer time. Coarse 5-degree-resolution atmospheric models can be integrated with a time step of one hour using the spectral method and semiimplicit time differencing.[6] Such models require several hours of computer time to simulate one year of atmospheric weather using current supercomputer technology, and doubling the resolution to 2.5 degrees increases the computation time by a factor of eight or more.

[6]Washington and Parkinson (1986).

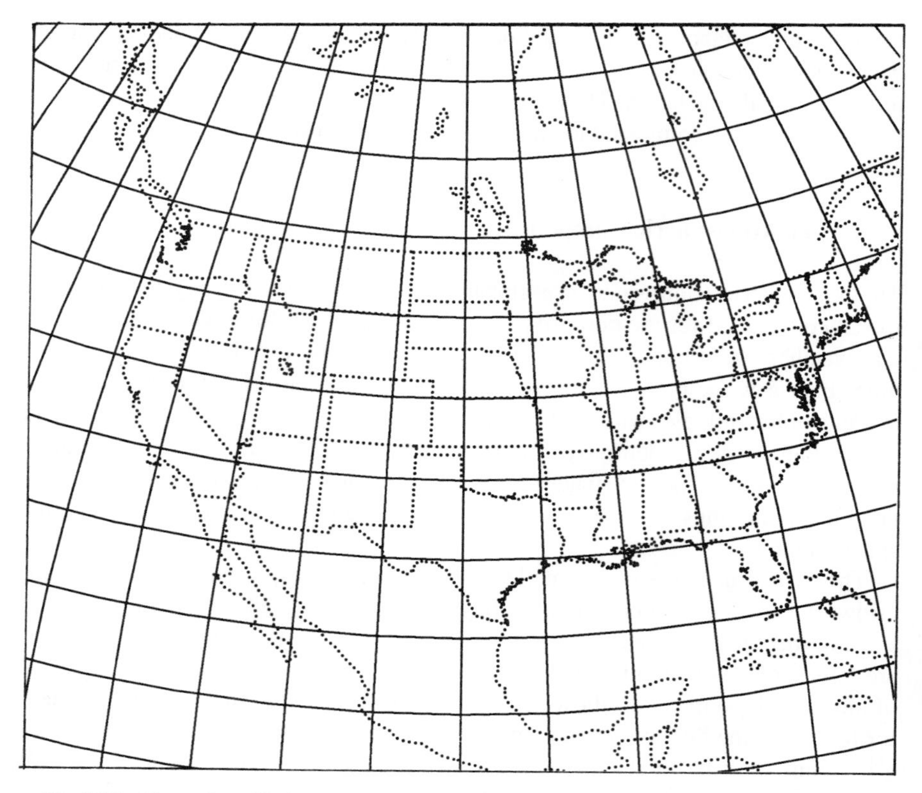

Fig. 10.2 Illustration of 5-degree latitude–longitude resolution typical of some current climate models intended for long-term integrations. Orthographic projection over North America.

Vertical resolution is also important to the quality of an atmospheric simulation. Climate models currently have typically about 10 levels, while weather prediction models have 20 or more levels. The levels are not equally spaced in height or pressure, but tend to be most closely spaced near the lower boundary and near the tropopause, where rapid changes take place and greater resolution is needed (Fig. 10.3). The total number of latitude, longitude, and height grid points represented in an atmospheric model may thus range from about 10^4 to 10^6.

10.3.1 The Conservation Equations

The basic dynamical framework of a GCM includes the solution of the equations describing the conservation of momentum, mass, and energy for a fluid. In addition, a water conservation equation for air is needed to predict the amount of water vapor in the atmosphere. Water vapor is critical for the formation of clouds and rain, and is

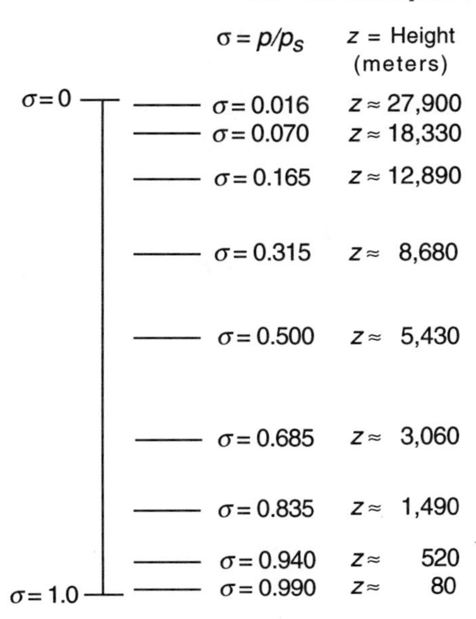

Fig. 10.3 Example of the vertical grid spacing in sigma coordinates, and the corresponding average altitudes of the levels in a nine-level atmospheric general circulation model employed by Manabe *et al.* (1970).

one of the most important variables in the calculation of radiative heating. Numerical models of the atmosphere use the equations of motion in a simplified form called the *primitive equations.* Use is made of the fact that the atmosphere is thin in comparison to its horizontal extent in order to approximate the vertical momentum balance with the hydrostatic balance. Earth is assumed to be spherical and some other small terms in the momentum equations are neglected. Because the hydrostatic balance is assumed, the vertical coordinate can be height or pressure. Since it facilitates the inclusion of topography and makes it easier to construct a numerical solution that conserves mass, the *sigma coordinate,* which is pressure normalized by surface pressure, is often used as the vertical coordinate in atmospheric primitive equation models.

$$\sigma = \frac{p}{p_s} \tag{10.1}$$

Here p is pressure and p_s is surface pressure. The primitive equations in sigma coordinates can be expressed in vector notation as follows.

Momentum conservation:

$$\frac{D\vec{V}}{Dt} + f\,\vec{k} \times \vec{V} + \vec{\nabla}\Phi + RT\,\vec{\nabla}\ln p_s = \vec{P}_V \tag{10.2}$$

Energy conservation:

$$\frac{DT}{Dt} - \frac{RT\omega}{c_p\, p_s \sigma} = P_T \tag{10.3}$$

Water vapor conservation:

$$\frac{Dq}{Dt} = P_q \tag{10.4}$$

Mass conservation:

$$\frac{Dp_s}{Dt} - p_s\left(\vec{\nabla} \bullet \vec{V} + \frac{\partial \dot{\sigma}}{\partial \sigma}\right) = 0 \tag{10.5}$$

Hydrostatic balance:

$$\frac{\partial \Phi}{\partial \sigma} = -\frac{RT}{\sigma} \tag{10.6}$$

In these equations D/Dt represents the time derivative following the motion of the fluid, which in sigma coordinates is written

$$\frac{D}{Dt} = \frac{\partial}{\partial t} + \vec{V} \bullet \vec{\nabla} + \dot{\sigma}\frac{\partial}{\partial \sigma} \tag{10.7}$$

where $\vec{V}$ is the horizontal vector wind with eastward component u and northward component υ, $\dot{\sigma}$ is the sigma component of velocity, f is the Coriolis parameter, $\vec{k}$ is the vertical unit vector, $\vec{\nabla}$ is the two-dimensional gradient operator on a surface of constant sigma, Φ is the geopotential field (height times gravity), R and c_p are respectively the gas constant and specific heat of dry air, and P_x denotes the rate of change of some variable x as a result of the parameterized processes.

The sources and sinks of momentum and heat are associated with phenomena that occur at scales that are much smaller than the grid resolution and so cannot be explicitly simulated. These are often called sub-grid-scale phenomena, and their effects must be specified from knowledge of the state of the atmosphere at the grid scale. This process of including the effect of unresolved phenomena according to the large-scale conditions is called *parameterization*.[7] Key parameterizations in atmospheric models include radiation, the effects of unresolved turbulence and gravity waves, and the effects of convection on the heat, moisture, and momentum budgets. The behavior of the climate model is critically dependent on these parameterized processes.

[7]The word parameterization is used because one or more adjustable constants or parameters must be introduced to relate sub-grid-scale processes to large-scale conditions.

10.3.2 Radiation Parameterization

Radiative heating from the sun and infrared cooling to space provide the basic drive of the climate system (Chapters 2 and 3). Radiative transfer models that are incorporated in current climate models do not consider each individual absorption line but rather treat groups of lines or band systems in a statistical or empirical manner. They consider the atmosphere and the embedded clouds to be horizontally homogeneous within a grid box. Despite these simplifications, it is generally believed that the transmission of radiation, at least under clear sky conditions, is accurately treated in climate models. While there are some discrepancies between various parameterization schemes, the basic physical processes and the methods necessary to treat them are reasonably well understood. The largest uncertainties in radiative flux calculations are associated with clouds and the manner in which the amount and nature of the clouds in the model are determined. Early general circulation models typically specified cloud radiative properties as externally determined quantities that were often zonally invariant with a fixed seasonal variation. More recent climate models include some form of crude cloud prediction, and models are being developed in which cloud amounts and optical properties are predicted and allowed to interact with the other elements of the climate system. These newer models attempt to treat cloud microphysics in some parameterized way and carry cloud water and ice as predicted variables. The predicted cloud water and ice provide a means of connecting the water budget, radiation calculation, and heat budget in a manner that is consistent between the parameterized processes and the large-scale conservation equations.

10.3.3 Convection and Cloud Parameterization

Clouds affect radiative transfer, and the convective motions associated with clouds produce important fluxes of mass, momentum, heat, and moisture. The spatial scales at which cloud properties are determined are generally much smaller than the grid size at which the atmosphere is represented in a climate model, yet the vertical fluxes of heat and moisture associated with sub-grid-scale convection are often greater than those of the large-scale flow. The timing and intensity of precipitation in the tropics and over land areas during summer are controlled as much by unresolved mesoscale phenomena as by the large-scale flow. The atmospheric state averaged over the area of a typical climate model grid box may be stable to moist convection even when intense convection is occurring somewhere in the grid box. Cloud and convection parameterization seek to address the mismatch between the spatial resolution of climate models and the spatial scale of convective motions and clouds.

At least three important effects in a climate model are associated with the formation of clouds: (1) the condensation of water vapor and the associated release of latent heat and rain; (2) vertical transports of heat, moisture, and momentum by the motions associated with the cloud; and (3) the interaction of the cloud particles with radiation. A cloud parameterization in a climate model should treat each of these effects consistently, but this has not been the case with every climate model. For example, vertical transports by cloud-scale motions are often not parameterized separately, but are instead subsumed into a general parameterization for all types of sub-grid-scale mixing. Coupling between the location and intensity of moist convection and the local cloud radiative properties is a relatively recent development in climate modeling, since many early models simply specified invariant cloud radiative properties. The release of latent heat during intense convection drives the Hadley circulation and other important components of the general circulation, so that the coupling of deep convection to the large-scale flow is critical to a proper simulation of climate.

Most climate models include at least two types of clouds: convective clouds and large-scale supersaturation clouds. Large-scale supersaturation clouds occur when the relative humidity in a grid box at some model level exceeds a critical value. This can be implemented by assuming that condensation occurs in the grid box when the relative humidity reaches some threshold like 80%.[8] Another alternative is to assume that sub-grid-scale temperature variability occurs within the grid box and that the fraction of the grid box where the temperature variability causes the relative humidity to reach 100% is cloud covered.[9]

Convective clouds are associated with the buoyant ascent of saturated air parcels in a conditionally unstable environment. The simplest convective parameterization is moist adiabatic adjustment.[10] If the lapse rate exceeds the moist adiabatic lapse rate (Appendix C), the moisture and heat are readjusted in a vertical layer such that the air in the layer is saturated, the lapse rate equals the moist adiabatic lapse rate, and energy is conserved. Excess moisture is assumed to rain out, but momentum is not transported. In this parameterization the entire grid box is assumed to behave like a convective element, while in reality convection occurs at much smaller spatial scales. The Kuo (1974) parameterization considers the effect of large-scale convergence in supplying moisture for cumulus convection. This connection between large-scale moisture convergence and cumulus convection is supported by observations, but the assumption that the convection heats by mixing cloud air and environmental air seems unjustified, since the heating appears to be produced primarily by subsidence between the cloud elements. The Arakawa–Schubert (1974) parameterization includes a more comprehensive treatment of the interaction of a cumulus ensemble with the large-scale environment. The parameterization includes a model for how a population of cumulus clouds interacts with its environment through entrain-

[8]Manabe *et al.* (1965).

[9]Hansen *et al.* (1984).

[10]Manabe *et al.* (1965).

ment of environmental air during ascent, detrainment of air and moisture near the cloud top, and the generation of subsidence in the near environment. A quasiequilibrium state is hypothesized and clouds are dispatched at a rate sufficient to keep the atmosphere near equilibrium in the face of large-scale destabilization. Recent modifications of the Arakawa–Schubert approach have added the effect of downdrafts to the formulation.

10.3.4 Planetary Boundary-Layer Parameterization

In the atmospheric boundary layer, rapidly fluctuating phenomena with vertical and horizontal space scales much smaller than the grid spacing of climate models determine the fluxes of heat, momentum, and moisture between the surface and the atmosphere. These phenomena include turbulence, gravity waves, and rolls or other coherent structures that cannot be resolved by climate models and must therefore be parameterized. The simplest and most often used parameterizations are the bulk aerodynamic formulas and similarity theories briefly described in Section 4.4. These formulas allow the boundary layer to be treated as a single layer through which the fluxes can be calculated using the mean variables resolved by the model. They can be made dependent on vertical stability and surface roughness. These models may be elaborated by adding a prediction equation for the boundary-layer depth.

If the model has sufficient resolution to have several levels within the boundary layer, then eddy diffusion formulations may be used in which the vertical eddy fluxes are assumed to be proportional to the vertical derivative of the mean for the model grid box.

$$\overline{w'T'} = -K_T \frac{\partial \overline{T}}{\partial z} \tag{10.8}$$

The simplest approach is to make the flux coefficient, K_T, a constant, but more realistic parameterizations include height and stability dependencies.

Higher-order closure schemes include prognostic equations for the turbulent kinetic energy in the boundary layer and more complex equations than (10.8) for the vertical eddy fluxes.[11] These models require more than a few computational levels in the planetary boundary layer (PBL) and carry a heavy computational burden. In addition, it can be argued that they do not describe the essential physics of the coherent structures that appear to produce much of the vertical flux in the boundary layer and control the response of this flux to mean conditions.[12] In many cases the boundary layer contains clouds, and in such cases moist processes are critical to the behavior of the boundary layer. Some planetary boundary-layer parameterizations incorporate

[11] Mellor and Yamada (1974).

[12] Brown (1991).

a separate parameterization for boundary-layer clouds.[13] Oceanic stratocumulus and tradewind cumulus clouds are important examples where the interaction of the boundary layer with low clouds is essential. In regions of deep convection the boundary-layer properties and the cloud fluxes interact very strongly, so that the PBL parameterization and the convection parameterization must be compatible. The PBL parameterization must also be compatible with the land surface parameterizations that may incorporate the effect of plant canopies on the fluxes of momentum, heat, and moisture between the surface and the atmosphere.

10.4 The Land Component

The land component of a climate model must contain the surface heat balance equation described in Chapter 4 [equation (4.1)] and a surface moisture equation that is at least as sophisticated as the bucket model described in Section 5.6.1, and includes a model for snowcover. Land topography can readily be incorporated in a sigma-coordinate model, but the fidelity with which Earth's surface topography can be included is limited by the resolution of the model. Many important mountain ranges are much narrower than the grid spacing of a GCM, so that very smooth representations of topography must be used.

Experiments with GCMs that include these basic prescriptions for the land surface indicate a substantial climate response to the soil moisture and to the surface albedo. Low soil moisture generally results in a warmer surface and less local evaporation and rainfall. These changes are consistent with the observation that much of the precipitation in such areas as the central United States during summer and the Amazon Basin during the rainy season is reevaporated rainfall rather than water vapor that has been advected into the region. The changes in the surface fluxes of latent and sensible heat affect the thermal forcing of atmospheric flow and may thereby cause weather and climate anomalies at remote locations.

In land areas the storage of precipitation in the soil and the subsequent release of this moisture to the air are strongly dependent on the soil type and the vegetative cover. In addition, the absorption of solar radiation and the emission of longwave radiation are sensitive to the geometry and physical state of the vegetative cover. Turbulent transports of heat and moisture between the atmosphere and the soil are also affected by the physical structure of the plant canopy, particularly for forested areas. Parameterizations of surface processes that include all of these effects have been developed in recent years and are being incorporated in climate models.[14] These parameterizations treat the vegetative cover as a variable resistance that moderates the flow of moisture from the soil to the atmosphere.

[13] Suarez *et al.* (1983).

[14] Sellers *et al.* (1986); Dickinson (1983).

10.5 The Ocean Component

The oceanic component of a global climate model is also built on a framework of the equations of motion describing the general circulation of the ocean, although most climate simulation experiments have been conducted with a general-circulation model of the atmosphere coupled with a much simpler ocean model in which currents are not computed. Ocean general-circulation models differ from atmospheric models in that the fluid is water, rather than air, and the geometry of the ocean basins is more complex. The fundamental driving mechanisms for the ocean circulation include wind stress at the upper surface, radiative and sensible heat fluxes through the surface, and density variations caused by changes in salinity. Salinity changes are induced at the surface by precipitation, evaporation, runoff, salt rejection during sea ice formation, and the addition of freshwater during sea ice melting. These salinity variations are carried to the interior of the ocean by fluid motions, which are forced in part by the density variations associated with salinity and heat content.

The spatial resolution with which the oceanic circulation can be resolved is again limited by the computing time required to complete a simulation. A smoothed representation of the bathymetry of an ocean model used in a coupled climate model is shown in Fig. 10.4. This model has 12 vertical levels and 4.5° latitude × 3.75° longitude resolution. When you consider that critically important features like the Gulf Stream and Kuroshio currents are less than a degree in width, the horizontal resolution currently used in the ocean component of climate models is very crude. Actual turbulent diffusion in the ocean is believed to be very small and yet,

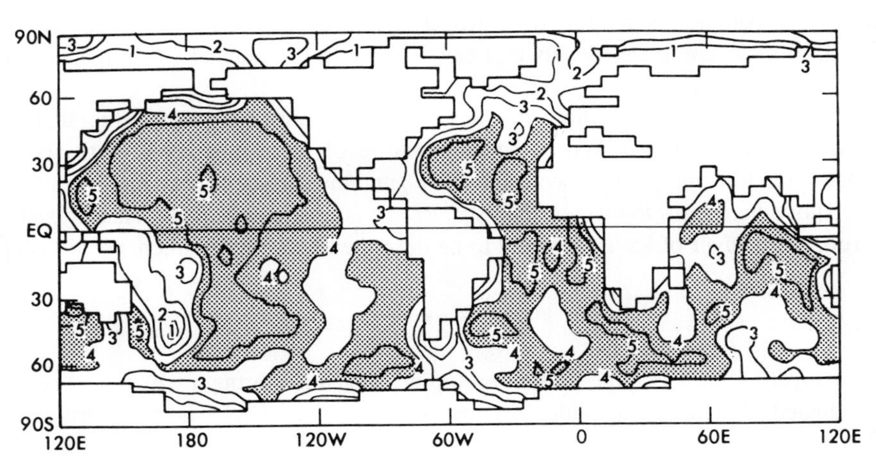

Fig. 10.4 Smoothed bathymetry of the global ocean as represented in the coupled ocean–atmosphere model of Manabe *et al.* (1990). Depths are given in kilometers. (Reprinted with permission from the American Meteorological Society.)

for numerical reasons, large turbulent diffusion coefficients must be used in the conservation equations for momentum, heat, and salt. As a consequence, ocean models currently used in coupled climate models produce only weak and diffuse western boundary currents, and cannot resolve the eddies that develop on these currents.

Global ocean models with spatial resolution as fine as half a degree of latitude and longitude have been used to simulate the ocean circulation in response to fixed atmospheric forcing, but have not been operated in a coupled mode with an atmospheric model. These high-resolution simulations are better able to simulate many key aspects of the oceanic circulation such as the western boundary currents, the Antarctic circumpolar current, and the narrow equatorial currents of the Pacific Ocean.[15] It is estimated that global ocean simulations with a resolution of about one-eighth of a degree will be required to properly simulate the eddies that develop on the Gulf Stream. Climate simulations with such fine spatial resolution are beyond the capabilities of current computers, but it is unclear whether such detail is necessary for an adequate simulation of climate.

10.5.1 Sea Ice Models

Sea ice plays a critical role in climate by increasing the albedo of the ocean surface, inhibiting the exchanges of heat, moisture, and momentum between atmosphere and ocean, and altering the local salinity during sea ice freezing and melting (Fig. 10.5). Thermodynamic processes lead to freezing, and melting of seawater and dynamic processes, such as driving by winds and currents, cause mechanical deformation and transport of sea ice. In most climate models at least a thermodynamic model of sea ice is employed. Transport of sea ice by winds and currents is treated with varying degrees of complexity, ranging from mixed-layer ocean models with no ice transport at all, to more complete models that calculate the movement of ice in response to both winds and currents.

The simplest model predicts the thickness of a layer of sea ice based on thermodynamic considerations. When the temperature of the upper layer of the ocean reaches the freezing point of seawater (–2°C) and the surface energy fluxes continue to remove heat from the water, then surface ice is assumed to form. Heat transport through this layer of ice is assumed to be described by a flux-gradient relationship.

$$F_I = -k_I \frac{\partial T}{\partial z} \tag{10.9}$$

A layer of snow may be present on top of the sea ice through which heat is also conducted. The thermal conductivity for ice, $k_I \approx 2\ \mathrm{W\ m^{-1}\ K^{-1}}$, is much larger than the thermal conductivity of snow, $k_s \approx 0.3\ \mathrm{W\ m^{-1}\ K^{-1}}$, so that a layer of snow greatly increases the thermal insulation provided by the sea ice. Snowcover also in-

[15] Semtner and Chervin (1992).

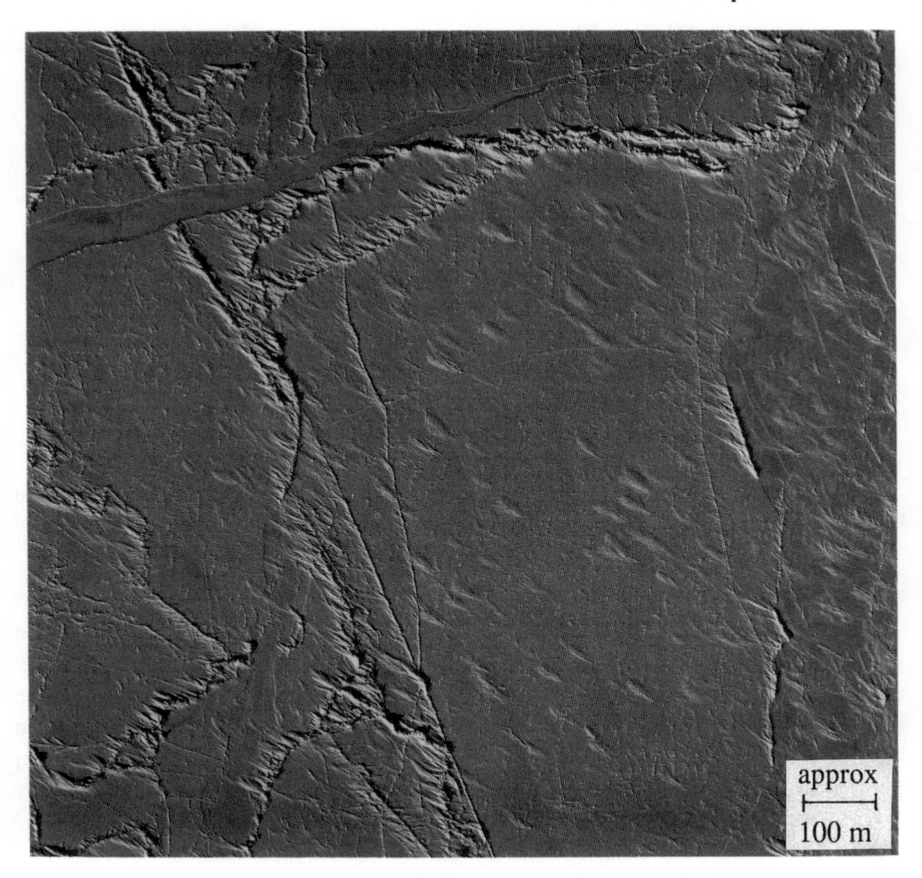

Fig. 10.5 Aerial photograph of sea ice in the Beaufort Sea. Note the refrozen lead at the upper left, the pressure ridges, and the barchans (bumps that are caused by windblown snow accumulation). (Photo: NASA, March 1972.)

creases the albedo. If all heat fluxes, including solar heating, are assumed to be applied at the surface rather than distributed over some depth, then in steady state the flux through the snow and ice would be equal at every depth. Constant flux implies a linear temperature profile, so that the heat flux becomes proportional to the temperature difference across the ice layer. Under these steady conditions the temperature profile within the layer of snow and ice would be as shown in Fig. 10.6. The fluxes across the ice and snow layers can then be written in terms of the temperature difference across the layers.

$$F_I = k_I \frac{T_B - T_i}{h_I}, \qquad F_s = k_s \frac{T_i - T_s}{h_s} \tag{10.10}$$

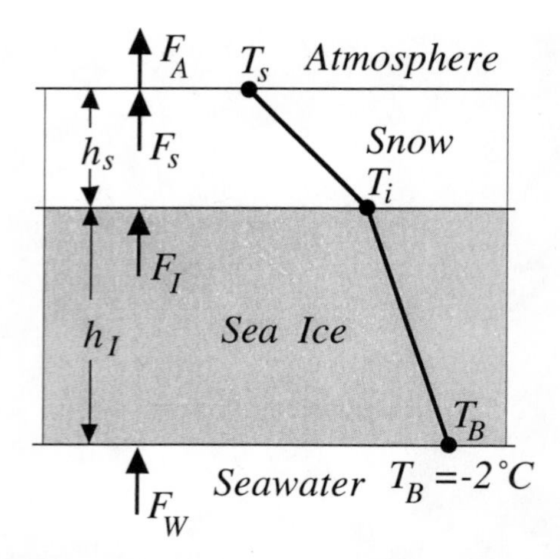

Fig. 10.6 Schematic diagram of a simple sea ice model with a layer of snow over a layer of ice and the fluxes of heat through the layers. The temperature profiles in the two layers are linear if penetration of solar radiation is ignored and a steady state is assumed. Symbols are as defined in the text.

where T_s is the surface temperature, T_B is the temperature at the base of the sea ice layer, T_i is the temperature at the interface between ice and snow, and h_I and h_s are the depth of the ice and snow layers, respectively. The interface temperature can be eliminated by applying the requirement that in a steady state the heat flux must be the same through the snow layer and the ice layer. The flux through the sea ice can then be expressed in terms of the temperature difference across the snow-ice layer and the depth of each layer.

$$F_I = \frac{k_I\, k_s}{k_I h_s + k_s h_I}\left(T_B - T_s\right) = \gamma_{\mathrm{SI}}\left(T_B - T_s\right) \tag{10.11}$$

Here γ_{SI} is the combined thermal conductance of the snow–ice layer. For a thick layer $\gamma_{\mathrm{SI}} \approx 1\ \mathrm{W\ m^{-2}\ K^{-1}}$, and for a thin layer $\gamma_{\mathrm{SI}} \approx 20\ \mathrm{W\ m^{-2}\ K^{-1}}$. The surface temperature can be determined from the requirement that the net heat flux at the surface be zero, which is consistent with our previous assumption of no heat storage in the ice. The net heat flux includes the conductive flux through the ice, and the latent, sensible, and net radiative heating of the surface. The depth of the snow layer is determined by the balance of snowfall, sublimation–evaporation, and melting. If no snow is present, then the ice layer may be reduced in thickness by melting and sublimation at the top. Growth of the ice layer occurs at the bottom, where it is assumed to thicken at a rate such that the heat of fusion will just balance the net heat flux at the bottom of the ice layer and thus keep the lower edge of the ice at the freezing point. Similarly, if the equilibrium surface temperature would exceed the freezing point, ice is assumed to melt at a rate sufficient to keep the temperature at the freezing point until all of the ice is melted.

Sea ice grows rapidly when it is thin and more slowly as thickness increases; when subjected to similar thermal forcing, ice a few centimeters thick grows nearly a hundred times faster than ice that is 2 or 3 m thick. The reason for this is because of the insulating effect of the ice itself. Consider a sheet of sea ice that is growing as a result of an imposed temperature difference between the ice surface and the seawater. The thickness of the sea ice increases at a rate necessary to balance the heat flux through the ice with the latent heat of fusion needed for freezing seawater at the base of the ice.

$$\rho_I L_f \frac{\partial h_I}{\partial t} = \frac{k_I}{h_I}\left(T_B - T_s\right) - F_w \tag{10.12}$$

Here ρ_I is the density of ice, L_f is the latent heat of fusion for seawater, and F_w is the rate at which heat is supplied to the sea ice by ocean fluxes. It is notable that for fixed surface temperature, the growth rate of the ice thickness is inversely proportional to the thickness itself. Integrating (10.12) over time assuming $F_w = 0$, we find that the ice thickness is proportional to the square root of the integral over time of the temperature difference.

$$h_I^2(t) - h_I^2(t_0) = \frac{k_I}{\rho_I L_f} \int_{t_0}^{t} \left(T_B - T_s\right) dt \tag{10.13}$$

Very simple models of sea ice like the one described above are used in many current climate models. Important thermodynamic processes, such as the heat capacity of the sea ice and the brine incorporated in the ice, are ignored in this model and these processes may have important effects on the amount and seasonality of sea ice and its response to climate forcing.[16] Horizontal transport of sea ice is also important in many situations and plays a significant role in heat and salt transport in the high-latitude oceans. In the Arctic Ocean the mean ice thickness is about 3–4 m, while in the Antarctic the average thickness is 1–2 m. The thinner Antarctic sea ice may be related to a greater supply of heat to the sea ice by ocean fluxes in the Antarctic. Also, more mechanical deformation of the ice in the closed basin of the Arctic Ocean may contribute to the greater average thickness there compared to the Antarctic sea ice, which is less constrained by shorelines and simply spreads equatorward until it melts (Fig. 10.7). Where sea ice is driven together by winds and current, pressure ridges can be formed where the ice thickness exceeds 10 m. As part of the same process small openings in the ice (leads) or larger areas of open water (polynyas) can be produced. In these regions of open water, enhanced heat loss and ice formation occur in the season of ice growth, and more absorption of solar radiation occurs in the melting season. These processes are not explicitly treated in most current climate models, but are clearly important.

[16]Maykut and Untersteiner (1971); Semtner (1976, 1984).

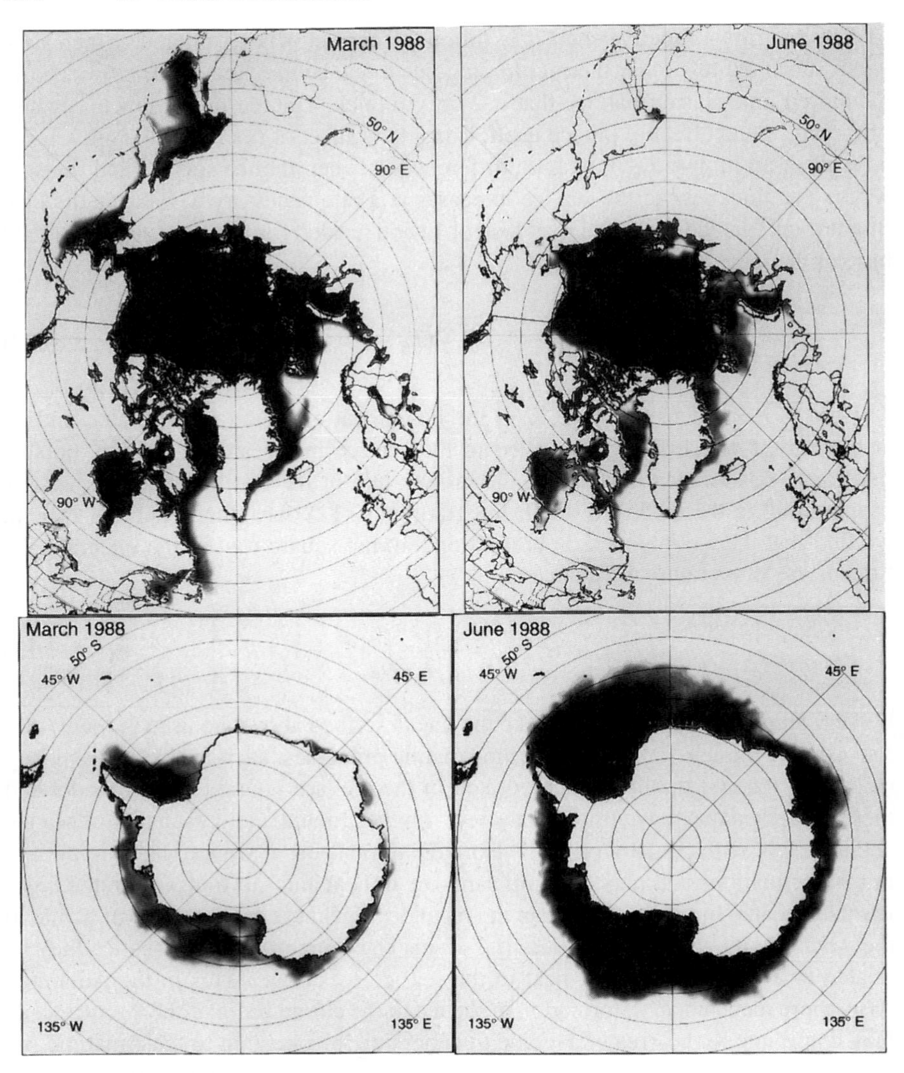

Fig. 10.7 Sea ice concentration observed by satellite microwave imaging in the Arctic and Antarctic for 4 months during 1988. Shaded areas indicate sea ice concentration in excess of 15%. See scale. (Figure courtesy of J. Comiso, NASA Goddard Space Flight Center.)

10.6 Validation of Climate Model Simulations

The simulations obtained with current climate models are in reasonable accord with observations of the present climate for a variety of key dynamic and thermodynamic climate variables. This agreement is highly encouraging, but it does not

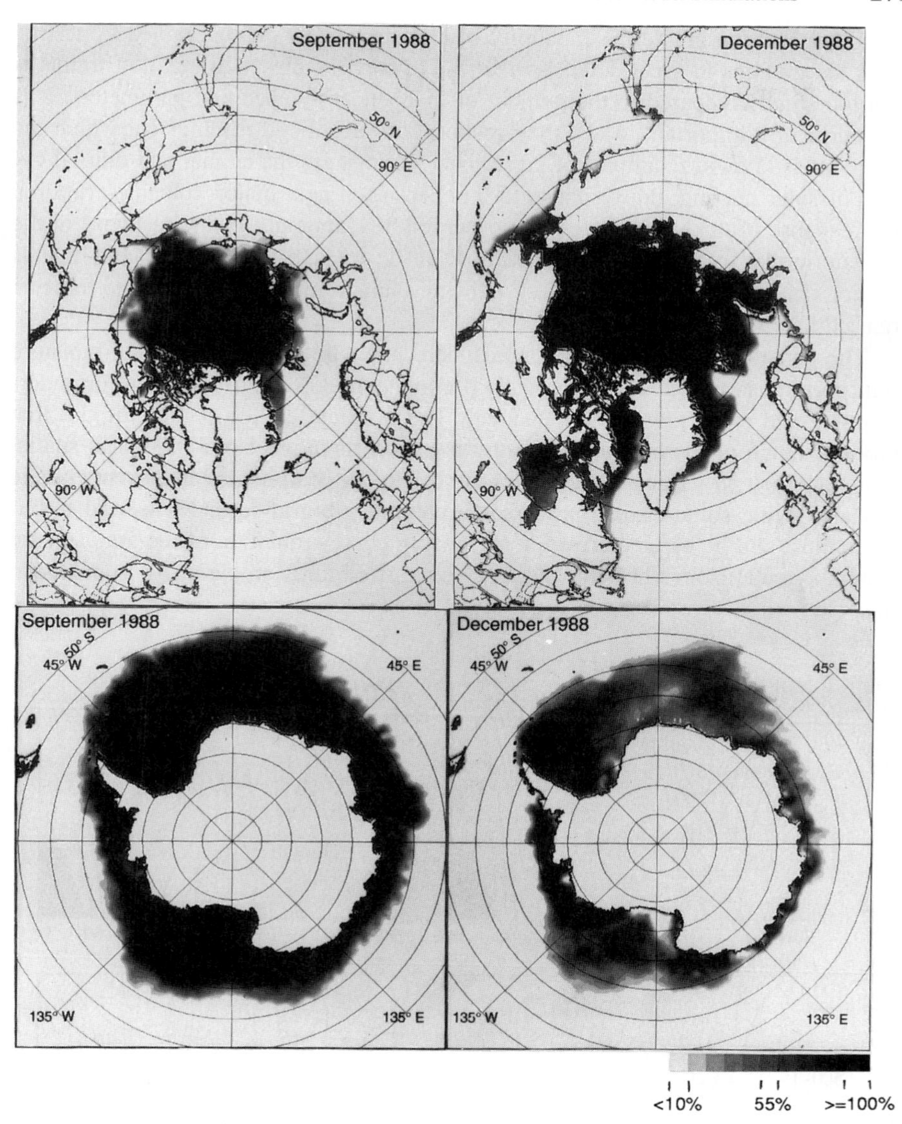

Fig. 10.7—*Continued*

by itself mean that climate models are capable of accurately predicting the response of climate to a natural or human-induced perturbation. The reason for skepticism is that a large number of adjustable constants are introduced in the parameterizations for sub-grid-scale phenomena and processes and these parameters often cannot be determined on the basis of fundamental principles, but rather are set to values that give the most realistic-looking simulation of the current climate. Real confidence in the predictive capability of climate models can be gained by testing their simulations in great detail, so that the number of observational constraints is not too small compared to the number of independently specified parameters. It is also critically important to test their response to prescribed forcings for which the response is known. Examples of prescribed forcings are the annual and diurnal cycles of solar heating, the response to an event such as a major volcanic eruption, the response of the atmosphere to an observed SST anomaly, or the response of the climate model to the boundary conditions of an ice age (see Chapter 11). In this section some examples of how well existing climate models can simulate the current climate will be presented.

The general circulation of the atmosphere, including winds, heat, and moisture transport, must be simulated accurately if climate models are to be useful for understanding and predicting climate changes. The transient eddies of midlatitudes that produce much of the heat and moisture transport there are large enough in spatial scale that their structure and evolution can be explicitly simulated in current climate models. While some critical processes involving sub-grid-scale waves and turbulence must still be parameterized, a reasonably good simulation of the atmospheric general circulation can be achieved, as characterized for example by the zonal mean flow (Fig. 10.8).

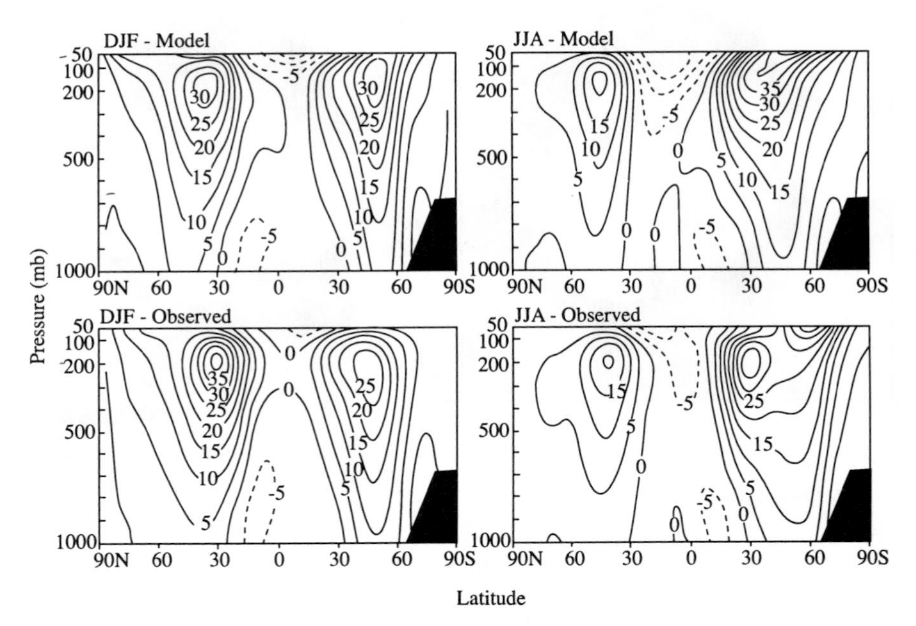

Fig. 10.8 Zonal cross sections of zonal-mean wind simulated in a GCM (top) and observed (bottom) during the solstitial seasons of December–February (left) and June–August (right). [From McFarlane *et al.* (1992). Reprinted with permission from the American Meteorological Society.]

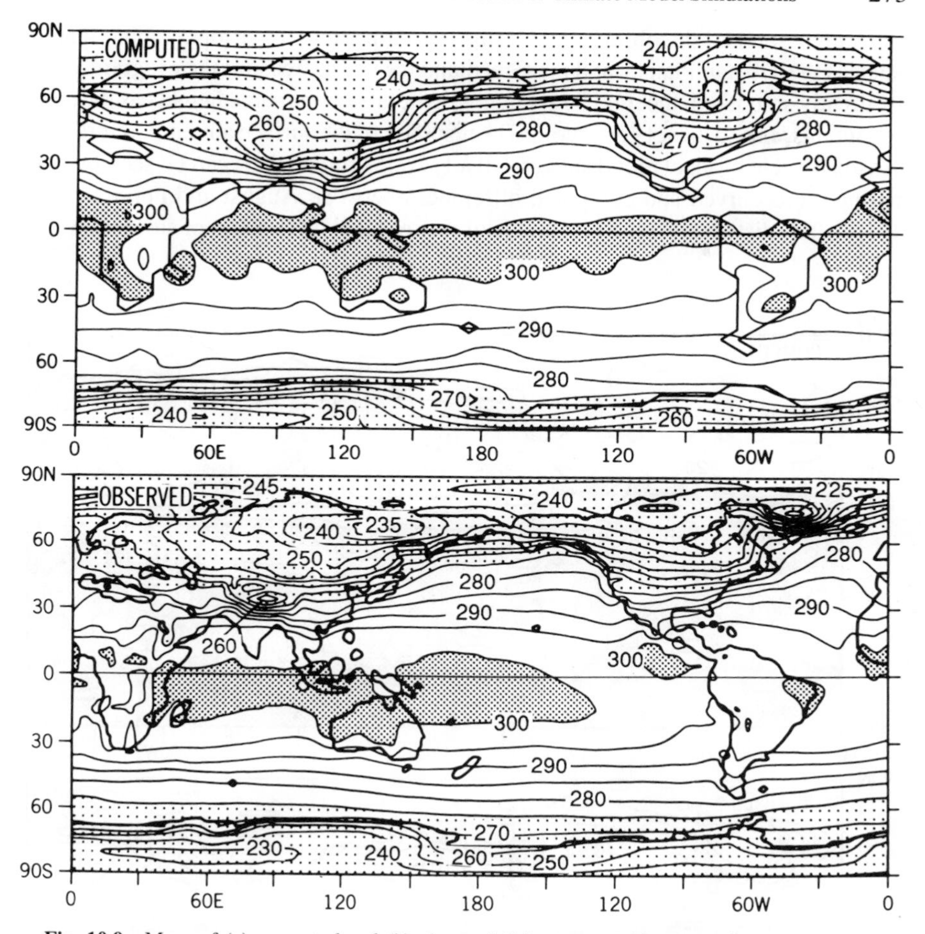

Fig. 10.9 Maps of (a) computed and (b) observed February monthly mean air temperatures (K). [Top computed distribution from Manabe and Stouffer (1980); bottom observed distribution from Crutcher and Meserve (1970) and Taljaard *et al.* (1969) as printed in Manabe and Stouffer (1980), © American Geophysical Union.]

A key climatic variable is the surface air temperature, and at least the broad outlines of the geographic and seasonal variation of air temperature can be simulated in current climate models. An example of the surface air temperature simulation from a model with a mixed-layer ocean and a horizontal grid spacing of about 600 km is shown in Fig. 10.9. One apparent shortcoming of this simulation is the failure to produce an east-to-west temperature gradient in the equatorial Pacific Ocean. This is because the ocean model used does not transport heat, so that the upwelling of cold water in the eastern Pacific is not included in the simulation. Another obvious difference between the simulation and the observations is the air temperature over topographic features such as the Tibetan Plateau and the Greenland ice sheet. Because the orography must be greatly smoothed, the model mountains do not reach the same heights as the actual mountains, and so do not get as cold. The response of the surface air temperature to the seasonal cycle of insolation is also relatively well simulated, as can be seen by comparing the observed and computed differences between August and February (Fig. 10.10). The distribution and magnitude of the seasonal differences of surface air tem-

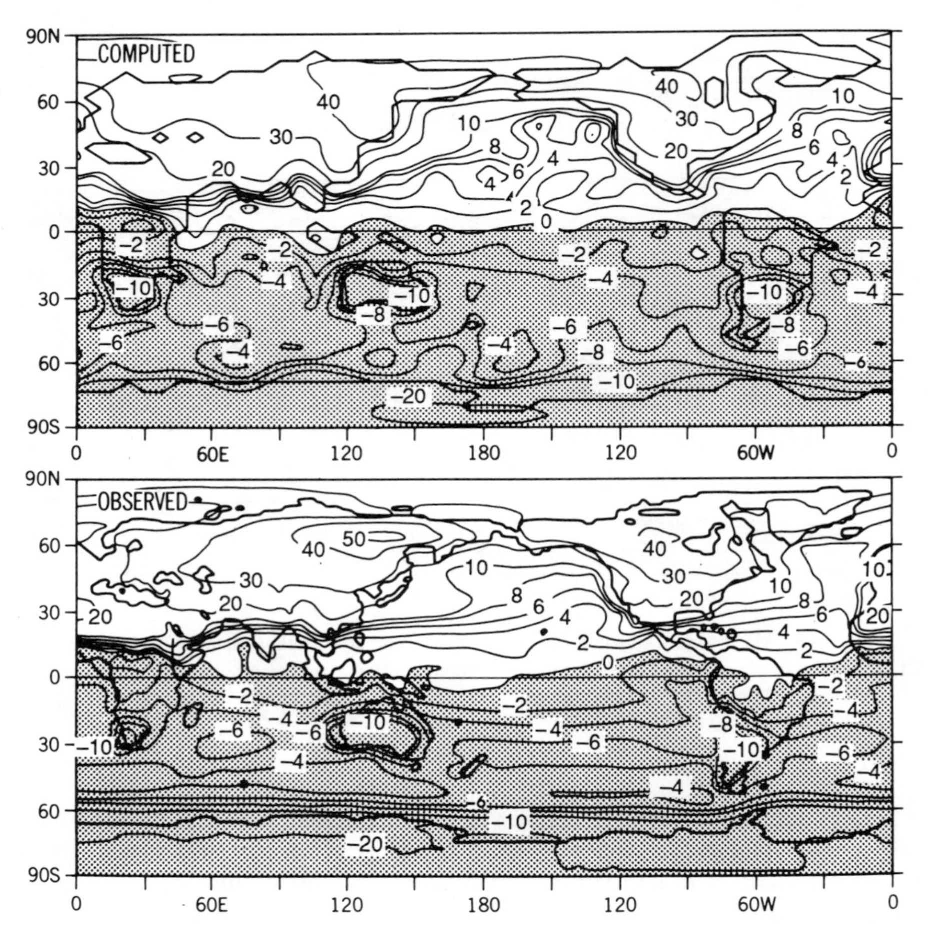

Fig. 10.10 Maps of (a) computed and (b) observed temperature difference (K) between August and February. The contour interval is 2 K when the absolute value of the temperature difference is less than 10 K and is 10 K when it is more than 10 K. [Top computed distribution from Manabe and Stouffer (1980); bottom observed distribution from Crutcher and Meserve (1970) and Taljaard *et al.* (1969) as printed in Manabe and Stouffer (1980), © American Geophysical Union.]

perature are reasonably well matched. The smaller seasonal differences in temperature over the ocean are of course a result of the larger heat capacity of the ocean, which is accounted for in the model by its mixed-layer ocean.

Precipitation is also a critically important climate variable, and it is one that is difficult to simulate accurately. Comparisons between observed and computed precipitation patterns for two seasons are shown in Figs. 10.11 and 10.12. The climate model simulates the position and strength of the major precipitation centers and their seasonal movement relatively well. The strength and narrowness of the Intertropical Convergence Zone are underestimated, especially in the eastern Pacific Ocean where

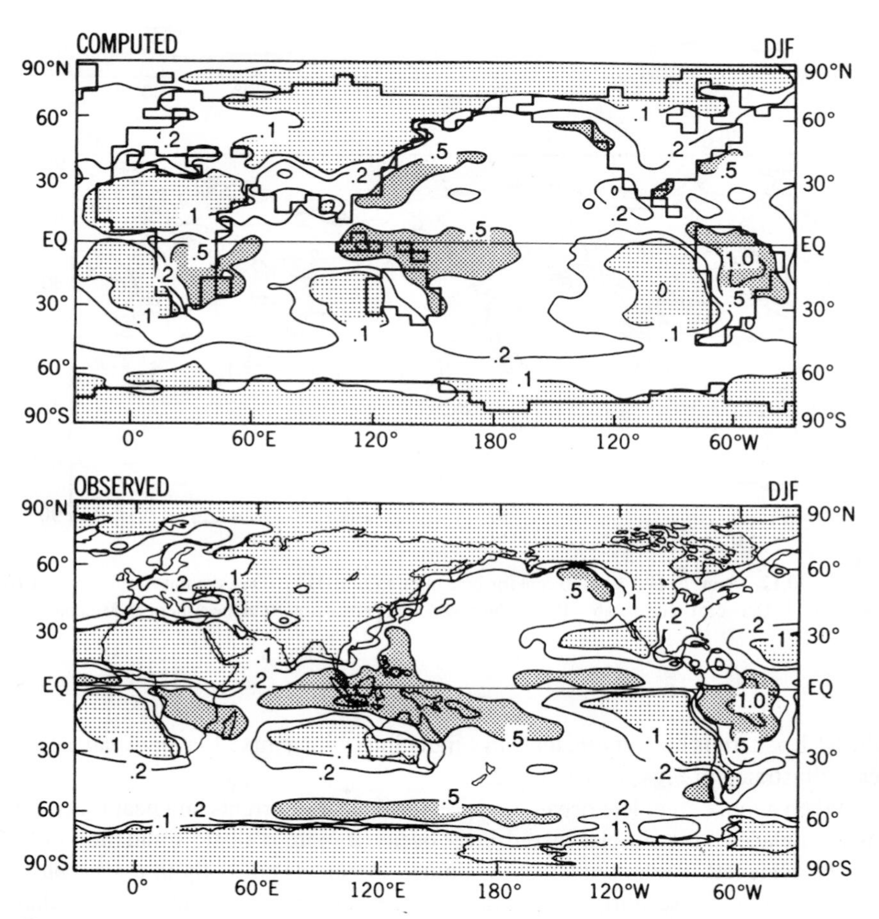

Fig. 10.11 Map of mean precipitation rate (a) computed and (b) observed for December–February (DJF). Heavy stippling indicates areas where the precipitation is >0.5 cm day^{-1}, and light stippling indicates where it is <0.1 cm day^{-1}. Both maps are smoothed. [Observations from Jaeger (1976) as printed in Delworth and Manabe (1988). Reprinted with permission from the American Meteorological Society.]

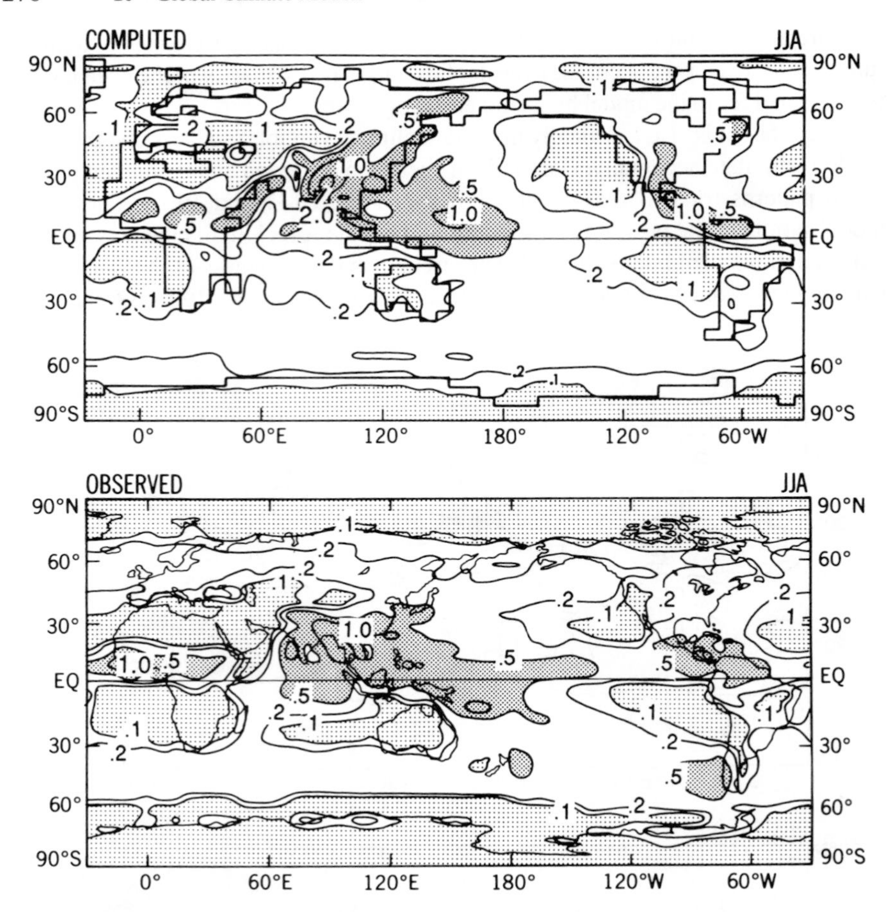

Fig. 10.12 Same as Fig. 10.11 except for June through July (JJA). [Observations from Jaeger (1976) as printed in Delworth and Manabe (1988). Reprinted with permission from the American Meteorological Society.]

the SST pattern is not very realistic in simulations with a mixed-layer ocean, as is the case illustrated in Figs. 10.10–10.12.

When a more complete ocean model that calculates currents and heat transport is used in place of the simpler mixed-layer ocean, a more realistic simulation of SST can be achieved. The improvement is limited in models currently in use because of the relatively coarse horizontal grid resolution. Figure 10.13 shows a simulation from a coupled climate model including an ocean with a horizontal grid spacing of 4.5° longitude × 3.75° latitude and 12 vertical levels. The SST is still too zonally uniform and the gradients are too weak compared to the observations.

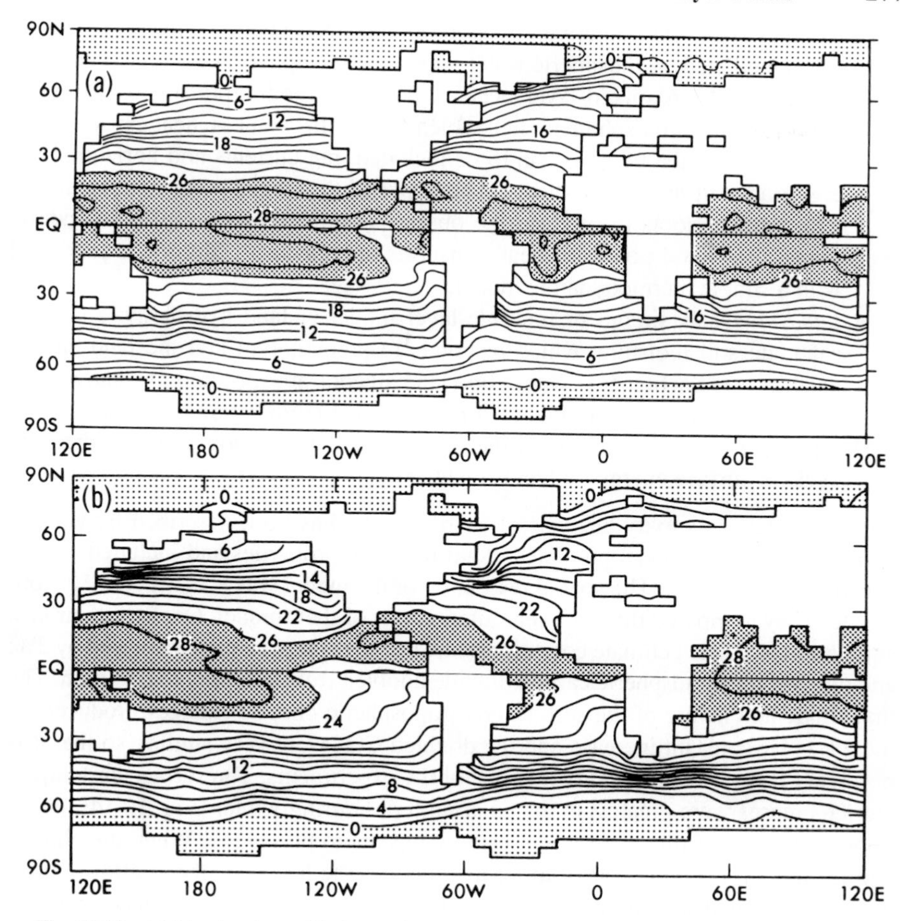

Fig. 10.13 (a) Simulated and (b) observed annual mean sea surface temperature (SST) in °C. [From Manabe *et al.* (1990); bottom panel data from Levitus (1982). Reprinted with permission from the American Meteorological Society.]

10.7 Sensitivity Estimates from Climate Models with Mixed-Layer Oceans

Global climate models can be used to estimate how various processes interact to determine the sensitivity of climate. The first class of models for which extensive GCM sensitivity experiments have been conducted is the type of model that was prevalent in the 1980s. These models have the following basic characteristics.

- *Atmosphere model.* Approximately 500-km horizontal resolution and 7–12 vertical levels are used. A complete parameterized treatment of radiative transfer is included, which incorporates the effects of clouds. Convection is parameterized, but radiative effects of clouds are sometimes fixed and sometimes predicted. Atmospheric water vapor and precipitation are predicted.

- *Ocean model.* A mixed-layer ocean with a specified depth of 50–100 m is coupled to the atmosphere model, which allows the wetness, heat capacity, and low albedo of the upper ocean to be included, but currents and horizontal energy transport are not predicted. The depth of the mixed layer may be varied with latitude so as to produce the observed annual variation of temperature. Sometimes fixed poleward heat transports are specified. Sea ice is predicted with a simple thermodynamic model.
- *Land model.* A single- or multiple-layer model of land hydrology is used to calculate soil moisture, which responds to precipitation and the land energy balance (see Section 5.5). Snow-free land albedos are specified, but snowcover and its albedo effect are predicted from surface temperature and precipitation rate. No explicit simulation of the active role of vegetation on the hydrologic cycle or surface energy budget is included.

A thorough sensitivity analysis of such a model has been described by Hansen *et al.* (1984). Their model predicted cloud radiative properties and included a fixed oceanic heat transport. The spatial resolution of the model was 8° latitude × 10° longitude. They compared three 35-year integrations of their model. One was their best simulation of current climate conditions; one had the solar constant increased by 2%; and one had the atmospheric carbon dioxide doubled from 315 ppm to 630 ppm. The magnitude and nature of the surface and tropospheric climate changes produced by the 2% solar constant increase and the doubled CO_2 were very similar, so that it is necessary to show results from only one of the experiments. The surface air temperature increases were greatest in the polar latitudes during winter [Fig. 10.14(a)]. An important cause of this polar amplification appears to be sea ice. When the climate warms, the sea ice melts back farther in summer. The lowered albedo allows much more heat to be absorbed in summer, but this heat does not result in a large increase in surface temperature since it is spread over the large heat capacity of the mixed layer. During the winter this stored heat is returned to the atmosphere and results in a large wintertime warming.[17] Strong polar temperature inversion during winter and the reduction of the strength of this inversion in a warmed climate also contribute to the polar amplification.

The temperature increase in the upper tropical troposphere is much larger than the increase at the surface, so that the average lapse rate of the atmosphere decreases [Fig. 10.14(b)]. The additional solar heating at the surface is largely used to evaporate water rather than increase the surface temperature, and the increased supply of water vapor enhances deep convection, which heats the upper troposphere and supports the reduction in lapse rate in the tropics.[18]

The relative contributions of various radiative feedback mechanisms to pro-

[17] Manabe and Stouffer (1980).

[18] See Fig. C.1 and the related discussion in Appendix C.

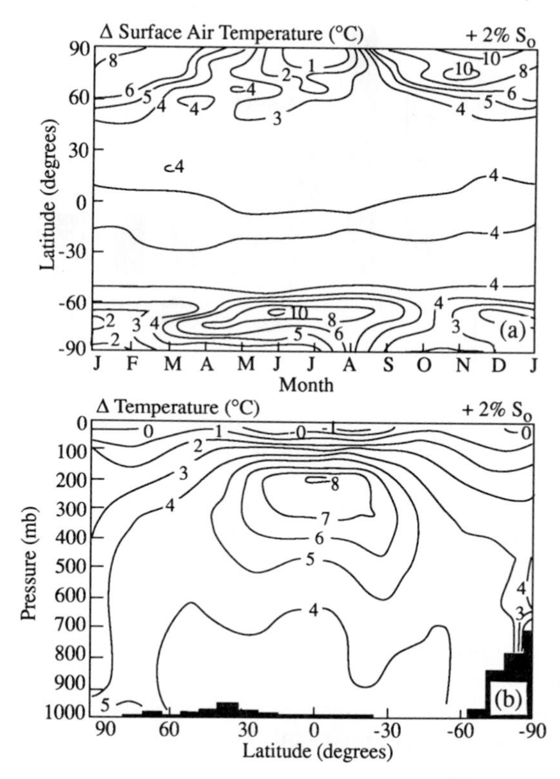

Fig. 10.14 (a) Latitude versus season contour plot of the zonally averaged surface air temperature difference and (b) pressure–latitude contour plot of zonally averaged air temperature difference resulting from a 2% increase of the solar constant. [From Hansen *et al.* (1984). © American Geophysical Union.]

ducing the observed global-mean warming can be evaluated with the help of a one-dimensional radiative–convective model (Chapter 3). The changes in surface albedo, water vapor amount, water vapor distribution, lapse rate, cloud cover, and cloud height are globally averaged from the three-dimensional model simulation and then inserted individually into a one-dimensional model to compute their effect on surface temperature. The results of these calculations are shown graphically in Fig. 10.15. The direct effect of increasing the solar constant by 2% is about 1.3°C, compared to a total warming of 4°C, so that most of the surface warming appears to be contributed by the feedback processes. Among these the most important is the increased water vapor and its changed vertical distribution. The water vapor resides at a higher altitude on average and would therefore emit at a lower temperature. This effect is more than offset by the decreased lapse rate, which makes the infrared opacity of this water vapor less effective in reducing the escaping longwave radiation. The surface albedo decrease associated with decreased surface ice, the decreased cloud amount, and the increased cloud altitude each contribute about 0.5°C of warming.

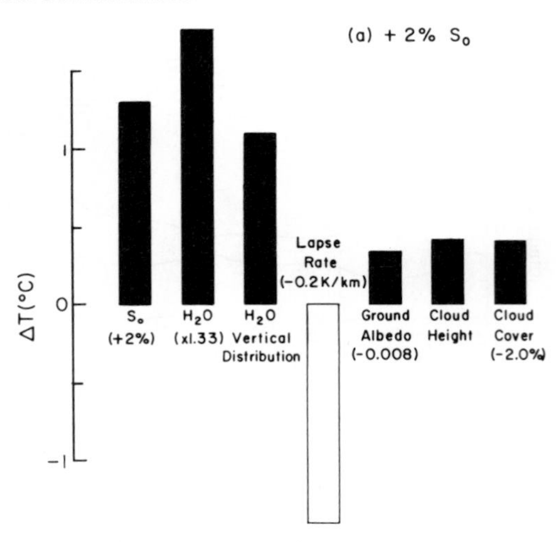

Fig. 10.15 Estimated contributions by the direct forcing of a 2% increase of solar constant and various feedback mechanisms to the surface temperature increase in a climate model. [From Hansen *et al.* (1984), © American Geophysical Union.]

The strong positive water vapor feedback is consistent with the fixed relative humidity assumption often made in simple climate sensitivity arguments and seems to be supported by observations.[19] Among these feedbacks the positive cloud feedback is the most controversial. The cloud radiative feedback in climate models varies greatly from model to model, even for climate models in which cloud optical properties are fixed.[20] Cloud albedo is specified in many models, but is potentially sensitive to the mean climate and could be the basis for a strong feedback. For example, a modest sensitivity of cloud optical depth to mean surface temperature could provide a strong feedback.

10.8 Coupled Atmosphere–Ocean Processes and the Thermohaline Circulation

Fully coupled climate models simulate the atmosphere and the ocean jointly, including the exchanges of momentum, heat, and water between them. An example of the mean meridional mass circulation in a coupled model is shown in Fig. 10.16. The Hadley, Ferrel, and polar cells of the atmospheric circulation are shown, along with a representation of the deep thermohaline circulation of the ocean. A large poleward cell exists in the Northern Hemisphere ocean with sinking to intermediate depths at about 60°N. Most of this circulation occurs in the Atlantic. Shallow circulations occur near the equator. Upwelling occurs in midlatitudes of the Southern Hemi-

[19] Raval and Ramanathan (1989).
[20] Cess *et al.* (1990).

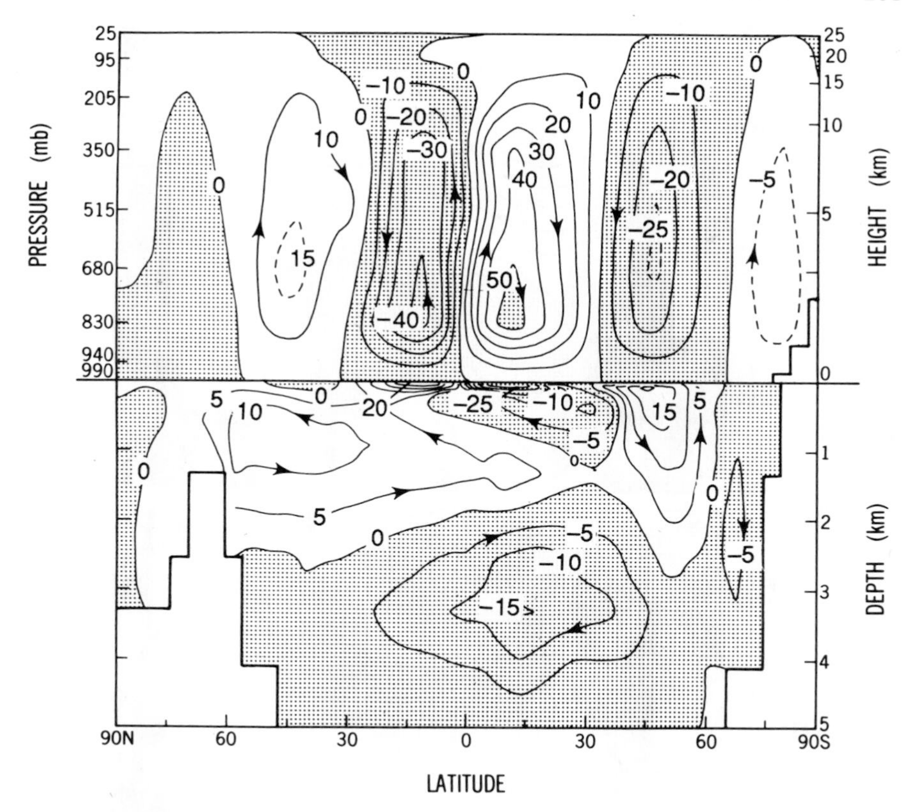

Fig. 10.16 Zonally integrated mass transport of the atmosphere and ocean for an equilibrium climate with an active thermohaline circulation. Units are megatons s^{-1} (10^9 kg s^{-1}). [From Manabe *et al.* (1990). Reprinted with permission from the American Meteorological Society.]

sphere at about 50°S, and some deep water is formed in Antarctic waters poleward of 60°S. A deep circulation cell is simulated, which is centered near the equator and at a depth of about 3500 meters.

The deep thermohaline circulation of the Atlantic Ocean and its interaction with the atmosphere seem to provide a mechanism for variability within the climate system. Numerical experiments suggest that the coupled ocean–atmosphere climate system may possess at least two stable equilibria: one with a thermohaline circulation in the North Atlantic as in today's climate, and one in which the thermohaline circulation is weak or absent.[21] This model result is consistent with paleoclimatic data discussed in Chapter 8, which suggest that the thermohaline circulation in the north Atlantic may have slowed down or ceased to operate for a time about 11,000 years ago and during the last glacial maximum.

To demonstrate that two different equilibrium states are possible: a single model

[21]Manabe and Stouffer (1988).

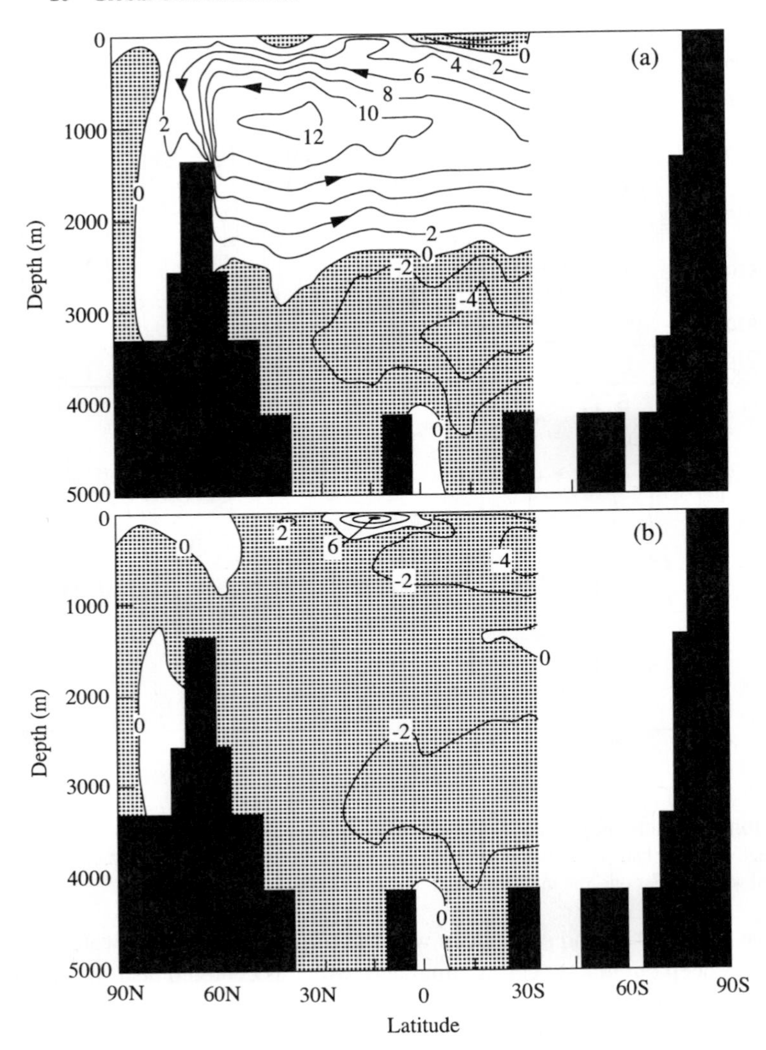

Fig. 10.17 Model-simulated meridional mass streamfunction in the north Atlantic Ocean for (a) equilibrium with a strong thermohaline circulation, (b) equilibrium without one. Units are megatons s^{-1}. [From Manabe and Stouffer (1988). Reprinted with permission from the American Meteorological Society.]

with identical external forcing conditions is integrated from two different initial conditions. One initial condition with saline water in the north Atlantic leads to an equilibrium state with a thermohaline circulation, and an initial condition with relatively fresh water in the north Atlantic does not. The meridional mass streamfunctions for the two equilibria are shown in Fig. 10.17. In one case poleward motion near the sur-

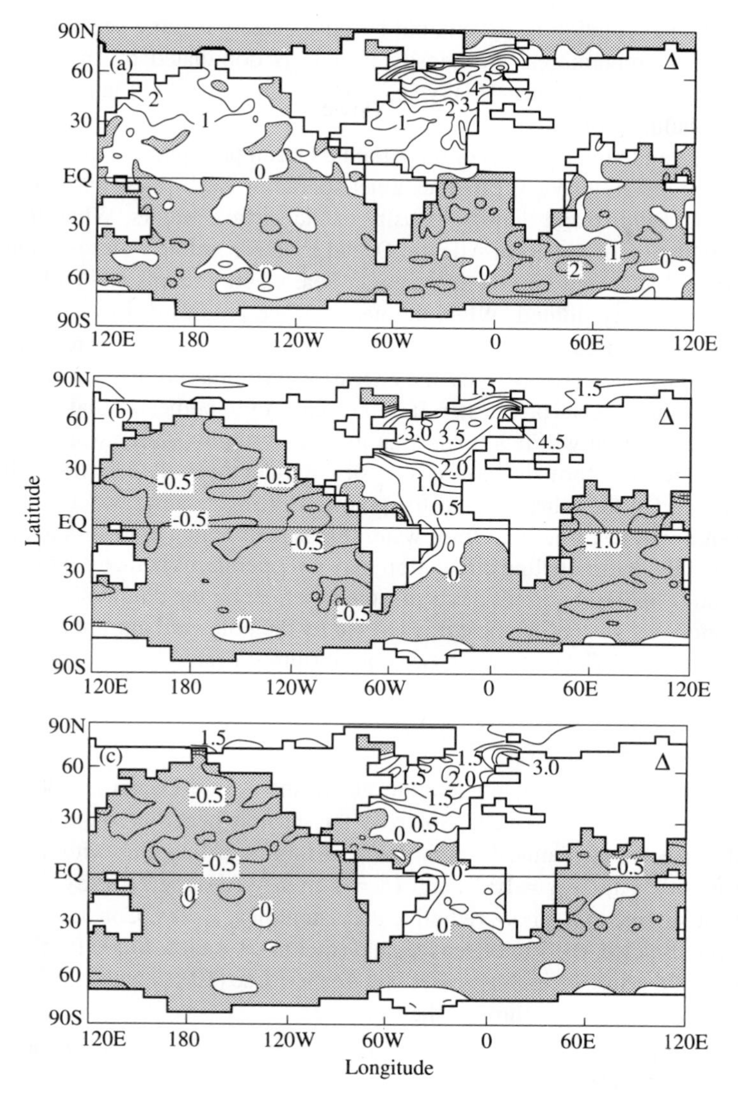

Fig. 10.18 Difference in (a) SST (K), (b) surface salinity (‰), and (c) surface water density (kg m^{-3}) between a model equilibrium with a thermohaline circulation and without one. [Panel (a) from the unpublished version of Manabe and Stouffer, 1988; panels (b) and (c) from published version of Manabe and Stouffer (1988). Reprinted with permission from the American Meteorological Society.]

face and sinking at high latitudes are evident, but in the other case the meridional circulation in the ocean is weak. The differences between the two equilibrium states as seen in SST, surface salinity, and surface density are shown in Fig. 10.18. Temperature and salinity are both increased in the high-latitude north Atlantic when the thermohaline circulation is present, but because density is more sensitive to salinity at these low temperatures, the density change is dominated by the effect of the greater salinity. This greater density sustains the sinking at high latitudes that feeds the thermohaline circulation.

In middle and high latitudes of the Atlantic Ocean precipitation exceeds evaporation, so that the atmosphere provides a net source of freshwater that would tend to decrease the salinity and thus the density of the surface waters. When the thermohaline circulation of the north Atlantic Ocean is active, saline water is brought from the south and advected into the Norwegian Sea, where it cools and sinks before its salinity is seriously diluted. When the near-surface poleward flow associated with the thermohaline circulation slows down, the source of salty water is reduced and the surface salinity decreases because the excess of precipitation over evaporation continues to freshen the surface water. It is thus easy to hypothesize that a strong thermohaline circulation will tend to sustain itself by bringing saline water to the polar regions where it can form the basis of very dense water that will then sink to the depths of the ocean. On the other hand, if the poleward flow of saline water is never initiated, then it will be difficult to form water that is dense enough to sink to substantial depths, and the thermohaline circulation may not operate. Whether the thermohaline circulation is operating or not has a substantial effect on the climate in the north Atlantic region, because of the heat transported by the poleward current and the resulting difference in SST between the two equilibrium states.

Exercises

1. Suppose that the air temperature above the Arctic Ocean is –30°C and the water temperature is –2°C. Solve for the sensible heat flux to the atmosphere if the bulk aerodynamic formula for heat flux (4.26) applies with surface pressure 10^5 Pa, $C_{DH} = 10^{-3}$, and $U_r = 5 \text{ m s}^{-1}$. Assume a steady state with no heat storage and ignore radiative and latent heat fluxes. Consider three cases: (a) no sea ice, (b) 1 m of sea ice, and (c) 3 m of sea ice. Solve for the temperature of the surface in those cases with sea ice. Use (10.10) with $k_I = 2 \text{ W m}^{-1}\text{K}^{-1}$ to solve for the flux through the ice.
2. Repeat problem 1 for the case of 1 m of sea ice with 10 cm of snow on top. Use $k_s = 0.3 \text{ W m}^{-1}\text{K}^{-1}$.
3. Calculate the surface temperature and sea ice thickness in equilibrium under the following conditions: polar darkness, air temperature –40°C, surface pressure 10^5 Pa, $C_{DH} = 10^{-3}$, $U_r = 5 \text{ m s}^{-1}$, and $k_I = 2 \text{ W m}^{-1}\text{ K}^{-1}$. The downward longwave flux is 100 W m^{-2}, and the surface has unit longwave emissivity. Consider cases where the ocean supplies heat to the base of the ice at the rate of (a) 10 W m^{-2} and (b) 20 W m^{-2}. *Hint:* Use the energy balance at the surface and at the base of the ice and linearize the Stefan–Boltzmann emission about the air temperature. Ignore latent cooling.

4. What would be the equilibrium sea ice thickness for the two cases of problem 3 if the sea ice were covered by 20 cm of snow with $k_s = 0.3$ W m^{-1} K^{-1}?
5. Use Fig. 10.16 to estimate the magnitude of the zonally averaged meridional wind in the lowest 200 mb of the atmosphere at 15°N and of the zonally averaged northward current in the upper km of the ocean at 45°N.
6. Using Fig. 7.2, estimate the individual contributions to the density change shown in the lower panel of Fig. 10.18 from the SST change and the salinity change at 60°N in the mid-Atlantic Ocean shown in the upper panels. Use the mean SST shown in the lower panel of Fig. 10.13.
7. Explain how the effect of lapse-rate changes on emitted longwave radiation might be offset by humidity changes if the relative humidity remains constant.

Chapter 11 Natural Climate Change

11.1 Natural Forcing of Climate Change

The geologic record indicates that dramatic changes in climate have occurred in the past. These changes occurred in the absence of humans, for the most part, and we can call them natural climate changes. Understanding these natural climate changes is a challenging and important problem that will help us to understand and predict future natural and human-induced climate changes.

In this chapter we will evaluate several mechanisms whereby past climate changes might have been forced to occur. The distinction between a forcing mechanism and a feedback mechanism is somewhat arbitrary, and depends on what you include as part of the climate system and what is considered external to it. We will regard the atmosphere, the ocean, and the land surface as internal to the climate system, and we will consider Earth's interior and everything extraterrestrial as being external to the climate system. Therefore solar constant variations, variations in Earth's orbit about the sun, and volcanic eruptions are all considered external forcing mechanisms. Another reason for making this distinction is that the solar constant, Earth's orbital parameters, and probably volcanic eruptions are not influenced by the climate, so interaction is one-way. Continental drift should also be considered an external forcing to climate, since continental movements are driven by convection in Earth's mantle, which is presumably not sensitive to the modest variations of Earth's skin temperature that are of interest to the climatologist. During the Holocene period when we have the most information about climate, the continental positions have been very nearly fixed, so that discussion of the effect of continental position on climate will not be expanded beyond what was presented in Chapter 8.

11.2 Solar Luminosity Variations

The total energy output of the sun is a central determinant of Earth's climate. A very direct way of forcing climate change would be to vary the luminosity of the sun. Theories of stellar evolution suggest that the luminosity of the sun has increased steadily by about 30% since the formation of the solar system. This increase is associated with the conversion of hydrogen to helium, leading to concomitant increases of solar density, solar core temperature, rate of fusion, and energy production. If the luminosity of the sun were suddenly decreased by 30%, Earth would quickly become substantially cooler. Geologic evidence of sedimentary rocks 3.8 billion years old suggests that liquid water was present on Earth at this time. Other evidence seems to suggest that Earth was actually warmer early in its history, since no evidence for glaciation exists prior to 2.7 billion years ago. This combination of a less luminous early sun and warm surface climate of the early Earth can be called the *faint young sun problem*.[1] The most likely resolution of this paradox seems to be that the greenhouse effect was much stronger early in Earth's history, perhaps associated with elevated levels of carbon dioxide or some other biogeochemically controlled trace gas. Analysis of ancient soils suggests that 440 million years ago atmospheric CO_2 concentration may have been ~16 times greater than its current value.[2] Carbon cycle models suggest that increasing solar constant and the development of land plants may have accelerated the rate of CO_2 uptake by weathering and caused the apparent decline of atmospheric CO_2 during the mid-Paleozoic (400–320 million years ago).[3] This CO_2 decline may have set the stage for the late Paleozoic glaciation. Beyond the theoretical inference that the luminosity of the sun has increased over its history, we know little about variations in solar luminosity on climatic time scales. Although the sun is a very dynamic entity, it is widely assumed to be a very steady source of energy.

Most of the energy received from the sun originates in the photosphere, which has an emission temperature of about 6000 K. The dominant features seen in the photosphere are dark spots called *sunspots*, which can be seen in both visible light and in the sun's total broadband emission. The center of a typical sunspot has an emission temperature about 1700 K colder than the average for the photosphere, so that the energy emission is only about 25% of average. The darkness of sunspots is produced by a disruption of the normal outward energy flow by the strong magnetic field disturbances that are associated with sunspots. Sunspots are transient features and range in scale from those a few hundred kilometers in diameter and lasting a day or two, to some that are tens of thousands of kilometers wide and last several months. On average sunspots persist for a week or two. The area of the visible disk of the sun that is covered by sunspots ranges from zero to about 0.1%. Dark sunspots are accompanied by bright regions called faculae that cover a much larger fraction of the area of the solar disk than the sunspots with which they appear to be associated. Faculae are about 1000 K hotter than the average photosphere and emit about 15% more energy.

Because sunspots are so easily observed,[4] a long record of sunspot occurrence

[1] Kasting and Grinspoon (1991).

[2] Yapp and Poths (1992).

[3] Berner (1993).

[4] Sunspots are visible with the naked eye through a hazy sky at sunset, and regular reports of sunspots can be found in the Chinese literature beginning in the fourth century. Galileo observed sunspots with his small telescope in 1609. He and his contemporaries also discovered faculae and named them with the Latin word for "torches."

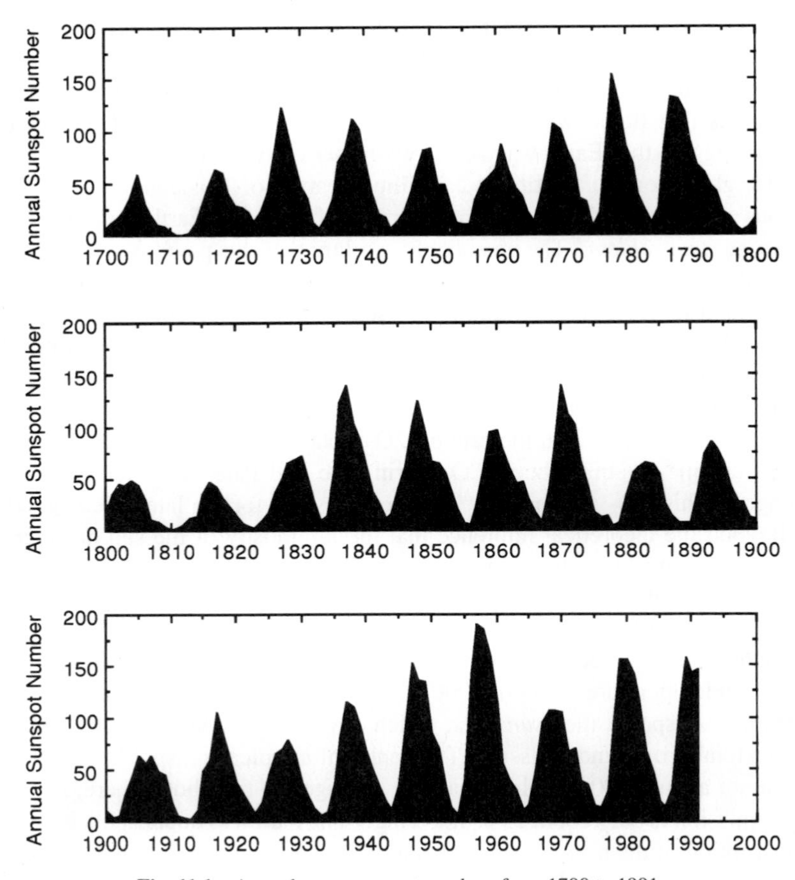

Fig. 11.1 Annual mean sunspot numbers from 1700 to 1991.

exists and an irregular cycle in sunspot abundance with a period of about 11 years has been documented. The number of sunspots varies from none to several hundred. The magnetic polarity of sunspot pairs alternates between sunspot maxima, so that a complete cycle of solar activity has a period of about 22 years. Reliable records of the annual mean sunspot number are available from about 1848, when the Wolf[5] sunspot number index was defined (Fig. 11.1). Variations in the magnitude of the solar cycle are evident, with a significant reduction in sunspot activity at the beginning of the nineteenth century.[6] A collection of scattered and somewhat less reliable

[5]Johann Rudolf Wolf (1816–1893) defined the sunspot number as the number of sunspots on the visible hemisphere of the sun plus 10 times the number of sunspot groups. Later, when better techniques of observation became available, it was shown that his index is approximately proportional to total sunspot area. The minimum size of the sunspots counted was limited by optical distortion by the atmosphere.

[6]These variations are associated with an 80–90 year cycle in sunspot activity (Gleissberg, 1944).

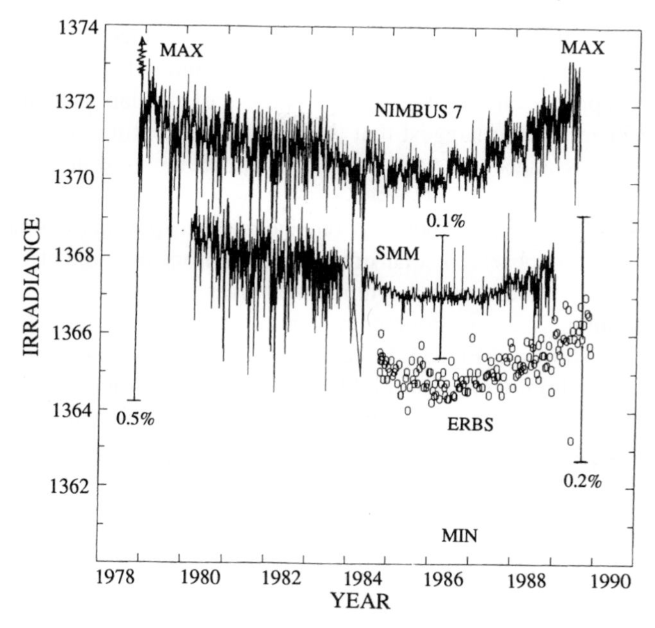

Fig. 11.2 Total solar irradiance measurements from 1978 to 1989 taken from Nimbus-7, Earth Radiation Budget Experiment (ERBS), and Solar Maximum Mission (SMM) satellites. The bars indicate the estimated absolute accuracy of the measurements. The units are W m^{-2}. [From Lee *et al.* (1991).]

data indicates that sunspots were virtually absent during the period 1645–1715, which is called the *Maunder minimum.*[7] It is intriguing that this corresponds roughly with the period of the Little Ice Age in Europe.

Solar flares and eruptive prominences are spectacular features of the sun's corona and cause significant increases in ultraviolet, X-ray, and gamma-ray radiation. The variation of high-energy radiation and particles associated with solar flares has a significant influence on the upper atmosphere. The effect of flares and prominences on the total energy output of the sun is negligible, however, and their influence on the surface climate is presumed to be small.

Very precise measurements of the solar irradiance can be made from Earth-orbiting satellites, and measurements from several space platforms extending over one solar cycle have been taken. These measurements indicate an average solar irradiance of about 1367 W m^{-2} at the mean Earth–sun distance. They indicate a small, less than ~0.1% or 1.5 W m^{-2}, increase in solar output from sunspot minimum (~1985) to sunspot maximum (~1980 and ~1990) (Fig. 11.2). The precision of solar irradiance measurements is generally greater than their absolute accuracy, so that different estimates of the magnitude of the variation over a solar cycle

[7]Eddy (1976).

agree better than the estimates of the mean solar irradiance. It is interesting that the largest solar irradiance occurs at sunspot maximum, when the greatest number of dark spots appears on the photosphere. Indices of faculae area such as Lyman-α flux have been used to suggest that the facular brightening dominates the darkening by sunspots and causes the solar constant to increase at the maximum of solar activity.

The statistical relationship between faculae and sunspot number can be used in conjunction with the observed solar irradiance variations over the last solar cycle to estimate the variation of the solar luminosity over the last century. Estimates of sunspot dimming, facular brightening, and solar constant over the period 1874–1988 are shown in Fig. 11.3. According to these estimates the sunspot dimming and facular brightening are highly correlated and strongly compensate each other, but the facular effect is slightly larger, especially in recent decades. According to this analysis the maximum solar constant observed at the peak of solar ac-

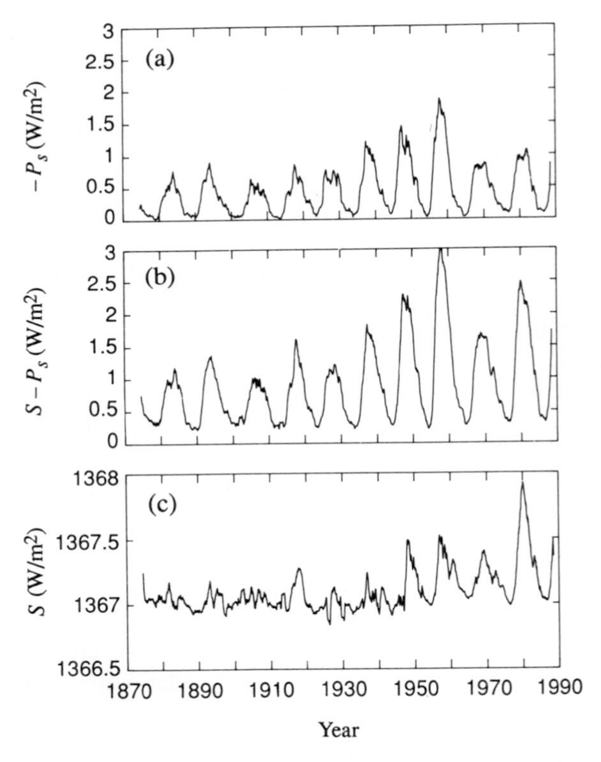

Fig. 11.3 Reconstruction of (a) sunspot dimming (plotted as a positive quantity), (b) facular brightening, and (c) solar irradiance. All are given in W m^{-2} and have been smoothed with a 12-month running mean. [From Foukal and Lean (1990), © by the AAAS.]

tivity in 1980 was caused by an anomalously large ratio of facular brightening to sunspot dimming.

The variations of solar irradiance associated with the 11-year sunspot cycle are of great interest, but have only a very small influence on climate. A change in solar irradiance of 1 W m^{-2} translates into about a 0.175 W m^{-2} climate forcing, which, using a climate sensitivity of λ_R = 0.5 K/(W m^{-2}), yields an equilibrium climate response of <0.1°C. Moreover, the 11-year cycle has a time scale significantly shorter than the relaxation time scale of climate, which is determined by the large heat capacity of the oceanic mixed layer, so that the magnitude of the transient response is much less than the response of a steadily applied forcing of the same magnitude, and would be too small to be observed. It is consistent with this analysis that no evidence of a significant 11-year cycle in surface climate has been discovered.[8]

In summary, what we have been able to directly observe of total solar irradiance suggests that luminosity variability will have a negligible effect on climate variations. We have precise observations from only a little over a decade, however, and it is possible that solar variability is an important factor on time scales of the 80-year Gleissberg cycle and longer.[9] To study other mechanisms of climate change, one can adopt the hypothesis that the luminosity of the sun is effectively constant and attempt to understand past variations on that basis.

11.3 Natural Aerosols and Climate

An aerosol is a suspension of liquid or solid matter in air. According to this definition, clouds are composed of water aerosols, but we will generally use the word aerosol to denote all noncloud aerosols. Since most aerosols contain some water, and cloud droplets generally form from growth by condensation on a tiny aerosol particle, the distinction between cloud and noncloud aerosol may seem arbitrary. A distinction is made in that cloud aerosols are at equilibrium in air with relative humidities in excess of 100%, whereas noncloud aerosols can be in equilibrium at relatively low humidities. Aerosols range in size from only slightly larger than individual air molecules to radii of several tens of micrometers. Aerosols larger than about 20 μm are removed quickly by gravitational settling to the surface. The smallest aerosols dominate the number of aerosols in the atmosphere, but their mass and surface area are so small that they have little direct effect on climate. The surface area of aerosols is important both for their role as cloud condensation nuclei (CCN) and also for their direct effect on radiation. The largest contribution to the total surface area of atmospheric aerosols comes from aerosols with radii in the

[8]But see, e.g., Kelly and Wigley (1992); Schlesinger and Ramankutty (1992).

[9]Baliunas and Jastrow (1990) and Lean et al. (1992).

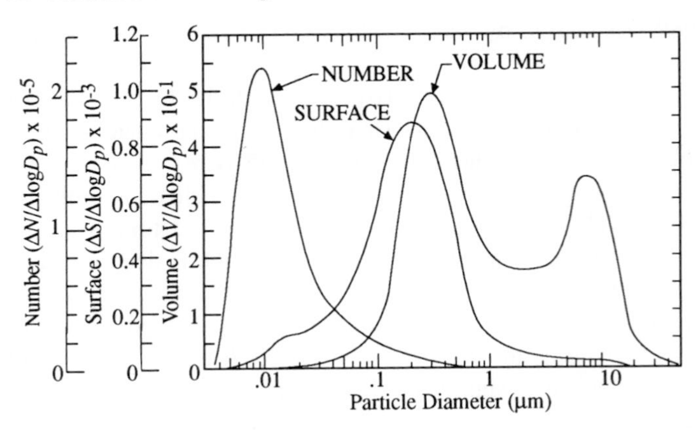

Fig. 11.4 Normalized plots of the number, surface area, and volume distributions for a tropospheric aerosol. [From Charlson (1988). Reprinted with permission from Wiley and Sons, Inc.]

range of ~0.1–1 μm (Fig 11.4). These are the aerosols that are most important for climate. Aerosol number concentrations typically range from 10^3 cm^{-3} in clean air over the oceans, to 10^5 or 10^6 cm^{-3} in polluted urban air.

Aerosols are produced by a large number of processes and can have varied chemical compositions. Aerosols can be injected directly into the atmosphere or be formed by the condensation or chemical transformation of vapors. Except for a small contribution from meteoritic debris, the sources of aerosols are at the surface. Direct emissions of aerosols come from the bursting of bubbles on the ocean surface, which produces small droplets that evaporate to leave a sea-salt aerosol. Other direct sources of aerosols include the elevation by wind of mineral dust from dry land surfaces, injection of ash and rock by volcanoes, soot from forest fires, meteoritic debris, and biological emissions such as spores and pollen. The largest contributions by mass to direct production of aerosols come from sea-salt and windblown dust (Table 11.1), but a large fraction of the mass of particles produced in these ways is contained in a relatively small number of large particles. Because they represent a comparatively small fraction of the total number of aerosol particles, sea-salt and mineral aerosols are of less importance for climate than their large mass would suggest.

Aerosols formed by gas-to-particle conversion consist mainly of sulfates, nitrates, and hydrocarbons, with sulfates making the largest contribution. Recent estimates indicate that industrial production of sulfur gases, mostly through SO_2 release during fossil fuel combustion, is larger than the natural sources of sulfur gases, although estimates of natural sulfur emissions are very uncertain (Table 11.2). The climatic effect of sulfate aerosols produced by fossil fuel burning appears to be significant, as will be discussed in Chapter 12.

Over the unpolluted oceans, the majority of aerosol particles are produced by

Table 11.1

Sources of Aerosols by Mass (Tg yr^{-1})

Source type	Peterson and Junge (1971)		SMIC (1971)
	All sizes	$r < 2.5$ μm	
	Natural sources		
Direct emissions			
Sea salt	1000	500	300
Mineral sources	500	250	100–500
Volcanoes	25	25	25–150
Forest fires	35	5	3–150
Meteorite debris	10	—	—
Biological material	—	—	—
Subtotal	1540	780	428–1100
Gas-to-particle conversion			
Sulfate	244[a]	220	130–200
Nitrate	75	60	140–200
Hydrocarbons	75	75	75–200
Subtotal	394	355	345–1100
Total natural	1964	1135	773–2200
	Anthropogenic sources		
Direct emissions			
Transportation	2.2	1.8	
Stationary sources	43.4	9.6	
Industrial processes	56.4	12.4	
Solid-waste disposal	2.4	0.4	
Miscellaneous	28.8	5.4	
Agricultural burnings	—	—	
Subtotal	133.2	29.6	10–90
Gas-to-particle conversion			
Sulfate	220	200	130–200
Nitrate	40	35	30–35
Hydrocarbons	15	15	15–90
Subtotal	275	250	175–325
Total anthropogenic	408	280	185–415
Sum total	2372	1415	958–2615

[From Warneck (1988).]

[a]Now regarded as high.

conversion of sulfur-bearing gases generated by ocean biology. Sea-salt aerosols probably account for more mass than biogenic aerosols in the marine boundary layer, but sea-salt aerosols are generally large, and account for only about 10% of the aerosols by number. A large fraction of the CCN over the oceans is thus biogenic

Table 11.2

Global Emission of Sulfur Gases

Source	Global	Northern Hemisphere	Southern Hemisphere	Global (Andreae, 1990)
Fuel combustion and industrial activities	77.6	69.8	7.8	80±10
Biomass burning	2.3	1.3	1.0	>2.5
Volcanoes	9.6	7.6	2.0	9.6–13
Marine biosphere	11.9	5.3	6.6	35–57
Terrestrial biosphere	0.9	0.5	0.3	4.8–13
Total	102.2	84.5	17.7	128–148

Units in Tg S year^{-1}. [From Spiro *et al.* (1992), © American Geophysical Union and Andreae (1990), reprinted with permission from Elsevier Publishers.]

sulfur aerosols. The primary medium for carrying sulfur from the ocean to the atmosphere is dimethyl sulfide gas (DMS; CH_3SCH_3). DMS accounts for about 90% of the reduced volatile sulfur in seawater and is present in concentrations that are in excess of equilibrium with air, so that a DMS flux from the ocean to the atmosphere occurs. DMS flux is estimated by using measurements of DMS concentration in the ocean mixed layer and the air, together with a flux formulation not unlike the aerodynamic formulas for heat and moisture flux discussed in Section 4.5. In the atmosphere DMS is oxidized photochemically to form sulfate (SO_4^{2-}) in aerosol form.

DMS is produced in ocean surface waters by certain species of phytoplankton, particularly algae. DMS is formed by the enzymatic cleavage of dimethylsufonium propionate (DMSP), which produces equal parts of DMS and acrylic acid. DMSP has an osmoregulatory function in marine algae. The release of DMS from DMSP in algae occurs continuously, but the rate increases when the alga is under stress and as it ages. Algae also excrete DMSP directly to seawater, where it is cleaved to produce DMS. The biological functions of DMS and DMSP emission by phytoplankton and their relative contributions to the concentration of DMS in ocean waters are unknown at this time. By regulating the release of DMS, and hence the production of oceanic CCN, phytoplankton could influence the cloud albedo over the oceans and hence regulate to some extent the climate. When more CCN are present the cloud water tends to reside in more, smaller droplets than when CCN are less abundant. As illustrated in Fig. 3.14, if the mean particle radius decreases and all else remains the same, the cloud albedo increases.

11.4 Volcanic Eruptions and Stratospheric Aerosols

Volcanoes play an important role in climate by cycling atmospheric elements such as carbon and sulfur between the lithosphere and the atmosphere. Violent eruptions can inject dust and source gases for aerosols into the stratosphere, where they remain for many months and can have a significant influence on climate. Much has been learned about the effects of volcanic eruptions in the last several decades because the major eruptions of El Chichón in Mexico (1982, 17.3°N) and Pinatubo in the Philippine Islands (1991, 15.1°N) were observed with modern satellite and in situ techniques. These eruptions were spectacular manifestations of explosive outgassing from Earth's interior and caused major changes in the optical properties of the stratosphere. These two eruptions resulted in peak global-average stratospheric aerosol optical depths of 0.07 and 0.15, respectively. The time history of aerosol optical depth can be measured accurately from satellites and shows these two eruptions very clearly (Fig. 11.5).

Although ash and fine rock particles can be injected high in the stratosphere by a violent eruption, most of these aerosols are too large to remain long in the atmosphere. Particles that are small enough to not precipitate serve as condensation nuclei for sulfuric acid or water vapor. Most stratospheric aerosols consist of a mixture of 75% sulfuric acid (H_2SO_4) and 25% water. A nonvolcanic background aerosol level is maintained in the stratosphere through an upward flux of sulfur-bearing gases from the troposphere. For example, carbonyl sulfide (COS) is produced in soils and has a lifetime of several years in the troposphere. It is mixed upward into the stratosphere where it encounters ultraviolet radiation or atomic oxygen and is oxidized to form sulfur dioxide (SO_2). Sulfur dioxide is then further oxidized to sulfuric acid. Since the vapor pressure of sulfuric acid is generally above saturation in the lower

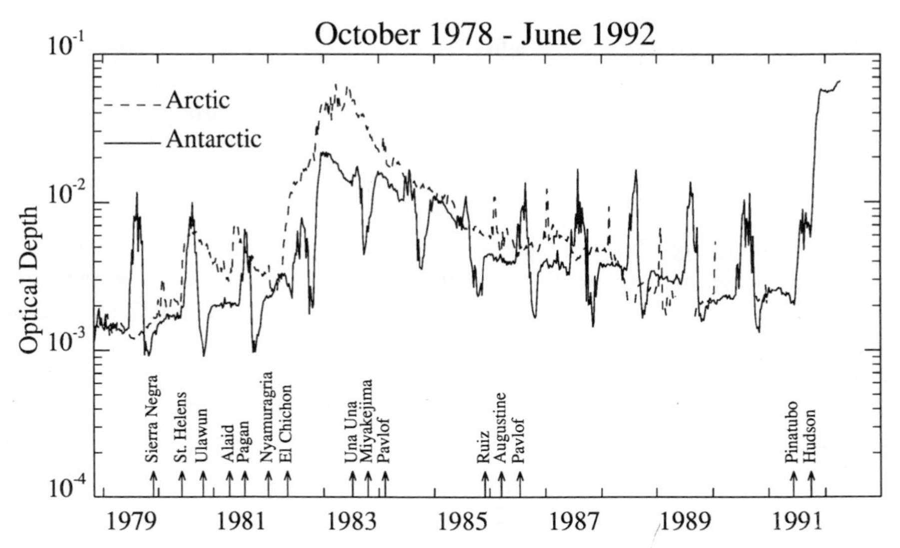

Fig. 11.5 Polar stratospheric optical depth versus time derived from SAM II and SAGE solar extinction measurements. Superimposed on the normal seasonal variations are major injections of aerosols associated with the El Chichón and Mt. Pinatubo eruptions. [From McCormick, *et al.* (1993).]

stratosphere, it condenses to form aerosols and is ultimately removed when the aerosols become large enough to precipitate or when the air in which they reside returns to the troposphere. Sulfuric acid is very hygroscopic and collects water to add to the mass of the aerosol. The enhancement of stratospheric aerosols after a volcanic eruption is caused by the direct injection of large amounts of SO_2 gas. The optical depth of the aerosol cloud peaks about 6 months after the eruption because of the time required to convert the SO_2 gas to sulfuric acid aerosols of the proper size to interact efficiently with solar and longwave radiation.

Stratospheric aerosols interact with both solar and terrestrial radiation and may both scatter and absorb radiation. Extinction is defined as the sum of scattering and absorption of the direct beam of radiation, so we may write the extinction cross section per unit mass (Section 3.6) as the sum of the scattering and absorption cross sections.

$$k_{ext} = k_{sca} + k_{abs} \tag{11.1}$$

The extinction coefficient is the ratio of the extinction cross section to the geometric cross section (shadow area) of an aerosol, and can similarly be written in terms of its scattering and absorption components.

$$Q_{ext} = Q_{sca} + Q_{abs} \tag{11.2}$$

The extinction coefficient for sulfuric acid aerosols is near 1 for solar radiation and decreases to less than 0.5 for terrestrial radiation [Fig. 11.6(a)]. The larger particles that characterize a volcanic aerosol cloud in the early phases after an eruption have a larger extinction coefficient and thus interact more efficiently with radiation, especially at longer wavelengths.

The single scattering albedo is defined as the ratio of the scattering coefficient to the extinction coefficient, and measures the ratio of extinction by scattering to total extinction during a single interaction of a photon beam with a particle.

$$\varpi = \frac{Q_{sca}}{Q_{ext}} \tag{11.3}$$

The single scattering albedo for sulfuric acid aerosols is nearly one for solar radiation, but decreases abruptly near 3 μm, so that very little absorption of solar radiation takes place, but a large fraction of the extinction of terrestrial radiation is absorption [Fig. 11.6(b)].

The probability that a scattered photon will depart in a particular direction relative to the direction of the incident beam of photons is given by the phase function, $\hat{P}$, which is normalized in the following way.

$$\frac{1}{4\pi} \int_{4\pi} \hat{P}\, d\omega = 1 \tag{11.4}$$

where $d\omega$ is the increment of solid angle (Section 3.3). The phase function can be characterized by the single scatter asymmetry factor $\overline{\cos\ \theta}$, which is defined

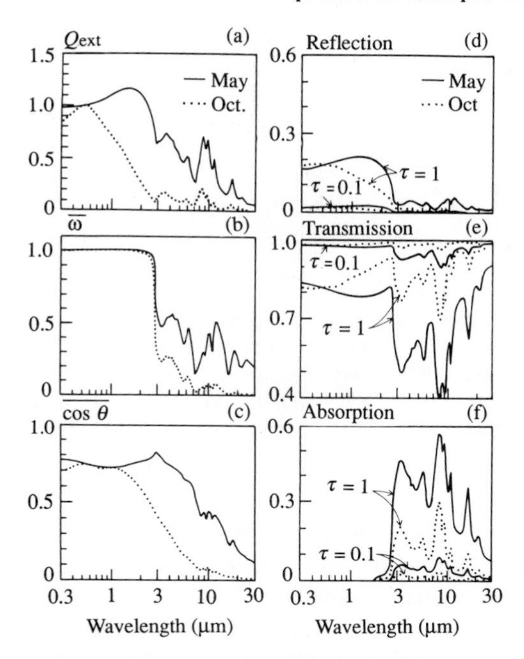

Fig. 11.6 Single scattering (left) and multiple scattering (right) by 75% sulfuric acid aerosols for two size distributions characterizing the early stages 1.5 months after the eruption of El Chichón 1982 ($r_{\text{eff}} \sim 1.4\ \mu$m, May) and the later stages when the effective particle radius was smaller ($r_{\text{eff}} \sim 0.5\ \mu$m, October). Multiple scattering calculations are given for a plane-parallel atmosphere with visible (λ = 0.55 μm) optical depth $\tau = 0.1$ and 1.0. Single scattering parameters are explained in the text. [From Lacis *et al.* (1992), © American Geophysical Union.]

as follows,

$$\overline{\cos\,\theta} = \frac{1}{4\pi} \int_{4\pi} \cos\,\theta\ \hat{P}\ d\omega \tag{11.5}$$

where θ is the angle between the direction of the incident beam and the scattered beam. The single scatter asymmetry factor varies between 1.0, which indicates forward scattering in the same direction as the incident beam, and −1.0, which indicates backscattering toward the direction of the source of the incident beam. Isotropic scattering, where scattering is equally probable in all directions, gives an asymmetry factor of zero. Stratospheric aerosols are strongly forward scattering for solar radiation and become increasingly isotropic scatterers for longer wavelengths [Fig. 11.6(c)].

As a result of the radiative properties of stratospheric sulfuric acid aerosols, the modest optical depths associated with recent large volcanic eruptions ($\tau \sim 0.1$) give rise to a modest reflection of solar radiation and a modest absorption of terrestrial radiation [Figs. 11.6(d–f)]. Because of their relatively efficient absorption of infrared

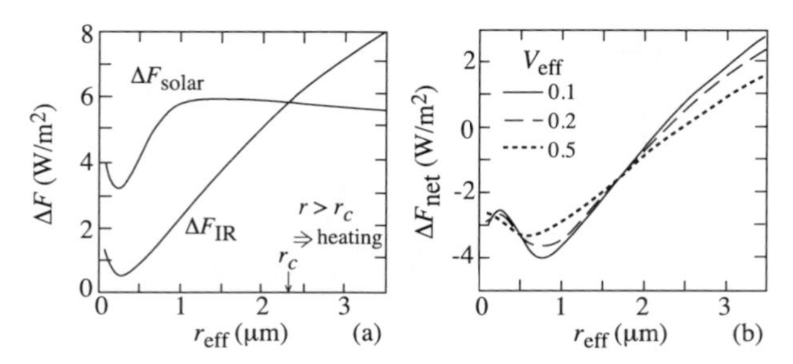

Fig. 11.7 (a) Decrease of incoming solar and outgoing infrared fluxes at the tropopause caused by adding a stratospheric aerosol layer (20–25 km) with visible optical depth $\tau = 0.1$ in a one-dimensional radiative–convective model with fixed surface temperature. Plotted as a function of effective radius, r_{eff}, with an effective variance of particle radius $v_{eff} = 0.2$. (b) Change of net radiation at the tropopause for three values of effective variance of particle radius. [From Lacis *et al.* (1992), © American Geophysical Union.]

radiation, volcanic aerosols generally heat the layer in which they occur (20–25 km). Their effect on the tropospheric climate is more complex, as they both reflect solar radiation and provide a small greenhouse effect. The balance of these two opposing effects is sensitive to the size of the aerosol particles. Smaller particles have a substantial influence on solar radiation, but not on infrared radiation, so that a reduction of net radiation of 3–4 W m^{-2} results for particles smaller than ~1 μm (Fig. 11.7). The greenhouse effect increases approximately linearly with particle radius, while the solar reflection becomes independent of particle radius when the radius is larger than the effective wavelength, so that a net warming results for particles larger than about 2 μm. This result is relatively insensitive to particle size dispersion [Fig. 11.7(b)]. A 1-μm spherical sulfuric acid particle at 20 km falls about 2 km per month, so that larger particles than this are relatively rare. Therefore, except for the brief period immediately after the eruption, when relatively large dust and ash particles may be abundant, we expect that volcanic aerosol clouds in the stratosphere will cause a significant reduction in net energy input to the climate system, with a large eruption like Mt. Pinatubo producing a 0.15 optical depth cloud and a –4 W m^{-2} climate forcing during its peak (Fig. 11.8).[10]

Because the aerosol cloud enhancement resulting from a single eruption is flushed out of the stratosphere in a year or two, and the thermal capacity of the ocean mixed layer gives it a thermal-response time scale of decades, the response of the climate system to a single volcanic eruption is rather modest, unless it is an exceptionally large eruption with a high sulfur content. A sequence of major volcanic eruptions each spaced about a year apart could cause a significant climate signal, however. The extent to which aerosol clouds from volcanic eruptions have

[10]Hansen et al. (1992).

Fig. 11.8 Ash and vapor cloud associated with the Mt. Pinatubo eruption on June 15, 1991.

influenced the climate of the past is largely unknown, but remains a subject of considerable interest.

The effect of a volcanic eruption on climate depends on the amount of long-lasting stratospheric aerosol produced, which depends both on the explosiveness of the eruption and the abundance of sulfur in the gases released. Evidence suggests that large eruptions in the past have had a significant effect on short-term climate anomalies (Table 11.3). One of the largest eruptions we know of was the 1815 eruption of Mt. Tambora on Sumbawa Island, Indonesia. Approximately 150 km^2 of ash and pumice was spread over a large area and the eruptive plume was estimated to have reached 50-km altitude. Within a few months the volcanic cloud had spread around the world and its optical effects were observed in Europe. The sun and stars were dimmed noticeably by the stratospheric aerosol cloud for nearly 2 years after the

Table 11.3

Estimates of Stratospheric Aerosols and Climatic Effects of Some Volcanic Eruptions[a]

Volcano	Latitude	Date	Stratospheric aerosols (Mt)	Northern Hemisphere τ_D	Northern Hemisphere ΔT (°C)
Explosive eruptions					
St. Helens	46°N	May 1980	0.3	<0.01	<0.1
Agung	8° S	March–May 1963	10	<0.05[b]	0.3
El Chichón	17°N	March–April 1982	20	0.15	<0.4
Pinatubo	15°N	June 1991	30	0.25	~ 0.5
Krakatau	6° S	August 1883	50	0.55	0.3
Tambora	8° S	April 1815	200	1.3	0.5
Rabaul?	4° S	March 536	300	2.5	Large?
Toba	3°N	– 75,000 years	1000?	10?	Large?
Effusive eruptions					
Laki	64°N	June 1783 to February 1784	~ 0	Locally high[c]	1.0?
Roza	47°N	– 14 million years	6000?	80?[d]	Large?

[From Rampino *et al.* (1988); reprinted with permission from the *Annual Review of Earth and Planetary Sciences,* Vol. 16, © 1988 by Annual Reviews, Inc. Mt. Pinatubo data added by the author.]

[a]Rampino and Self (1984). Optical depths (τ_D) are visual, direct beam.

[b]Southern Hemisphere $\tau_D \approx 0.2$.

[c]Aerosols were mostly tropospheric.

[d]If the aerosols were dispersed globally, the average Northern Hemisphere optical depth would have been about 40.

eruption. The following year (1816) was known as the year "without a summer," since it was much colder than normal both in Europe and North America. Unusual spring and summer frosts and a 6-in. snowfall in the second week of June caused crop failures in New England. It is unlikely that these climate anomalies can be completely attributed to the Tambora eruption, since the entire decade (1810–1820) was somewhat colder than normal.

11.5 The Orbital Parameter Theory of Ice Ages

11.5.1 Historical Introduction

The idea that Earth has experienced ice ages was first suggested in Europe in the eighteenth century. Continental ice sheets were offered as an explanation for the appearance of large boulders far from their source of origin and the presence of deep grooves on exposed bedrock. At the time many people believed that the erratic boulders were transported by a great flood, and the idea of great ice sheets was slow to be accepted. Credit for convincing the scientific community that the European Alps were once covered by a great ice sheet is usually given to the Swiss-born geologist Louis Agassiz, who argued the case forcefully and published a monograph on the subject in 1840, at about which time the argument was beginning to turn in his favor.

Among the first explanations for glacial ages was an astronomical theory by the French mathematician Joseph Alphonse Adhémar published in 1842. Adhémar clearly stated the hypothesis that the primary cause of the ice age succession was secular variation in Earth's orbit. His theory was based on the precession of the equinoxes, which in 1754 d'Alembert had shown completes a cycle every 22,000 years. Adhémar based his ice age timing on the number of days of daylight per year. Because Earth travels faster when it is closer to the sun (perihelion) than when it is far away (aphelion), and Southern Hemisphere summer currently occurs near perihelion, the Southern Hemisphere receives fewer hours of daylight than the Northern Hemisphere. He concluded that ice ages alternated between the hemispheres every 11,000 years, and that the Southern Hemisphere is currently in an ice age. Adhémar apparently did not understand that energy input and not hours of daylight is the critical consideration, and that annual-mean insolation in each hemisphere is independent of the precession cycle.

In 1864 the Scotsman James Croll extended the work of Adhémar to include the modulation by eccentricity variations of the effect of precession of the equinoxes. Eccentricity variations had been established by the work of LeVerrier in 1843, who also showed that the annual insolation is only weakly affected by eccentricity variations. Croll deduced that if eccentricity variations were to have an effect on ice ages, it must arise from the larger variations in seasonal insolation that accompany eccentricity changes. He adopted the idea that colder winters lead to growing glaciers. His theory predicts that an ice age will occur in one hemisphere or the other whenever two conditions occur simultaneously: a markedly eccentric orbit and a winter solstice near aphelion. According to his theory the last ice age occurred 80,000 years ago. He also hypothesized that ice ages would be more likely when the tilt of Earth's axis of rotation was less, because at these times the polar regions would receive less insolation. The timing of the obliquity variations was not known, however, so he could not incorporate this factor in his theory for past glaciations. In developing his theory Croll introduced the idea of ice albedo feedback in order to amplify the effect of the small insolation changes associated with changes in Earth's orbital parameters. He also included a feedback mechanism involving the tradewinds, ocean currents, and the geometry of the Atlantic Ocean that would amplify cooling in one hemisphere by diverting oceanic heat fluxes to the other hemisphere. Croll's theory met with a great deal of interest and approval, since by this time it was accepted that not one but many ice ages had occurred in the past, and the astronomical theory predicts cyclic ice ages associated with the cycles in Earth's orbital parameters. Although geologic dating techniques were very primitive at that time, what evidence could be gathered suggested that the last glaciation was much more recent than 80,000 years ago as Croll predicted. In addition, evidence came in from the Southern Hemisphere suggesting that glacial ages there occurred simultaneously with those in the Northern Hemisphere, which is inconsistent with the predictions of Adhémar and Croll that glacial ages would alternate between the hemispheres. By the end of the nineteenth century Croll's theory had been largely dismissed.

The orbital parameter theory of climate change was next taken up by the Serbian applied mathematician Milutin Milankovitch. Milankovitch began his work in 1911 with the benefit of Pilgrim's determination of all of the orbital parameters necessary to compute insolation as a function of time and latitude. His work on the orbital parameter theory extended over 30 years and was interrupted by three wars. In 1920 his work attracted the attention of the German climatologist Wladimir Köppen, who encouraged Milankovitch and engaged in a correspondence with him about the physical basis of climate. Milankovitch had developed the mathematical tools for calculating the insolation at any point and time as a function of the orbital parameters, but he was unsure at what latitude and season the ice sheets would be most sensitive to the insolation. Köppen indicated that a reduction of the summertime insolation in the Arctic would be the critical factor in producing growth of ice sheets. He reasoned that the temperatures were always cold enough for snow to form in the winter, but that a reduction in the summertime melting of glaciers would allow the ice sheets to expand. This was a critical departure from the work of Adhémar and Croll, who had focused on the coldness of winter as the critical parameter. With this advice Milankovitch set about to calculate the summertime insolation at 55°, 60° and 65°N for the last 650,000 years. After 100 days of calculation he sent his insolation curves to Köppen, who replied that they matched fairly well with what was known about the timing of the last several glacial epochs. The astronomical climate theory of Milankovitch was accepted to varying degrees until the 1950s when radiocarbon dating showed that the ages of many important glacial drift features on the land surface did not correspond to the times predicted by the radiation curves.

By the mid 1960s the early paleoclimatic evidence that had been used to support the Milankovitch theory had been dismissed and the theory itself was increasingly questioned. New evidence began to emerge that supported the theory, however. The sources of new information included new dating techniques that extended the 40,000-year time reach of radiocarbon dating. Coral terraces that could be associated with rising sea level were dated at 122,000, 103,000, and 80,000 years ago and these dates corresponded to times when Milankovitch had calculated higher than normal summertime insolation in midlatitudes.

The discovery of a reversal of Earth's magnetic field whose date could be fixed at 700,000 years ago was used to set the relationship between depth and time for ocean sediment cores by assuming a constant sedimentation rate. The oxygen $^{18}O/^{16}O$ ratio in these cores was found to be a good measure of the global ice volume. The continuous nature of the global ice volume record contained in ocean sediment cores allowed the use of time series analysis techniques. Using these techniques it was shown that the oxygen isotope record in ocean cores contains periodicities that correspond to variations in Earth's orbital parameters at periods of 100,000, 41,000, 23,000, and 19,000 years. It is somewhat puzzling that the record of the last 700,000 years is dominated by a 100,000-year periodicity, since the direct forcing of seasonal insolation by eccentricity variations at this long time scale is much less than the forcing associated with the obliquity and longitude of perihelion variations at 41,000 and about 20,000 years, respectively. Prior to 700,000 years ago, the available global ice volume records constructed from ocean cores appear to be dominated by the higher frequencies as the Milankovitch hypothesis would suggest (Fig. 8.6). The reason for the apparent amplification of the 100,000-year periodicity in the more recent era remains a mystery.

11.5.2 The Orbital Parameters and Their Influence

Early in the seventeenth century, Johannes Kepler discovered his first law, that the orbits of the planets are ellipses with the sun at one focus (Fig. 11.9). The degree to which the orbit departs from a circle is measured by the eccentricity of the ellipse. The point where the planet is closest to the sun is called *perihelion,* and the point on the orbit most distant from the sun is called *aphelion.* If the Earth–sun distance is d_p

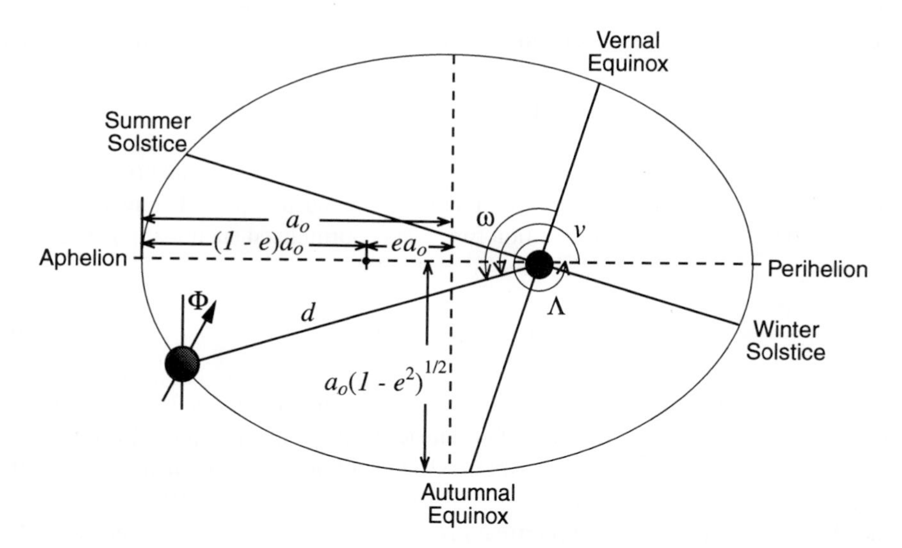

Fig. 11.9 Schematic diagram of Earth's elliptical orbit about the sun showing the critical parameters of eccentricity (e), obliquity (Φ), and longitude of perihelion (Λ) defined relative to the vernal equinox. The size of the orbit is defined by the greatest distance between the ellipse and its center point, which is called the semimajor axis length, a_0. The Earth–sun distance at any time (d), the angle between the position of Earth and perihelion that we call the true anomaly (ν), and the angle between the position of Earth and the vernal equinox (ω) are also shown.

at perihelion and d_a at aphelion, then the eccentricity is defined as the ratio of the difference to the sum of these two distances.

$$e = \frac{d_a - d_p}{d_a + d_p} \tag{11.6}$$

The maximum distance from the center of the ellipse to the orbit, a_0, occurs at perihelion and aphelion and is called the semimajor axis. From geometric considerations it can be shown that the minimum distance from the center of the ellipse to the orbit is given by $a_0 \sqrt{(1 - e^2)}$. The Earth–sun distance d as a function of the true anomaly v is given by

$$d(v) = \frac{a_0 \left(1 - e^2\right)}{1 + e \cos v} \tag{11.7}$$

The variation of the Earth–sun distance as Earth moves along its orbit contributes to the annual cycle of insolation and is one of the key factors in the orbital parameter theory.

Of even greater importance for the annual cycle of climate is the obliquity. The obliquity is the angle between the axis of Earth's rotation and the normal to the plane of Earth's orbit about the sun. The magnitude of the seasonal variation of insolation in high latitudes and the annual-mean insolation in high latitudes both increase with increasing obliquity. This can be seen most easily by considering a perfectly circular orbit with zero eccentricity. The daily insolation at any latitude and season, Q, can then be written

$$Q = \frac{S_0}{4} \ \tilde{s}(\Phi, x, t) \tag{11.8}$$

where S_0 is the solar constant at the mean Earth–sun distance, $\overline{d}$ and $\tilde{s}(\Phi, x, t)$ is the distribution function for a circular orbit and a particular obliquity Φ, sine of latitude x, and time of year t. This distribution function is normalized such that its area average is 1.

$$\frac{1}{2} \int_{-1}^{1} \tilde{s}(\Phi, x, t) \, dx = 1 \tag{11.9}$$

The distribution function for a circular orbit with an obliquity of 23.5° is shown in Fig. 11.10(a). Except for the effect of eccentricity, this is just the normalized form of Fig. 2.6. The sensitivity of this distribution to the obliquity can be illustrated by plotting the difference between $\tilde{s}$ for $\Phi = 24.5°$ and $\Phi = 22.5°$ [Fig. 11.10(b)]. Increasing the obliquity increases the summertime insolation in high latitudes and decreases the wintertime insolation in midlatitudes. The magnitude of the change of insolation during summer caused by an increase of obliquity from the minimum to the maximum experienced by Earth is about 40 W m^{-2}, or about 10% of that for the average obliquity. The annual average insolation at the pole is also sensitive to the obliquity.

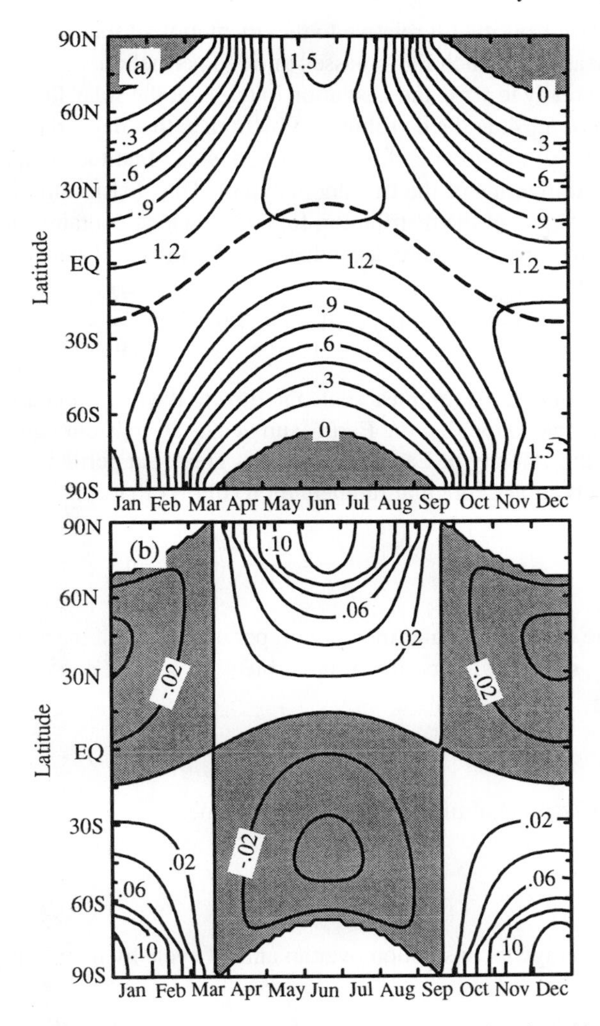

Fig. 11.10 (a) Insolation distribution function $\tilde{s}(\Phi,x,t)$ plotted as a function of latitude and season for an obliquity of Φ = 23.5°. (b) Sensitivity of the insolation distribution function to obliquity, $\Delta\Phi(\partial\tilde{s}/\partial\Phi)_{\Phi=23.5}$, evaluated for $\Delta\Phi$ = 2°.

If the obliquity is zero, the annual-mean insolation at the pole is zero. For a circular orbit the annual-mean value of $\tilde{s}$ at the pole varies as $4 \sin \Phi/\pi$.[11]

The longitude of perihelion is defined as the angle Λ, between the line from Earth to the sun at vernal equinox and the line from the sun to perihelion. It defines the

[11]Held (1982).

direction of Earth's orbital tilt relative to the orientation of the orbit, and thereby determines the season at which Earth passes through perihelion.

We are now ready to write the insolation in terms of the solar-flux density at some mean distance $\overline{d}$, $S_0/4$, and a distribution function, s, that depends on the eccentricity (e), obliquity (Φ), longitude of perihelion (Λ), sine of latitude (x), and position in the orbit defined in terms of the true anomaly (ν). This distribution function can be written as the product of the distribution function $\tilde{s}$ which contains the obliquity dependence, and the inverse square dependence on Earth–sun distance, d.

$$\frac{S_0}{4}\, s(e, \Lambda, \Phi, x, \nu) = \frac{S_0}{4}\, \tilde{s}(\Phi, x, \nu) \frac{\overline{d}^2}{d(\nu)^2} \tag{11.10}$$

To make use of (11.10) we must relate the time to the true anomaly ν. From Kepler's second law we know that the Earth–sun vector sweeps out equal area in equal time, and that the velocity of Earth in its orbit is greater at perihelion than aphelion. From the conservation of angular momentum it follows that

$$\frac{d\nu}{dt} = \frac{M_e}{d^2} \tag{11.11}$$

where M_e is the constant angular momentum per unit of Earth mass. If we define the unit of time, t', such that the period of the orbit

$$P_0 = \frac{2\pi}{M_e}\, \overline{d}^2 \sqrt{1 - e^2} \tag{11.12}$$

corresponds to 2π units of time, then (11.11) becomes

$$\frac{d\nu}{dt'} = \frac{\overline{d}^2}{d^2} \sqrt{1 - e^2} \tag{11.13}$$

We may now average the insolation over an annual cycle by integrating over t'.

$$\begin{aligned} \frac{1}{2\pi} \int_0^{2\pi} \frac{S_0}{4}\, \tilde{s}(\Phi, x, \nu) \frac{\overline{d}^2}{d(\nu)^2}\, dt' &= \int_0^{2\pi} \frac{S_0}{4}\, \tilde{s}(\Phi, x, \nu) \frac{\overline{d}^2}{d(\nu)^2} \frac{dt'}{d\nu}\, d\nu \\ &= (1 - e^2)^{-1/2} \int_0^{2\pi} \frac{S_0}{4}\, \tilde{s}(\Phi, x, \nu)\, d\nu \end{aligned} \tag{11.14}$$

From (11.14) we deduce that the annual-mean insolation at any point is independent of the longitude of perihelion. The annual-mean insolation at every latitude is proportional to $(1 - e^2)^{-1/2}$, which for small eccentricity is approximately $(1 + 0.5e^2)$. This dependence of annual insolation on eccentricity is very weak, since in going from a circular orbit to the most eccentric orbit Earth experiences ($e \sim 0.06$) causes a change of annual insolation of only ~0.18%.

The variations of annual-mean insolation associated with eccentricity changes are too small to produce a significant effect, if our estimates of climate sensitivity are approximately correct. It is for this reason that Milankovitch sought an explanation for the ice ages in the changes of the seasonal and latitudinal distribution that are caused by variations in the orbital parameters. Summertime insolation in high latitudes will be larger when the obliquity is larger, and summertime insolation in the Northern Hemisphere will be greater when the northern summer solstice occurs at perihelion. The effect of longitude of perihelion variations on insolation increases with the eccentricity of the orbit. To see this quantitatively, we can insert (11.7) into (11.10), and use Fig. 11.9 to deduce that $v = \omega - \Lambda$, so that we can write an expression for the insolation that shows its dependence on eccentricity and longitude of perihelion.

$$\frac{S_0}{4} s(e,\Lambda,\Phi,x,v) = \frac{S_0}{4} \tilde{s}(\Phi,x,v) \frac{(1+e\ \cos(\omega-\Lambda))^2}{\left(1-e^2\right)^2} \tag{11.15}$$

At northern summer solstice $\omega = \pi/2$, so that (11.15) becomes

$$\frac{S_0}{4} \tilde{s}(\Phi,x,v) \frac{(1+e\ \sin\Lambda)^2}{\left(1-e^2\right)^2} \approx \frac{S_0}{4} \tilde{s}(\Phi,x,v)(1+2e\sin\Lambda) \tag{11.16}$$

where the approximation holds for small values of e. The critical parameter that describes the influence of eccentricity and longitude of perihelion on northern summer insolation is $e \sin \Lambda$, which we will call the precession parameter.

A summary of the effects of the orbital parameters on insolation reveals their importance for the seasonal and latitudinal distribution of insolation.

1. The annual-mean insolation varies only as a result of eccentricity in proportion to $(1 - e^2)^{-1/2}$, which, for the observed eccentricities $e < 0.06$, produces changes that seem too small to be important.
2. Obliquity controls the annual-mean equator-to-pole gradient of insolation and, together with the eccentricity and longitude of perihelion, the amplitude of the seasonal variation of insolation at a point. Obliquity variations within the observed range of 22.0°–24.5° can produce ~10% variations in summer insolation in high latitudes.
3. The precessional parameter $e \sin \Lambda$ controls the modulation in seasonal insolation due to eccentricity e and longitude of perihelion Λ variations. The combined effects of eccentricity and longitude of perihelion can result in ~15% changes in high-latitude summer insolation.
4. The combined effects of the three orbital parameters can cause variations in seasonal insolation as large as ~30% in high latitudes.

The time histories of Earth's orbital parameters can be constructed very accurately from celestial mechanics calculations assuming that Earth and other planets orbit the sun without perturbations from any large bodies passing through the solar

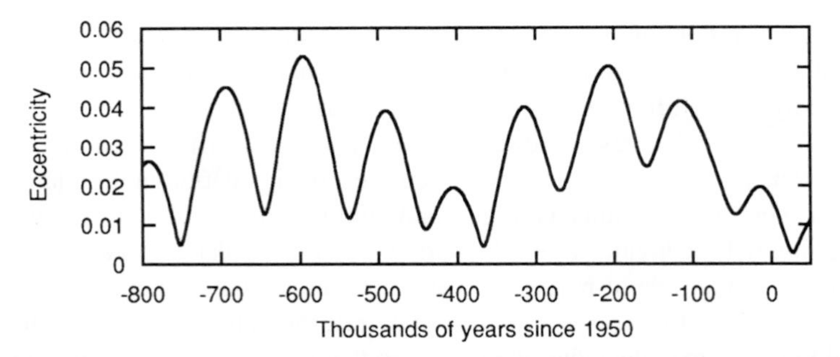

Fig. 11.11 Eccentricity of Earth's orbit as a function of time from 800,000 years before present to 50,000 years into the future. [Calculated using the series expansions of Berger (1978).]

system. The time variation of the eccentricity of Earth's orbit is shown in Fig. 11.11 for the period from 800,000 years ago to 50,000 years into the future. The dominant periodicities of about 100,000 and 400,000 years can be clearly seen. The present eccentricity of about 0.015 is relatively small compared to maximum values of about 0.055 attained about 200,000 and 600,000 years ago.

The obliquity and the precession parameter, $e \sin \Lambda$, are shown in Fig. 11.12 for the period from 150,000 years ago to 50,000 years in the future. The obliquity has a dominant periodicity of about 40,000 years and is currently near the middle of its range between about 22.5° and 24.5°. The longitude of perihelion cycle has a period around 20,000 years, but its effect on the precession parameter is modulated by the longer period variation of eccentricity. The precession effect will be small for the next 50,000 years because of the small eccentricity during this period. According to the Milankovitch theory, a relatively ice-free period will occur as a result of large Northern Hemisphere summertime insolation, which occurs when the obliquity is large

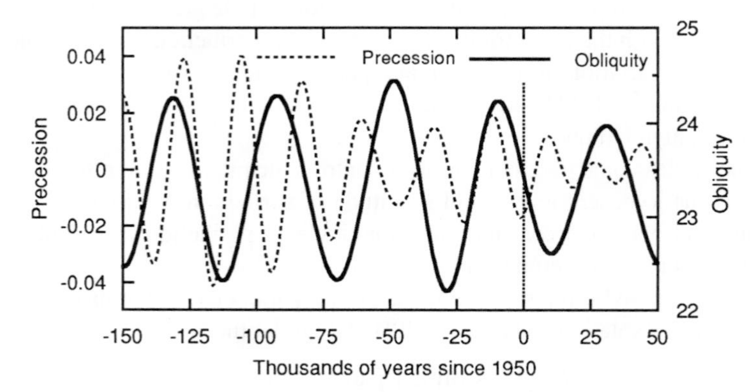

Fig. 11.12 Obliquity (Φ) and precession ($e \sin \Lambda$) parameters, as functions of time from 150,000 years before present to 50,000 years into the future. Units of obliquity are in degrees.

and the precession parameter is positive and large. This combination of events occurred about 10,000 years ago at a time when the climate was slightly warmer than today. About 23,000 years ago, when the ice sheets were growing rapidly toward the last glacial maximum, the precession parameter was large and negative and the obliquity was near its minimum value. Both the precession parameter and the obliquity were thus favorable for ice growth, according to the Milankovitch theory.

The combined effect of the orbital parameters can be seen by computing the insolation as a function of latitude for various times in the past and future. The temporal variation of daily averaged insolation at the solstices is shown in Fig. 11.13. Anomalies from the time average over the period from 150,000 years ago to 20,000 years in the future are shown. The upper panel shows the insolation variations at Northern Hemisphere summer solstice, which have amplitudes as large as 60 W m^{-2} near the pole. Large positive anomalies at 125,000 years ago and 10,000 years ago

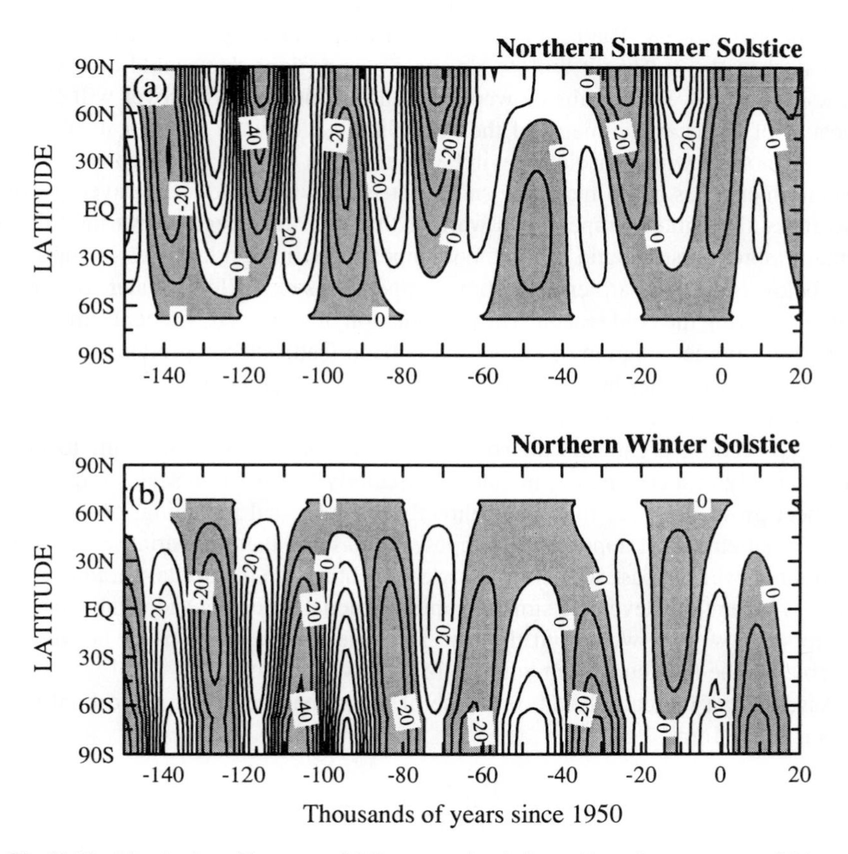

Fig. 11.13 Distribution of departure of daily average insolation at (a) northern summer and (b) northern winter solstices from average values for the period from 150,000 years ago until 20,000 years in the future. Contours are given in W m^{-2}.

correspond fairly well with the times of the last two interglacial periods. The last glacial maximum about 20,000 years ago was preceded by a relative minimum in northern summertime insolation.

11.5.3 Testing the Milankovitch Theory

Because we know Earth's orbital parameters for several million years into the past and future, and because we can deduce global ice volume for the past two million years or so from ocean cores, the Milankovitch theory is a testable hypothesis. It should be apparent from the discussion of the previous sections that the orbital parameter mechanism for producing secular changes in climate must depend heavily on the physics of the annual cycle of climate, and perhaps particularly in high latitudes during summer where insolation variations are large. The mechanism usually envisioned has to do with spring, summer, and fall snowmelt and the effect of insolation on that process. The idea is that if the insolation during summer is lower than normal, then the snowmelt will be slower. Some regions that would otherwise be snow-free by fall will remain snowcovered and the perennial icecover will expand. Because of the near symmetry of the perihelion cycle, when summer insolation is less than normal winter insolation will be greater than normal. This means that when insolation changes act to make the summer temperatures colder than average, they also make the winter temperatures warmer than average. Because warmer air can contain more moisture, and wintertime temperatures in high latitudes are normally well below freezing, warmer wintertime temperatures may allow greater snow accumulation during the cold season. The combination of increased winter accumulation, decreased summer ablation of ice, and ice albedo feedback can produce ice volume growth. When summer insolation is enhanced, all the arguments are reversed so that reduced ice volume and warming would result.

The preceding arguments all sound plausible, but are very difficult to model quantitatively in a convincing manner, particularly for the time scales required for ice sheet growth. Rather than jump directly into physical modeling, one may first consider an empirical approach.[12] Suppose we assume that the variations of global ice volume over the past 500,000 years are due entirely to the orbital parameter variations. We can then develop a simple empirical model with a few adjustable parameters to see how well we can fit the data and whether the relationship between the variables in the fit is consistent with the Milankovitch theory.

Assume that the equilibrium response of the global ice volume to orbital variations is of the form

$$I_{\mathrm{eq}} = I_0 - b_\Phi \frac{(\Phi - \Phi_0)}{\sigma(\Phi)} - b_e \frac{e}{\sigma(e)} \cos(\Lambda - \varphi_0) \tag{11.17}$$

[12]Imbrie and Imbrie (1980).

Here I_0 is the ice volume at zero eccentricity and some standard value of the obliquity Φ_0. The parameters b_Φ and b_e set the magnitude of the equilibrium response to obliquity variations and the perihelion cycle, φ_0 determines the phase of the perihelion cycle at which the ice is minimum, and $\sigma(\Phi)$ and $\sigma(e)$ are the standard deviations of obliquity and eccentricity over the past 500,000 years. Any linear response of ice volume to the orbital parameters would take a form such as (11.17).

Next assume that the ice volume relaxes toward its equilibrium value with a time scale that depends on whether the ice volume is increasing or decreasing

$$\frac{\partial I}{\partial t} = \frac{\left(I_{\mathrm{eq}} - I\right)}{\tau_I} \tag{11.18}$$

where

$$\tau_I = \begin{cases} \tau_c, & I < I_{\mathrm{eq}} \\ \tau_w, & I > I_{\mathrm{eq}} \end{cases} \tag{11.19}$$

Thus τ_c is the accumulation time scale and τ_w is the ablation time scale. The global ice volume record suggests an asymmetry between slow growth and rapid decay, and this is allowed for by using separate time scales for growth and decay of ice sheets.

A good fit to the ice volume record is obtained with τ_c = 42,000 years, τ_w = 10,600 years, $b_\Phi > 0$, $b_e / b_\Phi = 2$, and $\varphi_0 = 125°$. These optimal parameters, if taken at face value, imply that if the ice volume had time to adjust, the smallest ice volume would occur for large obliquity and for perihelion between northern summer solstice and autumnal equinox ($\varphi_0 = 90°$ would indicate minimum equilibrium ice volume when perihelion occurs at summer solstice), and with the obliquity and perihelion cycles being of comparable importance. These empirical determinations are in general accord with the original hypothesis of Milankovitch.

The empirical fit between the orbital parameter forcing and the global ice volume record indicates that the time scale for ice accumulation is 42,000 years. It is interesting to consider what processes in the climate system could account for this long time scale. Several candidates suggest themselves.

1. *The deep-ocean circulation.* As discussed in Chapter 7, the time scale for the deep ocean may be thousands of years because of its large heat capacity and slow overturning time. A liberal upper limit on the time scale would be a few thousand years, however, which is too short to explain the 10,000-year time scales required.
2. *Glacial and ice sheet dynamics.* The current precipitation in Green Bay, Wisconsin is about one inch of water for each of the months of December, January, and February. If we assume that all of the winter precipitation (3 in.) survives the summer melt season, it would take 39,000 years to grow a 3-km thick ice sheet. Glaciers and ice sheets also move around slowly and spread under their own weight, which would probably add to the time required.
3. *Isostatic adjustment.* The continents sink slowly under glacial loading and

bounce back slowly when ice sheets melt and the water flows into the ocean. The time scale for the response of the bedrock ranges from 5,000 to 20,000 years, depending on the size of the ice sheet and the character of the rock. The sinking of the continents under the ice sheets may allow the oceans to intrude over what would otherwise be continent and contribute to the rapid decay of the ice sheet via calving of icebergs from the ice sheet and heat and ice transport by the ocean currents. This has been offered as one mechanism to explain the more rapid decay than growth of the ice sheets. This asymmetry may also be because the precipitation rate limits the growth rate but no similar constraint applies to the melting rate.

4. *Geochemistry.* The carbon dioxide content of the atmosphere varies significantly between glacial and interglacial conditions, and the radiative effect of reduced atmospheric CO_2 contributes significantly to the cooling during an ice age. It can be argued that the CO_2 change must be the result of a change in ocean chemistry, since the CO_2 changes persisted for thousands of years and on this time scale the atmospheric CO_2 concentration is slave to the ocean. The ocean contains 60 times more carbon than the atmosphere. Ocean geochemistry and the budgets of certain key nutrients have time scales that are quite long and could contribute to the low-frequency response of the climate system.

11.6 Modeling of Ice Age Climates

Global climate models (GCMs) can be used to understand how past glacial climates were maintained and estimate the relative importance of the known feedback mechanisms. Paleoclimatic data can be used to estimate the extent and thickness of ice sheets, land vegetation, sea surface temperature, and atmospheric composition as a function of time since the last glacial maximum, approximately 18,000 years ago. Each of these factors can be specified individually in a climate model to determine its effect on the simulated climate, or they can be predicted independently to gauge the ability of the GCM to faithfully simulate the observed relationships among them. The orbital parameters can also be varied to determine how they affect the climate simulation. Currently, climate models of the most realistic type cannot be integrated for a long enough period to simulate ice sheet growth or changes in vegetation patterns, so these parameters have been specified in most simulations conducted to date. The quality of the simulation and the effect of these changes on ice age climate are best judged by comparing the simulated SST and air temperatures with estimates from paleoclimatic data.

11.6.1 Simulation of the Last Glacial Maximum

Broccoli and Manabe (1987) conducted a series of experiments with a GCM that shed light on the importance of the ice sheets, land albedo changes, and atmospheric

Table 11.4

Description of the Four Simulations Conducted by Broccoli and Manabe (1987) to Investigate Role of Ice-Age Boundary Conditions on the Climate of a GCM with a Mixed-Layer Ocean

Experiment	E1	E2	E3	E4
Land–sea distribution	P[a]	G[b]	G	G
Continental ice distribution	P	G	G	G
Atmospheric CO_2 concentration (ppm)	300	300	300	200
Snow-free land albedo distribution	P	P	G	G
Length of analysis period (years)	15	8	6	8

[© Springer-Verlag.]
[a]P = present conditions.
[b]G = conditions estimated for last glacial maximum.

CO_2 changes in forcing the temperature changes that existed 18,000 years ago. The orbital parameters at that time were not substantially different from today's, and so cannot contribute much to the calculation of instantaneous differences between ice age and current conditions. The model used in these experiments was a GCM with a 50-m-deep mixed-layer ocean, which allows the prediction of SST, but does not include the effect of ocean circulation. The model includes the seasonal cycle, but a fixed cloudiness distribution is specified that varies only with latitude. Snow cover and sea ice are predicted.

The sequence of experiments conducted is described by Table 11.4. The first experiment is a control experiment (E1) in which all conditions are for the present climate with a CO_2 concentration of 300 ppm. The next experiment (E2) changes the continental ice sheets and the sea level to those of the last glacial age, leaving everything else as in the control experiment. The third experiment (E3) adds the change in the snow-free surface albedo that would correspond to the land conditions estimated for the last glacial age. These snow-free albedo changes are associated with changes in the vegetation and soil inferred from pollen and soil studies. Finally experiment E4 incorporates all of the above ice age conditions plus a decrease of the atmospheric CO_2 concentration to its ice age value of 200 ppm.

The direct effect of these changes on the radiation balance of the planet can be estimated by changing each of them individually, keeping everything else fixed as in the control experiment for the current climate, and using the radiation model in the GCM to calculate the resulting changes in the radiative energy flux at the top of the atmosphere. The results show that the global radiative forcings are each on the order of 1 W m^{-2}, with the largest contribution from the CO_2 decrease and the smallest associated with the albedo of snow-free land areas (Table 11.5). The direct radiative effect of continental ice sheets is interesting because it is confined almost entirely to the Northern Hemisphere.

The response of the climate model to these radiative forcings can be evaluated by comparing the SST changes in the model with those inferred from paleoclimatic data

Table 11.5

Direct Forcing of Radiative Energy Balance at Top of Atmosphere as a Result of Including Boundary Conditions and CO_2 Changes Estimated for Last Glacial Maximum

Control experiment	Perturbation	ΔR Global	N. Hem.	S. Hem.
E1	LGM distribution of continental ice	–0.88	–1.71	–0.06
E3	Atmospheric CO_2 reduced to 200 ppm	–1.28	–1.24	–1.31
E2	LGM distribution of land albedo	–0.67	–0.77	–0.58

Values are in W m^{-2} (N. Hem., S. Hem. = Northern, Southern Hemispheres). [Adapted from Broccoli and Manabe (1987), © Springer-Verlag.]

by the Climate Mapping and Analysis Project[13] (Table 11.6). The global mean SST change of −1.9°C in the model compares favorably with the paleoclimatic estimate of −1.6°C. The similarity of the observed and simulated temperature changes suggests that the model sensitivity is not grossly in error. The continental ice sheets and the CO_2 decrease each contribute a little less than 1°C to the ice age cooling, and the land albedo changes make a smaller contribution. The temperature change associated with the ice sheets is confined almost entirely to the Northern Hemisphere. In this model the thermal communication between the hemispheres is very inefficient. Although it is possible that the thermal communication between the hemispheres would be more efficient in a model that included heat transport by ocean currents, it is tempting to conclude that the mechanism that makes ice ages simultaneous in the two hemispheres is associated with the CO_2 concentration and is therefore biogeochemical.

The latitudinal distribution of the changes in SST resulting from each additional ice age forcing shows the characteristic polar amplification of temperature change that characterizes Earth's climate (Fig. 11.14). The introduction of all of the ice age forcings produces a temperature change with a latitudinal structure that is similar in many respects to the paleoclimatic estimate. Two serious discrepancies are evident in the zonal averages. One is in the subtropical regions where the simulation cools more than the data indicate, and the other is in the region poleward of 60°N where the simulation cools less than the data suggest. The geographic distribution of the differences between current and ice age SST from the model simulation and the CLIMAP reconstruction is shown in Fig. 11.15. The model correctly predicts that the largest SST changes occur in the North Atlantic. The ocean area poleward of 60°N, where the large polar discrepancies occur, is a very small but important area of the Norwegian and Greenland Seas. The discrepancy in this region occurs because ocean currents are not included in the model, and the absence of ocean currents may also account for the discrepancies in the vicinity of the Kuroshio Current. Heat trans-

[13] CLIMAP Project Members (1981).

Table 11.6

Differences in Area-Averaged Annual Mean SST between Pairs of Experiments Conducted by Broccoli and Manabe (1987) in °C

		Global	N. Hem.	S. Hem.
E2–E1	(Ice sheet)	–0.8	–1.6	–0.2
E4–E3	(CO_2)	–1.0	–0.7	–1.1
E3–E2	(Albedo)	–0.2	–0.3	–0.2
E4–E1	(Combined)	–1.9	–2.6	–1.5
CLIMAP		–1.6	–1.9	–1.3

Only those grid points that represent oceans in all experiments are used in computing the differences. Differences between CLIMAP ice age estimates of SST and current values are shown for comparison. [© Springer-Verlag.]

port by ocean currents is critical to the sea ice distribution in the Norwegian Sea, and because these transports are absent in the model the sea ice extends too far south in the control simulation for the current climate. Below the ice the SST is at its minimum value of about –2°C and so cannot decrease further when the ice age forcings are applied.

The discrepancies between the model simulation and the CLIMAP reconstruction of SST changes in the subtropics are much more troublesome. The CLIMAP reconstruction indicates a warming in the subtropical Pacific Ocean of both hemispheres where the model indicates substantial cooling of more than 2°C. It is unclear where the resolution of this discrepancy lies. Geologic evidence indicates that the snowline

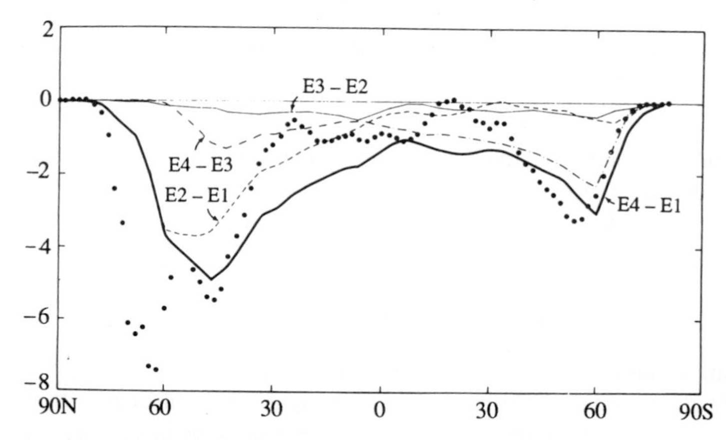

Fig. 11.14 SST changes (°C) induced by various ice age forcing mechanisms in the model of Broccoli and Manabe (1987) (various curves) compared with the SST inferred from ocean cores (dots). E2–E1, effect of ice sheets; E3–E2, effect of land albedo; E4–E3, effect of CO_2 concentration; E4–E1, difference between complete ice age simulation and simulation of current climate. [© Springer-Verlag.]

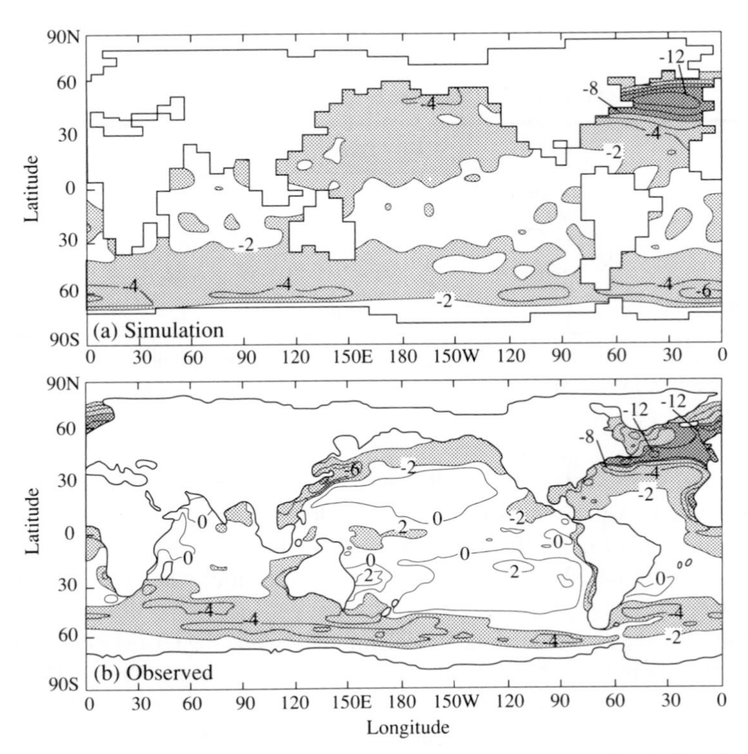

Fig. 11.15 SST differences (°C) between present and 18,000 years ago during the last glacial maximum derived (a) a climate model simulation (Manabe and Broccoli, 1985; ©American Geophysical Union) and (b) from paleoclimatic estimates (CLIMAP, 1981; permission granted from the Geological Society of America). The model simulation is the variable cloud simulation that is slightly more sensitive than the fixed cloud simulation presented in Fig. 11.14. Values less than −2°C are shaded; values less than −8°C are heavily shaded.

on tropical mountains, including those on the Hawaiian Islands, was about a kilometer lower during the ice age than at present.[14] It is difficult to reconcile this with the CLIMAP indications of warmer SST, which are based on assemblages of planktonic organisms reconstructed from ocean sediment cores.

11.6.2 Effects of Orbital Parameter Variations on Land Hydrology

According to the Milankovitch theory the rapid disappearance of the great continental ice sheets at the end of the last ice age was triggered by much greater insolation at high northern latitudes during summer. Northern summer insolation peaked about 10,000 years ago as a result of increased obliquity and because Northern Hemisphere summer occurred near perihelion. By 10,000 years ago, however, most of the

[14]See, e.g., Rind and Peteet (1985).

continental ice had already melted and the increased summertime insolation in the Northern Hemisphere had other important effects. Climate model simulations suggest that the moist period in the Sahara that occurred at about this time (see Section 8.5) was triggered by orbital parameter changes. The period around 10,000 years ago is characterized by insolation anomalies with a strong northward gradient from the equator to the North Pole in both seasons (Fig. 11.13). In the northern summer season this would tend to move the thermal equator northward. Recall from Fig. 6.25 and the discussion in Section 6.5.4 that the surface low-pressure belt and low-level wind convergence move northward and southward more or less following the sun over Africa. The strong solar heating of the surface during the summer seasons drives the low-level wind convergence deep into the Sahara during summer. About 10,000 years ago the increased summertime insolation must have intensified this summertime heat low over the Sahara to the extent that sufficient rising motion was generated to induce a prolonged summer rainy season that would extend to the center of the current Sahara Desert.[15] The moistening and greening of the surface would have amplified the effect of the insolation increase, since a moist vegetated surface would have a much lower albedo and would absorb a larger fraction of this insolation.

Paleoclimatic data and model simulations also suggest that increased summertime insolation in the Northern Hemisphere increases the intensity of the summertime monsoon over Asia. Greater summertime insolation results in increased heating of the land areas, increased low-level flow from ocean to land, and increased precipitation over land.[16]

Overall, there is much evidence to suggest that orbital parameter changes have a profound effect on climate. It is equally apparent that many other forcing mechanisms and the internal dynamics of the climate system also have a significant influence on climate variations. In the near future, it is likely that human actions will be the primary driver for climate change. This is the subject of the next chapter.

Exercises

1. Show that if the planetary albedo is 0.3, a change in solar constant of 1 W m^{-2} results in a global-mean climate forcing of 0.175 W m^{-2}.
2. The current magnitude of the greenhouse effect is measured by the difference between the emission from the surface and the emission from the top of the atmosphere, $G = \sigma T_s^4 - \sigma T_e^4 \approx 150$ W m^{-2}. What would be the required magnitude of the greenhouse effect to maintain the surface temperature at 288 K if the solar constant were reduced by 30%? By what distance would the effective emission level of the atmosphere need to rise if the lapse rate is approximately 6 K km^{-1}?
3. Calculate the percent change in annual mean insolation at a planet if the

[15]Kutzbach and Street-Perrott (1985); COHMAP Members (1988).
[16]Prell and Kutzbach (1987).

eccentricity is changed from zero to 0.2. How does this compare with the change in insolation at summer solstice if summer solstice occurs at perihelion?

4. Calculate the difference of insolation at 60°N at the summer solstice for conditions with perihelion at northern winter solstice and an obliquity of 22.5° and conditions with perihelion at northern summer solstice and an obliquity of 24.5°. Assume a solar constant of 1367 W m^{-2} at the mean Earth–sun distance and an eccentricity of 0.04. What are the relative contributions of precession and obliquity variations to this difference? *Hint:* Use (11.16) and Fig. 11.10.
5. What do the paleoclimatic data and their use in modeling studies such as that described in Section 11.6 imply about the energy balance albedo feedback models of the ice caps described in Section 9.4?
6. Discuss what adjustments to the Milankovitch theory described in Section 11.5 might be required by the results of the numerical experiments discussed in Section 11.6.

Chapter 12 Anthropogenic Climate Change

12.1 The Wings of Daedalus[1]

The greenhouse effect that warms the surface of Earth above its emission temperature results because a few minor constituents absorb thermal infrared radiation very efficiently. As a result of human activities, the atmospheric concentrations of some of these natural greenhouse gases are increasing, and entirely new man-made greenhouse gases have been introduced into the atmosphere. The increase in the atmospheric greenhouse effect will warm the surface of Earth. When the effects of feedback processes internal to the climate system are taken into account, it becomes clear that human activities are leading to a global climate change that may produce a mean surface temperature on Earth as warm as any for more than a million years. It is one of the great challenges of global climatology to predict future climate changes with adequate detail and sufficiently far in advance to allow humanity to adjust its behavior in time to avert the worst consequences of such a global climate change.

This challenge consists of several interlocking parts. We must first understand and predict the human-induced changes to the environment that are most important for climate, which may include atmospheric gaseous composition, aerosol amount and type, and the condition of the land surface. Then we must predict the climate change that will result from these changed climate parameters. These two steps are not independent, since climate change itself may feed back on the surface conditions and atmospheric composition through physical, chemical, and biological processes.

12.2 Humans and the Greenhouse Effect

Long-lived greenhouse gases that appear to be increasing as a direct result of human activities include carbon dioxide (CO_2), methane (CH_4), nitrous oxide (N_2O), and

[1] In Greek mythology Daedalus was an Athenian inventor, architect, and engineer. He was imprisoned with his son Icarus in a labyrinth, but managed to escape by constructing wings from feathers and wax. Icarus was young and imprudent and flew too close to the sun, where his wings melted and he fell into the sea. Daedalus flew on to safety.

Table 12.1

Characteristics of Some Key Greenhouse Gases That Are Influenced by Human Activities[a]

Parameter	CO_2	CH_4	CFC-11	CFC-12	N_2O
Preindustrial atmospheric concentration (1750–1800)	280 ppmv[b]	0.8 ppmv	0	0	288 ppbv[b]
Current atmospheric concentration (1990)[c]	353 ppmv	1.72 ppmv	280 pptv[b]	484 pptv	310 ppbv
Current rate of annual atmospheric accumulation	1.8 ppmv (0.5%)	0.015 ppmv (0.9%)	9.5 pptv (4%)	17 pptv (4%)	0.8 ppbv (0.25%)
Atmospheric lifetime[d] (years)	(50–200)	10	65	130	150

[From Watson *et al.* (1990). Reprinted with permission from Cambridge University Press.]

[a]Ozone has not been included in the table because of a lack of precise data.

[b]ppmv = parts per million by volume; ppbv = parts per billion by volume; pptv = parts per trillion by volume.

[c]The current (1990) concentrations have been estimated from an extrapolation of measurements reported for earlier years, assuming that the recent trends remained approximately constant.

[d]For each gas in the table (except CO_2), the "lifetime" is defined here as the ratio of the atmospheric content to the total rate of removal. This time scale also characterizes the rate of adjustment of the atmospheric concentrations if the emission rates are changed abruptly. Carbon dioxide is a special case since it has no real sinks, but is merely circulated between various reservoirs (atmosphere, ocean, biota). The "lifetime" of CO_2 given in the table is a rough indication of the time it would take for the CO_2 concentration to adjust to changes in the emissions.

halocarbons. Most of these gases have very long lifetimes in the atmosphere so that the amounts we release into the atmosphere today will remain in the atmosphere for up to two centuries, depending on the gas in question (Table 12.1). Most of these gases are naturally occurring, but most of the halocarbons are industrially created and have no sources in nature. Although water vapor is the most important greenhouse gas, we exclude it from this discussion because its atmospheric abundance is not under direct human control, but responds freely to the prevailing climate conditions and helps to determine them. We consider the changes in long-lived trace species to provide a forcing to the climate system, and we regard the changes in water vapor abundance that result to be a feedback process. Nonetheless, we must keep in mind that any temperature change associated with human activity will be composed of a large contribution associated with water vapor feedback and other feedbacks internal to the climate system. The net effect of all of these feedbacks is not known with great precision.

The contributions to the climate forcing from greenhouse gas changes during the 1980s are shown in Fig. 12.1. Carbon dioxide contributed more than half of the anomalous climate forcing, and chlorofluorocarbons (CFCs) contributed about one-quarter of the total for this period. Because of agreements to control CFC emissions to preserve the ozone layer, it is likely that the fraction contributed by CO_2 will increase in the future as the rate of CFC increase slows down. Carbon dioxide thus will

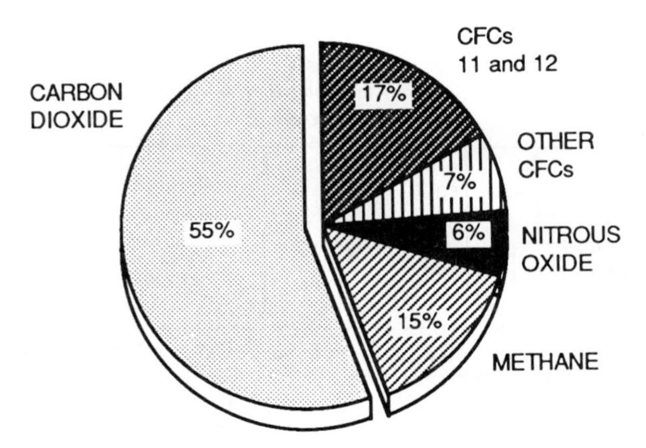

Fig. 12.1 Pie chart showing the contribution from each of the human-induced greenhouse gas changes to the change in radiative forcing from 1980 to 1990. The contribution from ozone is uncertain and has been ignored in this graph. [From IPCC Working Group I (1990). Reprinted with permission from Cambridge University Press.]

make the most important single contribution to greenhouse gas forcing of climate change, although the contributions from a number of other minor constituents add up to a substantial climate forcing.

12.2.1 Carbon Dioxide

Carbon dioxide is a naturally occurring atmospheric constituent that is cycled between reservoirs in the ocean, the atmosphere, and the land (Fig. 12.2). About 5 Gt C year^{-1} (gigatons of carbon per year) are currently released into the atmosphere from fossil fuel combustion, and roughly another 2 Gt C year^{-1} are released by deforestation. Some of this excess carbon is taken up by the ocean, but the atmospheric content is currently increasing by about 3 Gt C year^{-1} or about 0.5% year^{-1}. There are rapid exchanges of carbon between the atmosphere and the ocean, so that the particular CO_2 molecules in the atmosphere are changed in about 4 years. We may call this the *turnover time*. The time required for atmospheric CO_2 to achieve a new equilibrium in response to a perturbation such as fossil fuel burning is much longer, however, because of the slow rate at which carbon is exchanged between the surface waters and deep ocean. It requires 50–200 years for the atmospheric CO_2 concentration to achieve a new steady state. It is known that the 25% increase in atmospheric CO_2 during the industrial era is associated primarily with fossil fuel burning. Carbon dioxide estimates from ice cores indicate that the preindustrial atmosphere maintained a relatively constant CO_2 of about 280 ppmv for centuries. The recent rapid increase in atmospheric CO_2 concentration parallels very closely the known increase

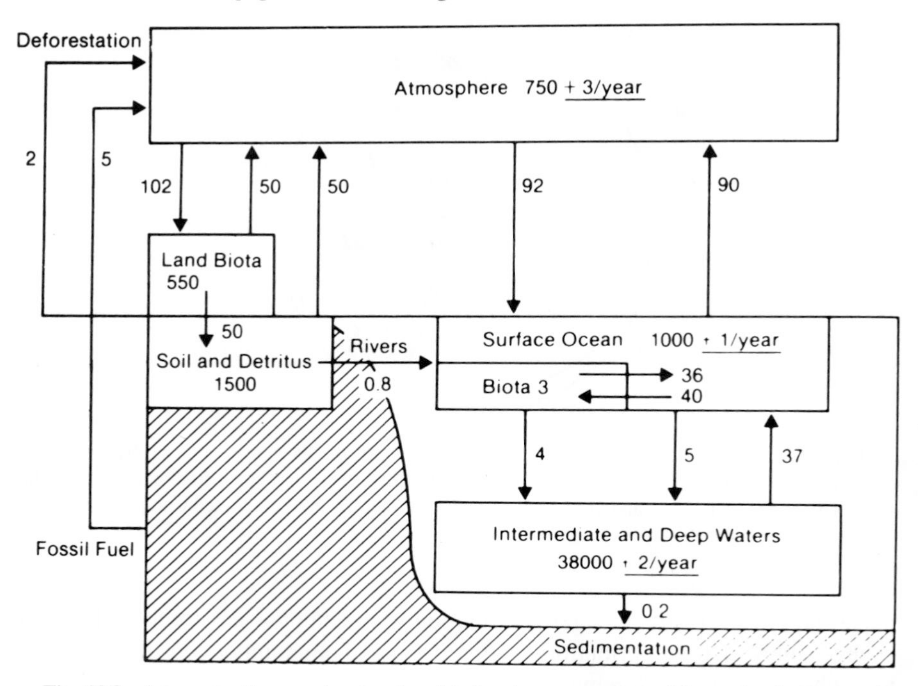

Fig. 12.2 Schematic diagram showing the global carbon reservoirs and fluxes for the present-day climate system. Fluxes are gross annual exchanges. Underlined numbers indicate net annual CO_2 accumulation because of human activities. Units are gigatons of carbon (Gt C; 1 Gt = 10^{12} kg) for reservoir sizes and Gt C $year^{-1}$ for fluxes. [From Watson *et al.* (1990). Reprinted with permission from Cambridge University Press.]

in fossil fuel combustion (Fig. 12.3). The fossil fuel origin of the recent CO_2 increase is further confirmed by changes in the isotopic abundance of ^{13}C and ^{14}C. Since fossil carbon has no ^{14}C isotope, the decrease of about 2% in atmospheric ^{14}C from 1800 to 1950 is consistent with a fossil carbon source.[2]

Future rates of CO_2 increase in the atmosphere are uncertain because the rates of release and the rates at which CO_2 will be taken up by the ocean and land biota are not known with great precision. It is estimated that during the period 1850–1986 195 ± 20 Gt C were released by fossil fuel burning and 117 ± 35 Gt C by deforestation and changes in land use, for a total airborne carbon production of 312 ± 40 Gt C.[3] About 41 ± 6% of this carbon has remained in the atmosphere. About 48 ± 8% of the total carbon release during the decade from 1980 to 1989 has remained in the atmosphere, but not all of the released carbon can be accounted for, as is illustrated in

[2]Stuiver and Quay (1981).
[3]Watson *et al.* (1990).

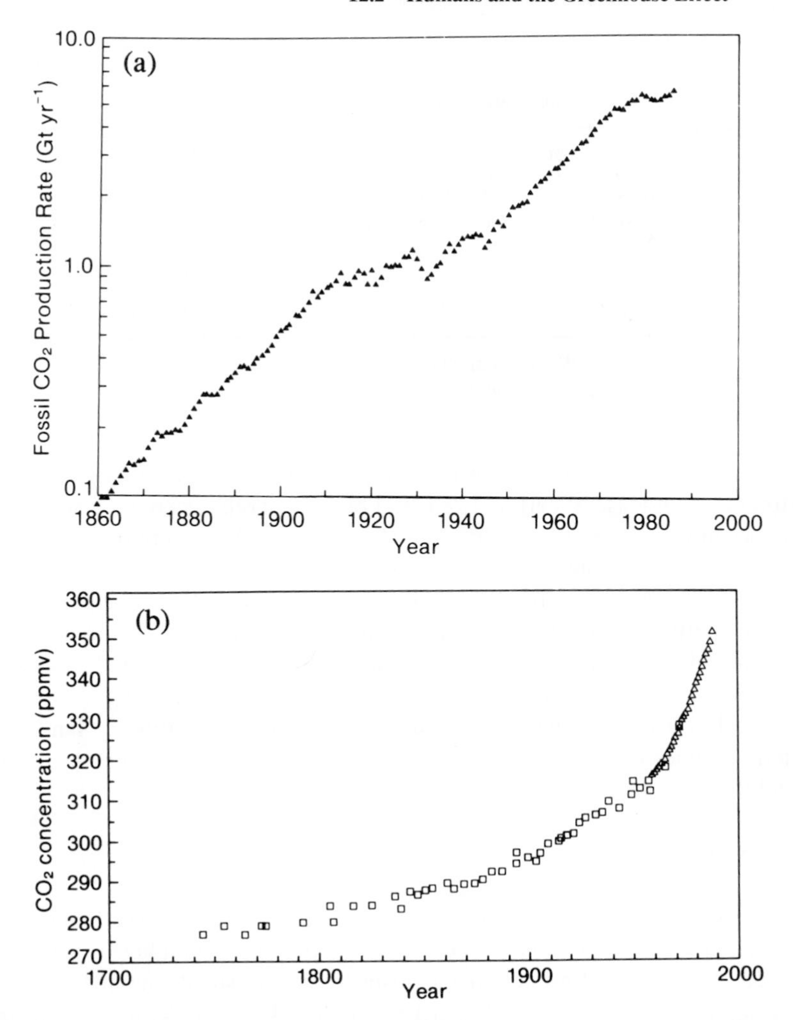

Fig. 12.3 (a) Global annual emissions of CO_2 from fossil fuel combustion and cement manufacturing expressed in Gt C year^{-1} (Rotty and Marland, 1986; Marland, 1989). The average rate of increase in emissions between 1860 and 1910 and between 1950 and 1970 is about 4% per year. Note the logarithmic scale. (b) Atmospheric CO_2 concentration for the past 250 years as indicated by measurements in air trapped in ice from Siple Station, Antarctica [squares, Neftel *et al.* (1985); Friedli *et al.* (1986), © Macmillan Magazines Limited] and by direct atmospheric measurements at Mauna Loa, Hawaii [Triangles, Keeling *et al.* (1989), © American Geophysical Union]. [From Watson *et al.* (1990). Reprinted with permission from Cambridge University Press.]

Table 12.2. The 1.6 Gt C year^{-1} imbalance in the carbon budget during the 1980s is relatively large and difficult to reconcile, although it is not much bigger than the uncertainty in the estimates. Emissions from fossil fuel combustion and the accumulation

Table 12.2

Estimated Human-Induced Perturbation to Global Carbon Budget for 1980–1989 Decade

Sources and sinks of carbon	GtC/yr
Emissions from fossil fuels into the atmosphere	5.4 ± 0.5
Emissions from deforestation and land use	1.6 ± 1.0
Accumulation in the atmosphere	3.4 ± 0.2
Uptake by the ocean	2.0 ± 0.8
Net imbalance	1.6 ± 1.4

[From Watson et al. (1990). Reprinted with permission from Cambridge University Press.]

in the atmosphere are known with relative precision. The uptake by the ocean and the storage and release of carbon from land biota are both uncertain. Carbon uptake by the ocean is estimated from differences in the partial pressure of CO_2 in the air and water combined with exchange coefficients that depend on the wind speed and the temperature. Both the observations of CO_2 partial pressure in the ocean and the exchange coefficients are somewhat uncertain. Studies using GCMs to estimate the distribution of atmospheric CO_2 from the distributions of sources and sinks suggest that a large sink of CO_2 in the Northern Hemisphere may be missing from these estimates. It has been hypothesized that the uptake of excess carbon by land biota in northern latitudes may be larger than expected and could account for the discrepancies in the carbon budget.[4]

12.2.2 Halocarbons

Halogens such as chlorine, bromine, and iodine have a wide variety of industrial applications, and in compounds with carbon they produce a number of useful gases (Table 12.3). Although present in the atmosphere in very small amounts, industrially produced gases such as CFC-11 (CCl_3F) and CFC-12 (CCl_2F_2) have a substantial influence on the greenhouse effect of Earth. The reason for the strong greenhouse effect of these gases is that they have strong absorption lines in the 8–12-μm region of the longwave spectrum where the surface emission is large. In this wavelength interval, naturally occurring gases do not absorb strongly, so the natural atmosphere is relatively transparent. These gases and other halocarbons are manufactured for use as the working fluid in refrigeration units, as foaming agents, as solvents, and in many other applications. Fully halogenated compounds are extremely unreactive and have very long lifetimes in the atmosphere. They are photodissociated by ultraviolet radiation in the stratosphere, where the chlorine and bromine thus released

[4]Tans *et al.* (1990).

Table 12.3

Concentrations and Trends of Selected Halocarbons

Halocarbon		Mixing ratio (pptv)	Annual rate of increase		Lifetime (years)
			pptv	%	
CCl_3F	(CFC-11)	280	9.5	4	65
CCl_2F_2	(CFC-12)	484	16.5	4	130
$CClF_3$	(CFC-13)	5			400
$C_2Cl_3F_3$	(CFC-113)	60	4–5	10	90
$C_2Cl_2F_4$	(CFC-114)	15			200
C_2ClF_5	(CFC-115)	5			400
CCl_4		146	2.0	1.5	50
$CHClF_2$	(HCFC-22)	122	7	7	15
CH_3Cl		600			1.5
CH_3CCl_3		158	6.0	4	7
$CBrClF_2$	(Halon 1211)	1.7	0.2	12	25
$CBrF_3$	(Halon 1301)	2.0	0.3	15	110
CH_3Br		10–15			1.5

[From Watson *et al.* (1990). Reprinted with permission from Cambridge University Press.]

participate in the catalytic destruction of ozone. Alternatives to fully halogenated compounds that can serve the same functions are being sought by replacing one of the halogens with a hydrogen atom and so making the molecules more reactive, giving them shorter lifetimes in the atmosphere, and producing less free chlorine and bromine when they dissociate.

12.2.3 Methane

Methane (CH_4) is produced in a wide variety of anaerobic environments, including natural wetlands, rice paddies, and the guts of animals (Table 12.4). It is also released during gas drilling and coal mining. The primary removal mechanism is oxidation by hydroxyl (OH) in the atmosphere. Methane oxidation by OH is the dominant source of water vapor in the stratosphere. Water in the stratosphere can be an important greenhouse gas, and both water vapor and ice can influence the photochemistry of ozone. The atmospheric concentration of methane has more than doubled since the preindustrial era, but its rate of increase has slowed in recent years. The reasons for the reduced rate of increase are not known.

12.2.4 Nitrous Oxide

Nitrous oxide (N_2O) is produced by biological sources in soils and water and has an atmospheric lifetime of about 150 years. Its primary sinks are in the stratosphere,

Table 12.4

Estimated Global Sources and Sinks of Methane for Current Conditions

Sources	Mean	Range
Natural		
Wetlands	115	(100–200)
Termites	20	(10–50)
Ocean	10	(5–20)
Freshwater	5	(1–25)
CH_4 hydrate	5	(0–5)
Anthropogenic		
Coal mining, natural gas and petroleum industry	100	(70–120)
Rice paddies	60	(20–150)
Enteric fermentation	80	(65–100)
Animal wastes	25	(20–30)
Domestic sewage treatment	25	?
Landfills	30	(20–70)
Biomass burning	40	(20–80)
Sinks		
Atmospheric (tropospheric + stratospheric) removal	470	(420–520)
Removal by soils	30	(15–45)
Atmospheric increase	32	(28–37)

Units are T_g CH_4 $year^{-1}$. [From Watson *et al.* (1992). Reprinted with permission from Cambridge University Press.]

where it is removed by photolysis and by reaction with electronically excited oxygen atoms. Evidence from ice cores indicates that N_2O concentration remained constant at about 285 ppbv for most of the past 2000 years and then began to increase at a rate of 0.2–0.3% $year^{-1}$ and reached a value of about 310 ppbv by 1990. Anthropogenic sources include the use of artificial fertilizers in cultivated soils, biomass burning, and a number of industrial activities.

12.2.5 Ozone

Ozone (O_3) is increasing in the troposphere and decreasing in the stratosphere. The tropospheric increase is related to industrial and automobile pollution that leads to the photochemical production of ozone near the ground. The climatic effect of the near-surface ozone is not great, but it is an environmental hazard because of its health effects on humans and plants. Ozone in the troposphere and lower stratosphere is an effective greenhouse gas, primarily because of the position of the 9.6-μm absorption band in the middle of the water vapor window. Recent evidence suggests that ozone concentrations in the lower stratosphere have declined during the past several decades, probably as a result of photochemical destruction of ozone by chlorine introduced into the atmosphere as CFCs. This ozone decline constitutes a reduction in the greenhouse effect that is estimated to have an amplitude that is comparable, for example, to the increase in the greenhouse effect directly contributed by the CFCs.[5] The reduction of ozone in the stratosphere also allows more solar energy to reach the troposphere, which partially offsets the effect of the decreased ozone greenhouse effect.

12.3 Anthropogenic Aerosols and Atmospheric Sulfur

Tropospheric aerosols have direct and indirect effects on the radiation balance. Aerosols directly influence solar and longwave radiative transmission. They also serve as cloud condensation nuclei (CCN), and the abundance of CCN influences the number, size, and atmospheric lifetime of cloud droplets or particles. Cloud abundance and radiative properties have a substantial influence on the net radiation balance of Earth. As discussed in Section 11.3, a substantial fraction of tropospheric aerosols is produced by conversion of SO_2 gas to sulfate aerosol, and more than half of the total sulfate aerosol production is anthropogenic and mostly related to fossil fuel combustion. Anthropogenic SO_2 emissions have increased from < 3 Tg S year^{-1} in 1860 to about ~80 in 1983.[6] These sulfur emissions have come mostly from the Northern Hemisphere and have led to a serious problem with acid rain. Northern Hemisphere emissions of sulfur have begun to decline during the last decade.

It is generally only a few days between the release of an aerosol precursor gas such as SO_2, its subsequent conversion to sulfate aerosol, and final precipitation in solution within a raindrop. Therefore, the aerosol burden of the troposphere is based on emissions from at most the previous 2 weeks, and if human production of aerosols were to cease, the aerosol burden of the atmosphere would return to its natural level within a similar period. Because of this short lifetime, the aerosol loading of the troposphere is highly variable, and tends to be highest near the sources of aerosols or their precursor gases. This is illustrated in Fig. 12.4, which shows a three-dimensional physical–chemical model simulation of the annual-mean sulfate concentration in the lower troposphere. The highest concentrations occur over the eastern portions of North America, Europe, and Asia, where they exceed the natural background by more than a factor of 10 and are therefore mostly anthropogenic. Since these aerosols are near the surface they will have a modest effect on longwave emission, but they may have a substantial effect on solar absorption when they overlie a dark surface in a relatively cloud-free and well-illuminated region.

[5]Ramaswamy *et al.* (1992).
[6]Ryaboshapko (1983).

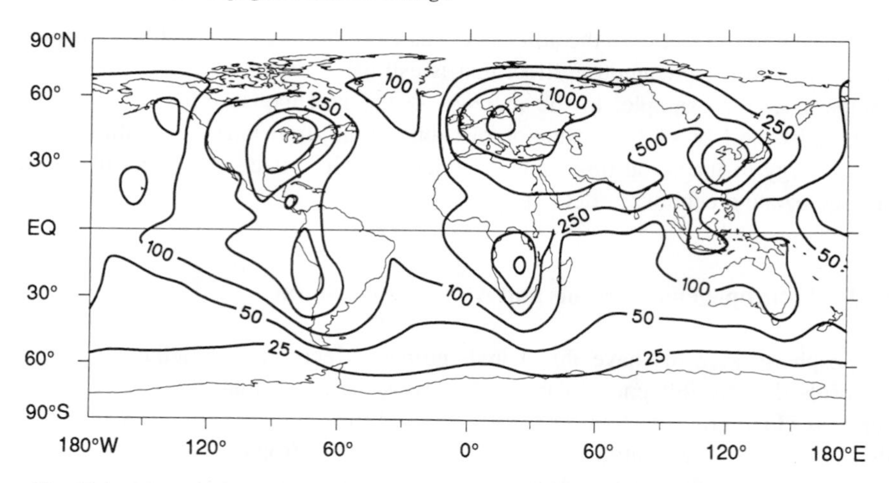

Fig. 12.4 Model simulation of the annual mean sulfate (SO_4^{2-}) concentration at 900 mb. Contours are shown at 25, 50, 100, 250, 500, 1000, and 2500 pptv. Concentrations over eastern North America and eastern Europe exceed natural (nonanthropogenic) levels by a factor of >10. [From Langner and Rodhe (1991). Reprinted with permission from Kluwer Academic Publishers.]

The direct effect of these anthropogenic aerosols on the reflection of solar radiation has been estimated and is shown in Fig. 12.5. They can cause a reflection of as much as 4 W m^{-2} in some regions, and when averaged over the globe provide an average climate forcing of about -1 W m^{-2}, with an uncertainty of a factor

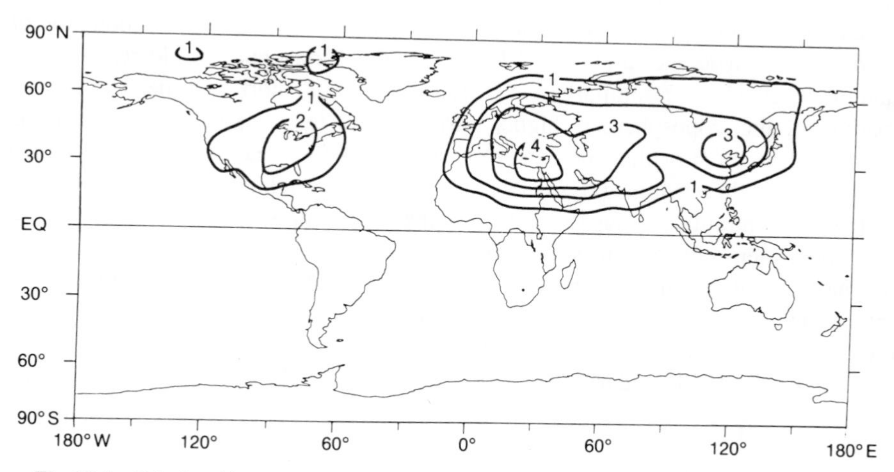

Fig. 12.5 Calculated increase of reflected solar flux caused by tropospheric sulfate aerosols derived from anthropogenic sources. Contour interval 1 W m^{-2}. [From Charlson *et al.* (1991). Reprinted with permission from Munksgaard International Publishers Ltd.]

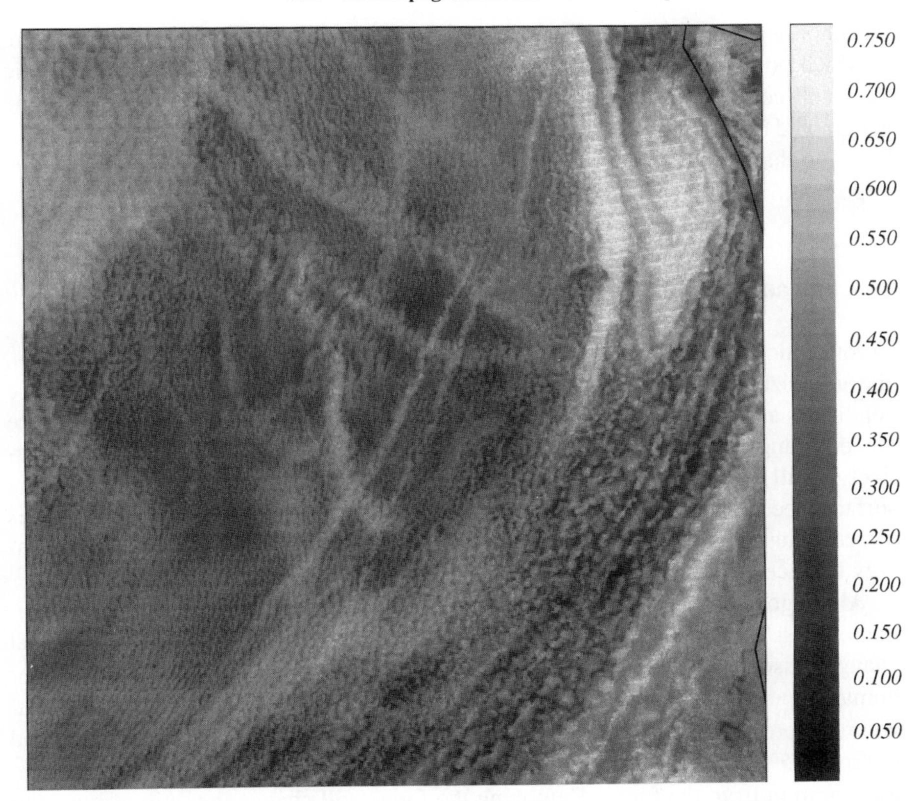

Fig. 12.6 GOES 1-km visible image of stratocumulus clouds over the Pacific Ocean on July 1, 1987 at 1615 UTC (Universal Time Coordinate). The scale of the image is approximately 5° square. The continental outline in the upper right is the Olympic Peninsula of Washington, and the one in the lower right is Cape Blanco on the Oregon coast. The linear features in the center of the image are ship tracks, enhancements of otherwise thin stratocumulus clouds caused by sulfur gas and particulate emissions of passing ships. The gray scale on the right indicates the visible albedo. (Image courtesy of Robert Pincus.)

of ~2.[7] This estimate includes only the direct radiative effect of the aerosols and does not include possible indirect effects that these aerosols might have on cloud albedos.[8] The enhanced cloud albedos would add to the direct aerosol cooling. The direct effect alone is only slightly less than the forcing currently provided by anthropogenic CO_2 (+1.5 W m^{-2}) and about equal to the effect of all the other greenhouse gas forcings (+1 W m^{-2}). If these estimates are correct, then a substantial fraction of the current anthropogenic greenhouse enhancement is being hidden by the cooling associated with anthropogenic sulfate aerosols (Fig. 12.6). This cancellation cannot continue, however, because of the very different atmospheric lifetimes of sulfur and

[7]Kiehl and Briegleb (1993).

[8]Coakley *et al.* (1987); Twomey *et al.* (1984).

greenhouse gases. Eventually the rate of increase of fossil fuel burning must slacken, at which point the sulfate aerosol concentration will level off, but the atmospheric CO_2 will continue to increase. Also, the aerosol cooling has a very characteristic geographic distribution, while the greenhouse warming is more geographically uniform, so that direct forcing of regional variations of climate change might be very important.

12.4 Changing Surface Conditions

Deforestation in midlatitudes and the tropics, expansion and contraction of deserts, and urbanization of the landscape can have significant effects on the local surface conditions and local climate. One of the most direct effects is that caused by surface albedo changes, which have a strong influence on the energy balance. When the relatively small fraction of land area on Earth and the effect of cloud cover in screening surface albedo changes are taken into account, however, land surface albedo changes directly caused by humans seem to have a very small effect on the global average energy balance, probably less than 0.1 W m^{-2}.[9]

Although land albedo changes may not be of first-order importance for global climate, most people live on continents and the regional climate and ecological changes associated with deforestation and urbanization can be quite important for human populations. It is estimated, for example, that complete removal of the Amazon rain forest would have a substantial influence on local surface temperature and hydrology.[10] About half of the rainfall over the Amazon basin is derived from evapotranspiration from the forest. Removing the forest will change the ratio of runoff to evapotranspiration, and the hydrological balance of the region may be seriously altered. The raised surface albedo also lessens the solar heating that ultimately drives upward motion and low-level moisture convergence. Although model simulations of complex interactions between vegetation and climate are at an early stage of development, some simulations suggest that removing the Amazon rain forest would result in a reduction of both moisture convergence and evapotranspiration, so that runoff would decrease significantly.[11]

12.5 Equilibrium Climate Changes

A standard experiment with a global climate model is to calculate the equilibrium climate for present and for doubled atmospheric carbon dioxide concentrations and study the differences between the two climate states. In such studies the transient na-

[9]Henderson-Sellers and Gornitz (1984).
[10]Dickinson and Henderson-Sellers (1988).
[11]Shukla *et al.* (1990); Nobre *et al.* (1991).

ture of the carbon dioxide increase and the climate response are ignored and only the state of the climate system when it is in balance with a particular concentration of CO_2 is considered. The response of the equilibrium climate to doubled CO_2 is much easier to compute than the transient response and gives useful information about the nature of the transient climate response to be expected. Doubled CO_2 is used as a surrogate for an equivalent climate forcing that may consist of contributions from many greenhouse gases. In an equilibrium calculation it is not necessary to choose a particular scenario for greenhouse gas release, and the thermal capacity of the deep ocean can be ignored. Most equilibrium calculations for GCMs are conducted using a mixed-layer ocean model.

12.5.1 One-Dimensional Model Results

A simple estimation of the response to changed greenhouse gases or other radiative forcings can be obtained from one-dimensional radiative–convective equilibrium models (Section 3.10). These models can be integrated very efficiently on the computer since only the globally averaged vertical structure is considered, so that a simple and quick assessment of the influence of radiative changes can be obtained. The results of such calculations for changed CO_2 concentration are shown in Fig. 12.7. These calculations were performed for average cloudiness and fixed relative humidity. A surface temperature increase of 2.4°C was calculated in response to a change from 300 to 600 ppmv atmospheric CO_2. It is interesting to note that while the surface

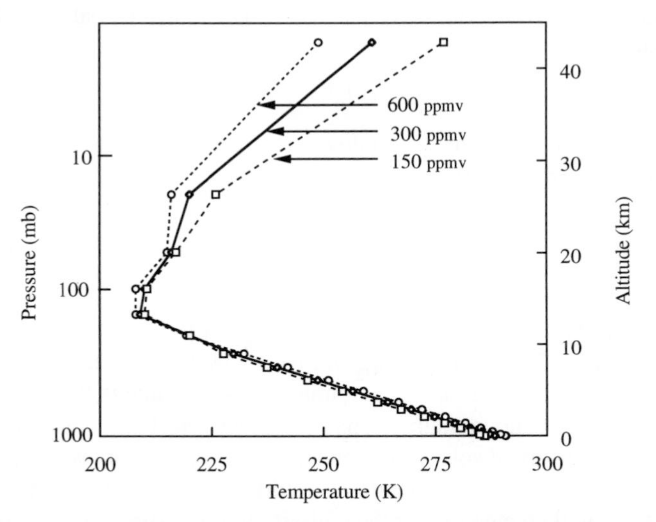

Fig. 12.7 Temperature profiles calculated with a one-dimensional radiative–convective equilibrium model for CO_2 at 150, 300, and 600 ppmv. [Data from Manabe and Wetherald (1967). Reprinted with permission from the American Meteorological Society.]

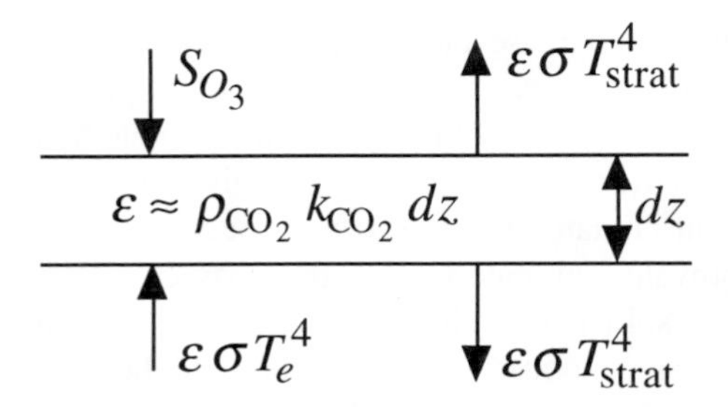

Fig. 12.8 Diagram showing the energy fluxes for a thin layer of stratospheric air in which ozone absorbs an amount of solar radiation S_{O_3}, and that absorbs and emits terrestrial radiation with an emissivity of $\varepsilon << 1.0$.

temperature increases with CO_2 concentration because of its enhancement of the total greenhouse effect, the stratospheric temperatures actually decline with increasing CO_2. The reasons for this can be easily understood by remembering that the approximate stratospheric energy balance is between heating through absorption of solar radiation by ozone and cooling by emission from CO_2 (Fig. 3.18).

We can illustrate the dependence of stratospheric temperature on its longwave emissivity through a very simple model (Fig. 12.8). Suppose that we consider a thin layer of the stratosphere with a longwave emissivity that is much less than 1 ($\varepsilon << 1.0$). This layer will absorb a fraction of the outgoing longwave radiation, and it will emit up and down according to its assumed emissivity. In addition, it will receive some amount of solar heating through the absorption by ozone S_{O_3}. The absorption by this layer of solar and outgoing longwave radiation (σT_e^4) must be balanced by its longwave emission.

$$S_{O_3} + \varepsilon\sigma T_e^4 = 2\varepsilon\sigma T_{strat}^4 \tag{12.1}$$

Solving for the temperature of the stratospheric layer shows the relationship of the stratospheric temperature to the emissivity of the layer.

$$T_{strat} = \sqrt[4]{\frac{\varepsilon^{-1} S_{O_3} + \sigma T_e^4}{2\sigma}} \tag{12.2}$$

Since CO_2 is the principal longwave absorber in the middle and upper stratosphere, and from (3.29) $\varepsilon \approx \rho_{CO_2} k_{CO_2} dz$, we know that the emissivity would increase when the CO_2 concentration is doubled. If we assume that the ozone amount and the planetary albedo remain unchanged when the CO_2 concentration is increased, then the absorbed solar radiation and the outgoing longwave radiation will be the same for a doubled CO_2 climate in equilibrium. Therefore nothing within the radical in (12.2) will change except the emissivity of the stratospheric layer, and the stratospheric temperature will decline with increasing CO_2.

12.5.2 Three-Dimensional Model Results

Equilibrium climate changes in response to doubled CO_2 have been calculated with more than 20 different GCMs. All of the models produce a surface temperature increase, but the amount of increase varies from about 2 to 5°C.[12] This large range is attributable to uncertainties in the feedback processes, and at this time it appears that cloud feedbacks produce the most uncertainty.[13] The seasonal and latitudinal variations of the zonal-mean warming in all of the models are very much like that shown in Fig. 10.14(a). The largest warming occurs in high latitudes in winter, mostly as a result of changes in the extent and thickness of sea ice. Summertime surface temperature changes in the areas with sea ice are much less than the global average because it is difficult to get the summertime SST values much above freezing. Tropical surface temperature changes are also somewhat less than the global average warming. Surface warming in the tropics is reduced because of the efficiency with which evaporation can remove excess heat (see Section 9.6).

The vertical and meridional distributions of zonal-mean air temperature changes in response to doubled CO_2 in calculations by two different GCM modeling groups are shown in Fig. 12.9. Both simulations show surface warming, stratospheric cooling, and strong surface warming in high latitudes of the winter hemisphere. The differences in the magnitude of the polar warming can be attributed to different amounts of sea ice in the simulation of the current climate in the two models. Greater amounts of sea ice in the simulation for current CO_2 concentration generally result in larger polar warming when the CO_2 is increased, since more sea ice in lower latitudes provides more scope for ice albedo feedback. The warming of the tropical upper troposphere is greater than the global-average surface warming and much more than the tropical surface warming. The greater evaporation that cools the tropical ocean surface results in more and deeper convection that warms and moistens the upper troposphere. The greater warming in the upper tropical troposphere than at the surface is fundamentally related to the moist adiabatic ascent of convective air parcels, whose lapse rate is decreased as the temperature is warmed (see Appendix C). The simulation by Wetherald and Manabe (1986) in the upper panel of Fig. 12.9 uses convective adjustment and produces a less pronounced upper-tropospheric warming than the simulation by Hansen *et al.* (1984). The differences in convection scheme may explain why the amount of tropical upper-tropospheric warming is different in the two simulations.

Many models also show a CO_2-induced summertime warming over the continents that is larger than the global-mean temperature increase. This enhanced continental warming is related to changes in the hydrologic cycle, which result in summertime drying of continental interiors in these simulations. Because of the warmer

[12]Mitchell *et al.* (1990); Schlesinger and Mitchell (1987).

[13]Cess *et al.* (1990).

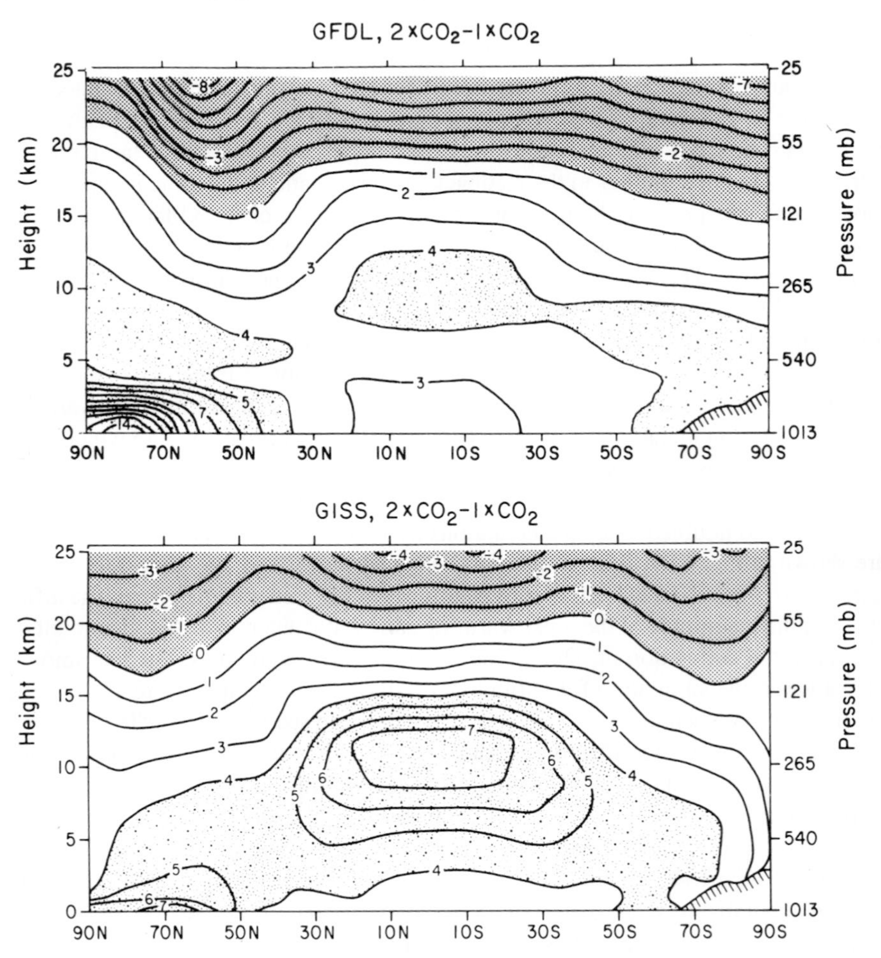

Fig. 12.9 Contour plot of the zonally averaged change in air temperature during DJF resulting from a CO_2 doubling in two models that each give a global-average surface temperature increase of 4°C. Cooling and warming greater than 4°C are shaded. [Top panel, Wetherald and Manabe (1986), reprinted with permission from Kluwer Academic Publishers; bottom panel, Hansen *et al.* (1984), © American Geophysical Union, as printed in Schlesinger and Mitchell (1987), © American Geophysical Union.]

surface air temperatures and consequent increase in saturation humidity, both the global-mean precipitation and evaporation are increased. The effect of this intensification of the hydrologic cycle on soil moisture depends on the location and season. In middle and high latitudes the increased precipitation during winter is not compensated by a similar increase in evaporation, so that the soil moisture generally increases in warming simulations. Because the water capacity of soil is basically fixed

(Section 5.6.2), the enhanced wintertime precipitation cannot easily be stored and results in enhanced runoff. In the simulations for the current climate, the soil moisture in the interiors of continents is generally greatest in the springtime after the snow melts and before the warm temperatures and high insolation of summer dry out the land. Although the maximum rates of precipitation often occur in the summer, the potential evaporation rate increases more than the precipitation and the soil gradually dries out over the course of a summer in many regions. In simulations of doubled CO_2 climates, the snowmelt and the commencement of the summer drying begin sooner and the evaporative capacity is greater. Because in these models the soil moisture is near its maximum capacity after the spring snowmelt even for the present climate, in a doubled CO_2 experiment the earlier snowmelt and earlier commencement of the drying season lead to reduced soil moisture during the summer months (Fig. 12.10). The reduced soil moisture leads to higher surface temperatures and less cloudiness, and both of these conditions encourage further drying of the surface.[14] Because of its critical dependence on the modeling of surface processes and their interaction with local climate, the prediction of summertime drying of midlatitude continental areas in a doubled CO_2 world is considered uncertain. In some simulations where the soil moisture is not near its maximum value during springtime in the $1 \times CO_2$ simulation, the springtime soil moisture in the $2 \times CO_2$ simulation is increased and the summertime drying is not significant.[15]

Most equilibrium doubled CO_2 simulations predict decreases in sea ice extent or thickness and these changes have a substantial influence on the high-latitude climate. In general, the sea ice models employed are the simple thermodynamic type (Section 10.5.1) and do not include changes in ocean circulation, ice dynamics, or salinity effects. The amount by which the sea ice is reduced varies from almost complete removal in summer to more modest changes.

A variety of changes in the circulation of the atmosphere have been noted in the doubled CO_2 equilibrium calculations. Associated with the reduced equator-to-pole temperature gradient are changes in the meridional pressure gradients and a slight weakening of the midlatitude storm tracks. On the other hand, the increased greenhouse gases lead to an increased radiative heating of the surface and increased radiative cooling of the atmosphere, and the increased surface temperature allows the surface air to retain more latent heat. All of these factors suggest an increase in the intensity of penetrative convection and possibly also in the intensity and frequency of tropical storms. It is likely that there would be regional shifts in weather patterns and stationary planetary waves in all seasons to accompany a global climate warming, but accurate prediction of regional climate changes seems to be beyond the capability of current climate models, since the predictions of regional climate shifts are not consistent between models.

[14]Wetherald and Manabe (1988); Delworth and Manabe (1988).
[15]Meehl and Washington (1988); Mitchell and Warrilow (1987).

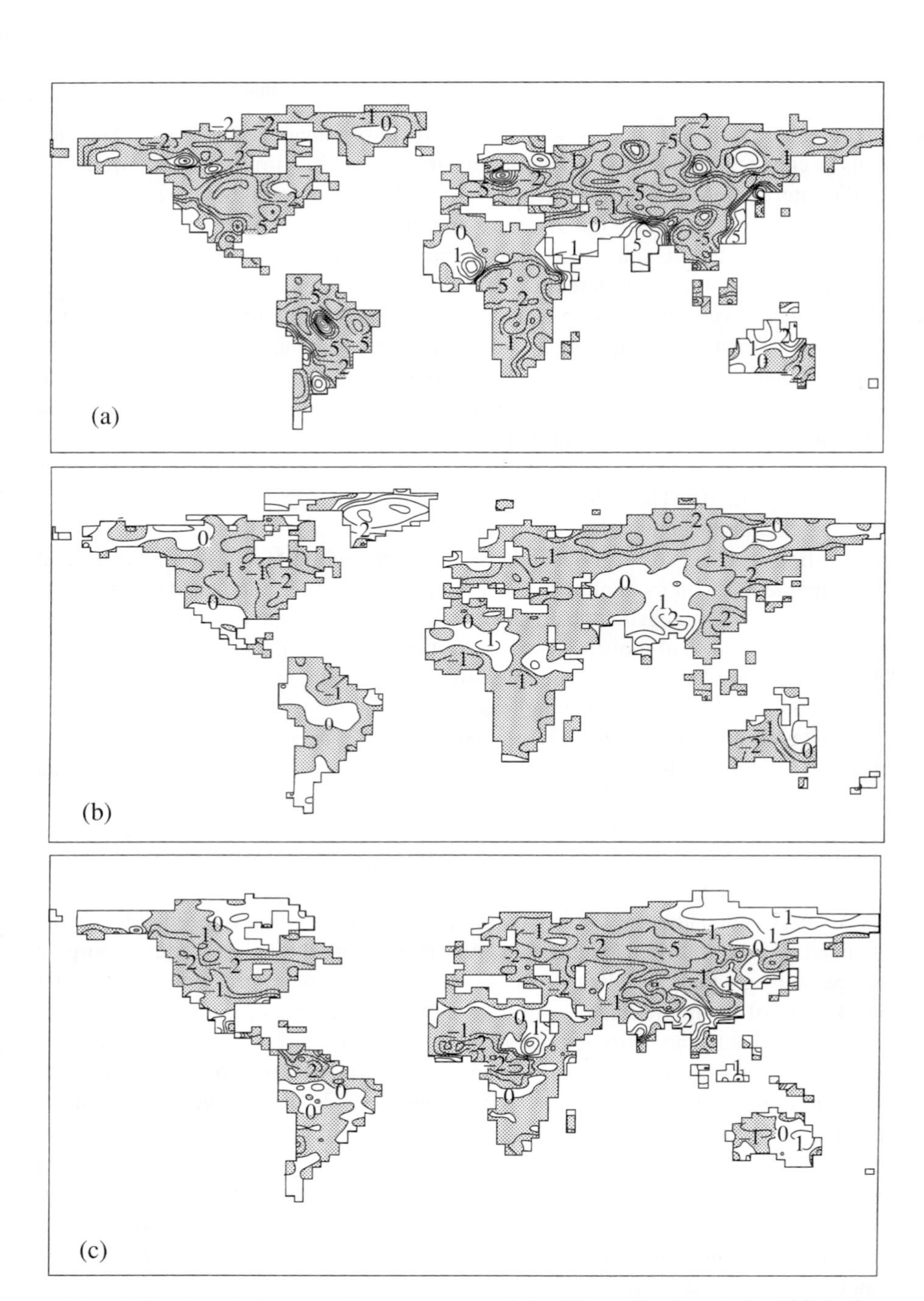

Fig. 12.10 Maps of change in soil moisture content during JJA resulting from doubled CO_2 in three GCMs. (a) Canadian Climate Center model; (b) NOAA GFDL model; (c) UK Meteorological Office model. Contour intervals are 0, ±1, ±2, and ±5 cm. Negative values are shaded. [From Mitchell *et al.* (1990). Reprinted with permission from Cambridge University Press.]

12.6 Time-Dependent Climate Changes

The changes in climate that result from human activities occur gradually in response to steadily increasing climate forcing by greenhouse gas increases, aerosol production, and land surface modification. The transient response of the climate system to the anthropogenic shift in climate forcing may be very different from the equilibrium response, and an equilibrium may never be achieved since it is unlikely that the anthropogenic forcing of climate will remain constant for the many centuries required for a steady state to be established. Moreover, the immediate need is to understand the history of climate that we are likely to experience over the next hundred years or so given a particular climate-forcing scenario. The response of the climate system to changed thermal forcing will be delayed by the large heat capacity of the ocean. In addition, the ocean currents and their slow response to this thermal forcing may yield geographic variations in the temperature change that are different from those of an equilibrium calculation. The ocean model is thus critical to the prediction of the transient response to climate forcing.

Transient climate simulations have been attempted with GCMs including both mixed-layer ocean models and ocean models with predicted currents and heat transports. The response of the mixed-layer ocean models can be understood using very simple conceptual models. In these models the ocean is represented by a wet surface with the heat capacity of the mixed layer of the ocean. Oceanic heat transports are not explicitly calculated and do not change with the climate. We may represent the global-mean transient temperature response T' to an imposed climate forcing, Q, with a first-order differential equation.

$$c\frac{\partial T'}{\partial t} = -\lambda_R^{-1} T' + Q \tag{12.3}$$

The time tendency term on the left represents the storage of energy in the ocean, c is the heat capacity of the ocean, and λ_R is the climate sensitivity parameter defined in (9.2). The solution to (12.3) for the case in which the temperature perturbation is initially zero is easily obtained.

$$T' = e^{-t/\tau_R} \int_0^t c^{-1}\, Q\, e^{t'/\tau_R}\, dt' \tag{12.4}$$

The response time, $\tau_R = c\,\lambda_R$, is proportional to the product of the heat capacity of the system times the sensitivity parameter. This makes it difficult to estimate the equilibrium climate change from the initial history of an induced warming, since both the sensitivity and the effective heat capacity are uncertain. Both the magnitude of the equilibrium response and the time required to achieve it are proportional to λ_R.

If the perturbation temperature is forced by an instantaneous switch-on of steady forcing

$$Q = \begin{cases} 0, & t \le 0 \\ Q_0, & t > 0 \end{cases} \tag{12.5}$$

where Q_0 is a constant, the solution (12.4) becomes

$$T' = \lambda_R Q_0 \left\{1 - e^{-t/\tau_R}\right\} \tag{12.6}$$

so that the equilibrium temperature perturbation $T' = \lambda_R Q_0$ is approached exponentially with an e-folding time scale of τ_R.

For an exponential increase of CO_2 the change of climate forcing is approximately linear with time, because the radiative forcing scales approximately as the logarithm of CO_2 concentration. If we apply a forcing that increases linearly in time

$$Q = \begin{cases} 0 & t \le 0 \\ Q_t\, t, & t > 0 \end{cases} \tag{12.7}$$

and insert it into the solution (12.4) for (12.3), we obtain

$$T' = \lambda_R Q_t \left\{t + \tau_R \left(e^{-t/\tau_R} - 1\right)\right\} \tag{12.8}$$

The exponential term within the brackets represents an initial transient, which for $t >> \tau_R$ is small compared to the other two terms, whence the solution becomes approximately

$$T' \approx \lambda_R Q_t \left\{t - \tau_R\right\} \tag{12.9}$$

The forcing at any time is $Q_t t$ to which the equilibrium response would be $\lambda_R Q_t t$. From (12.9) we infer that the transient response to linearly increasing forcing is just the equilibrium response, delayed by the response time τ_R. Choosing the thermal capacity to be that of 75 m of water [$c = 3.15 \times 10^8$ J K^{-1} m^{-2} from (4.5)] and choosing the sensitivity parameter to be such that doubling CO_2 leads to an equilibrium response of 4°C [λ_R = 1.0 K (W m^{-2})$^{-1}$], yields a response time of τ_R = 10 years. The response of this simple model to linearly increasing forcing is very similar to the global-mean response of a GCM with a mixed-layer ocean. The transient response is very much like the equilibrium response, except delayed by about 10 years.[16] The temperature responds sooner over the continental interiors where the effect of the ocean's heat capacity is less directly felt.

In GCM simulations with dynamically active oceans, the timing and spatial structure of the response are more complex. When ocean currents are included in the simulation the warming is much reduced around Antarctica and over the north Atlantic. Figure 12.11 shows the response of the surface air temperature to a 1% per year transient increase of CO_2 averaged over years 60–80, the equilibrium temperature response for doubled CO_2, and the ratio of the two temperature changes. A 20-year average centered on the 70th year of the transient simulation is chosen for comparison with the equilibrium calculation because a 1% per year increase of CO_2 will double its initial value in the 70th year. Larger temperature increases over the land areas and in north polar regions are observed in both the transient and equilibrium calculations. The transient temperature increase around Antarctica is less than 20% of the equilibrium response, while the tropical increases are about 80% of the equilibrium value. In the north

[16]Hansen *et al.* (1988).

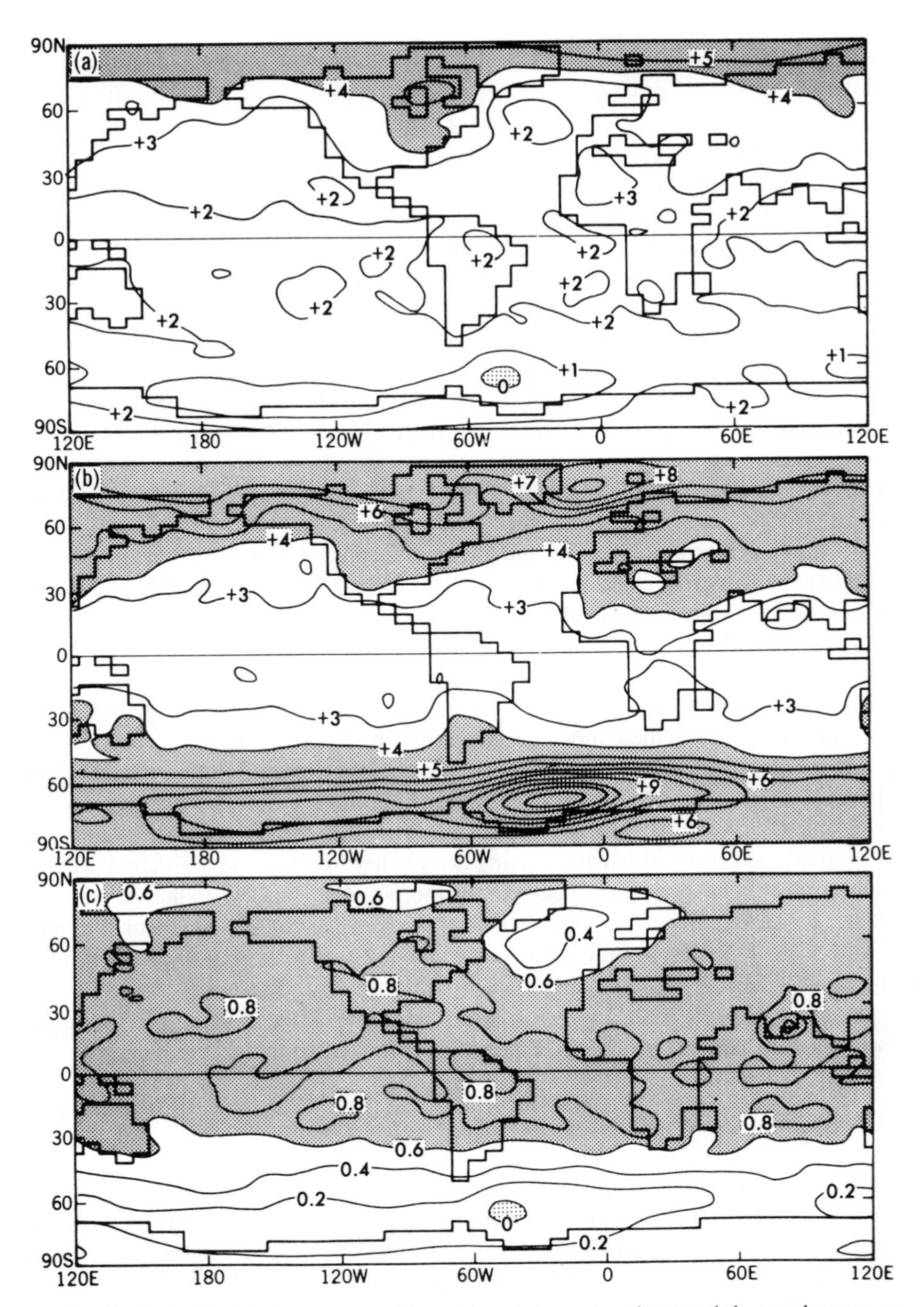

Fig. 12.11 (a) The transient response of the surface air temperature in a coupled atmosphere–ocean climate model to a 1% year^{-1} increase in atmospheric CO_2 concentration (°C). The response is the difference between the average for years 60–80 after the initiation of the CO_2 increase and the average of a 100-year simulation with current CO_2. (b) The equilibrium response of the surface air temperature to doubled CO_2. The response is the difference between two 10-year means from a doubled CO_2 equilibrium calculation and the simulation of the current climate. (c) The ratio of the transient response in panel (a) to the equilibrium response in panel (b). [From Manabe *et al.* (1991). Reprinted with permission from the American Meteorological Society.]

Atlantic Ocean region the transient response is only about 40% of the equilibrium response. According to our simple model for a mixed-layer ocean (12.9), if the response time is 10 years, the transient response to a linearly increasing forcing should be about 86% of the equilibrium response at year 70. The fact that much less than this response is observed in most regions except the tropics indicates a very important role for ocean currents in determining both the magnitude and the structure of the transient response.

The warming around Antarctica is suppressed because of the continuous upwelling of deep water around 60°S (Fig. 7.15). This upwelling is driven by the surface divergence forced by the strong zonal wind stress in the southern midlatitudes and is simulated by the coarse resolution ocean models used in climate change experiments (Fig. 10.16). The continuous replenishment of fixed temperature water from below makes it difficult for the changes in radiative energy fluxes to change the temperature at the surface, since the surface water is replaced by water from below before the surface fluxes can warm it up. Some initial experiments suggest that the upwelling of deep water around Antarctica is slightly enhanced in a doubled CO_2 climate.[17] This is because the surface temperature gradient in midlatitudes is enhanced by the tropical warming, which tends to increase the zonal winds and the driving of the surface divergence. The suppression of warming around Antarctica by this mechanism can be maintained for a very long time, since the time required to heat up the deep water is many centuries. Because of the suppression of SST increases around Antarctica, the sea ice there is not reduced, and so this feedback process does not contribute to the warming during the initial transient phase in these models with ocean currents.

In the north Atlantic the reason for the suppressed warming is more complicated. It appears that the modeled deep-water formation in the far north Atlantic Ocean is weakened, perhaps because enhanced precipitation freshens the surface waters and thereby decreases their density, making deep-water formation less efficient. The resulting decreased northward heat flux in the ocean offsets the radiative warming there. It is as yet unclear whether either of these ocean current effects on warming will operate in nature as they have been simulated in climate models. Natural interdecadal variability of the coupled ocean–atmosphere circulation seems to be substantial, so that at least the initial response to changed climate forcing could be overwhelmed by natural fluctuations, especially for limited geographic regions.

12.7 Comparison with Observed Temperature Trends

Simulations of climate change in response to enhanced greenhouse gas forcing suggest that the largest changes in surface temperature would be observed in high northern latitudes during winter and in the tropical upper troposphere during all seasons. Unfortunately, these are also regions with large natural variability, so that the unambiguous detection of a greenhouse gas warming signature is difficult. Moreover, the climate change in response to any forcing or internal mechanism will also have polar

[17] Bryan *et al.* (1988); Stouffer *et al.* (1989); Washington and Meehl (1989); Manabe *et al.* (1991).

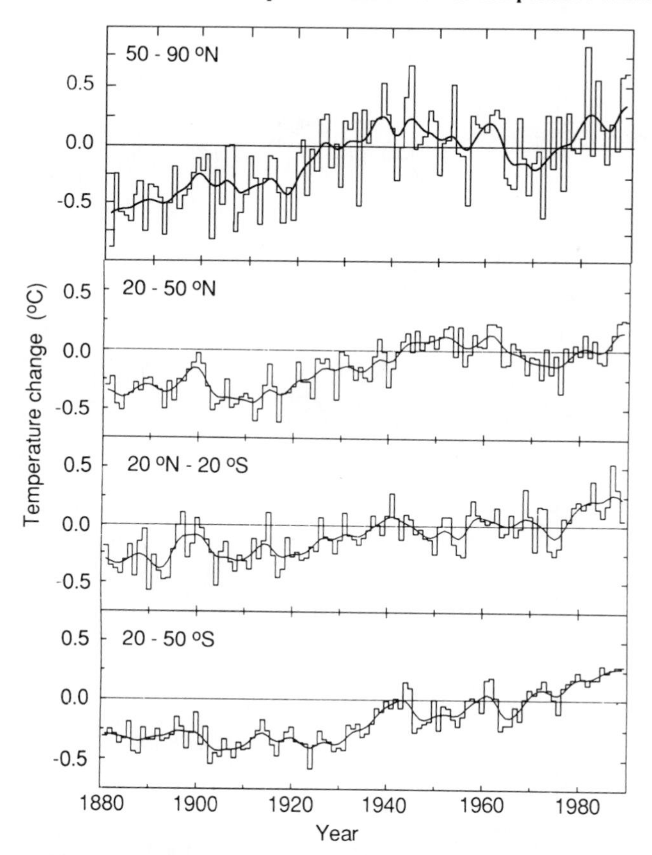

Fig. 12.12 Observed variations in annual-mean temperature for various latitude bands. The observations used are air temperature over the land areas and SST data for the oceans. The smoothed curves are provided to show decadal and longer trends more clearly. [From Wigley and Barnett (1990). Reprinted with permission from Cambridge University Press.]

and upper-tropospheric amplification, since this structure is caused by the feedback mechanisms and not by the forcing itself. The combination of stratospheric cooling and surface warming is characteristic of greenhouse gas forcing, and analysis of data suggests that such a signature is present in recent temperature trends.[18] The observation of decreased diurnal temperature range associated with nighttime warming is also consistent with an enhanced greenhouse effect.[19]

Observations also suggest that the warming of the past century has been larger in high northern latitudes than in the tropics, and larger in winter than in summer, as predicted by climate models (Figs. 12.12 and 12.13). The temperature measurements

[18] See, e.g., Angell (1988).

[19] Karl *et al.* (1984).

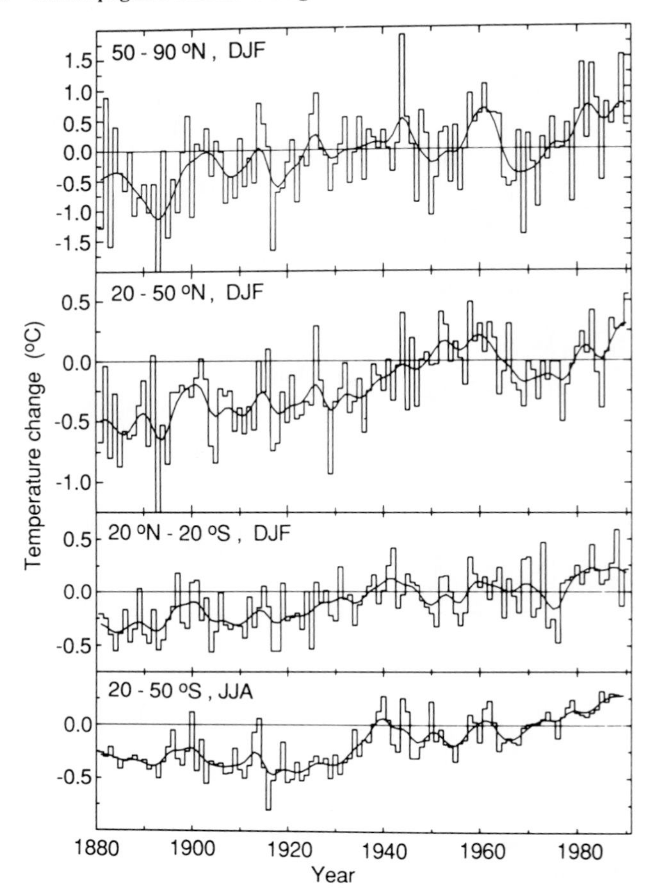

Fig. 12.13 Observed variations in winter temperature for various latitude bands. Otherwise, as in Fig. 12.12. Note the expanded temperature scale in the top panel. [From Wigley and Barnett (1990). Reprinted with permission from Cambridge University Press.]

around Antarctica are not abundant enough to determine whether the warming is suppressed there as simulations of transient climate change suggest.

12.8 Sea-Level Changes

Estimates from tide gauges suggest that the global sea level has risen by about 10 cm over the past century and continues to rise at a rate of 1–2 mm per year (Fig. 12.14). The estimates of sea-level rise are uncertain because of the quality of the tide gauge network and the necessity to correct for changes in continental elevation associated

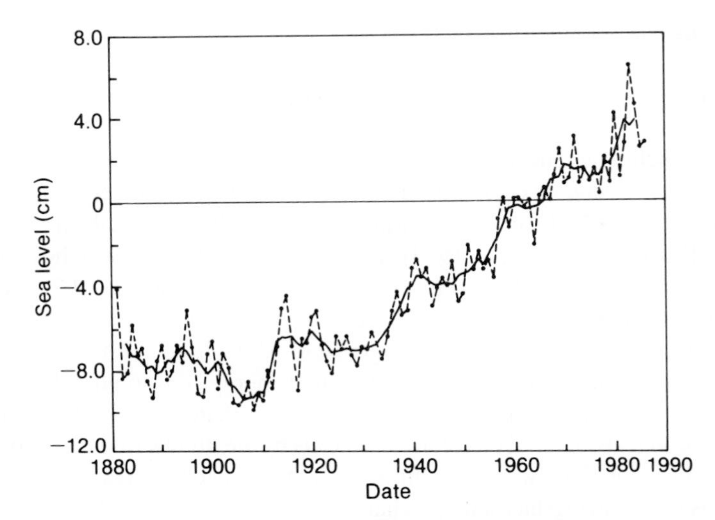

Fig. 12.14 Global-mean sea-level departures over the last century. The baseline is obtained by setting the average over the period 1951–1970 to zero. The dashed line represents the annual mean and the solid line the 5-year running mean. [From Barnett (1988).]

with geologic processes, particularly continuing rebound of the continents after the last ice age. This rise in sea level parallels the rise in temperature over this period (Fig. 8.1), and a causal relationship has been suggested between sea level rise and increasing temperatures.

The major contributors to sea-level rise over the last century are thermal expansion of the ocean (~4 ± 2 cm), melting of mountain glaciers and small ice caps (~4 ± 3 cm), melting of the Greenland Ice Sheet (~2.5 ± 1.5 cm), and possibly also changes in the Antarctic Ice Sheet (0 ± 5 cm).[20] The Greenland Ice Sheet seems to be shrinking, but the Antarctic Ice Sheet may actually expand in a warmed climate, since the warmer upper air temperatures will carry more moisture over the top of the ice sheet where it can precipitate as snow and add to the mass of the ice sheet.

If a substantial human-induced surface warming occurs, we may expect that there will be an accompanying rise in sea level. In the most extreme instance, the west Antarctic Ice Sheet may become unstuck from its submarine footing and slide into the southern ocean,[21] but this event is considered unlikely. Most estimates predict a sea-level rise of about 50 ± 30 cm over the next century.[22] Over the next 50 years the sea-level rise would come from thermal expansion and glacier melting in about the same proportion as the sea-level rise of the past century.

[20]Estimates from Warrick and Oerlemans (1990).

[21]Mercer (1978).

[22]Warrick and Oerlemans (1990).

12.9 Outlook for the Future

It is very clear that human activities have altered the atmosphere and that if we continue to release CO_2 and other greenhouse gases we can produce a significant change in the global climate. The magnitude and timing of this global climate change are both uncertain and depend on economic and social decisions that society has yet to make. Currently about 6 Gt C are released into the atmosphere annually from fossil fuel burning. Fossil fuel burning is closely tied to energy production, which is projected to increase by between 2 and 3% per year, which is comparable to the 2.2% per year increase over the past century. Economically recoverable fossil fuel reserves are uncertain, but estimated to be about 7500 Gt, most of which is coal.[23] These fossil fuel reserves are sufficient to sustain today's growth of energy consumption for many years into the future, but using coal to maintain future growth in energy consumption would probably lead to a doubling of atmospheric CO_2 by the middle of the twenty-first century. Doubling of CO_2, when combined with the probable increases in other greenhouse gases, would produce a major change in Earth's climate.

If we continue to alter the atmosphere in the future at an accelerating rate and the climate is as sensitive as some models suggest, then within a century we could produce a climate on Earth that would be warmer than any in more than a million years. Moreover, the rate of temperature increase would be very large by natural standards and would make it difficult for plants, animals, and humans to adapt. Because of the long lifetime of greenhouse gases in the atmosphere, delays in the response of the climate system, and natural variability, by the time a large climate change in response to greenhouse gases could be demonstrated from observations, we would be committed to that climate change for a long time into the future. For this reason long-term planning for global climate change will require accurate predictions of future climates that can withstand an extremely critical evaluation. A great deal of work by Earth scientists is needed to narrow the uncertainty in climate predictions such that they will become a broadly accepted basis for economic and environmental policy decisions. This is the challenge we face.

Exercises

1. Suppose that a volcanic eruption gives rise to a stratospheric aerosol cloud that changes the instantaneous Earth energy balance by 4 W m^{-2}. The cloud persists unchanged for 2 years. If the heat capacity of the climate system on this time scale is equivalent to a layer of water 50 m deep and the sensitivity of the climate is characterized by a sensitivity parameter $\lambda_R = 1\ K\ (W\ m^{-2})^{-1}$, what is the surface temperature anomaly produced by the aerosol cloud at the end of the 2-year period? What would the temperature response be if the aerosol cloud remained forever? What fraction of this long-term equilibrium response is achieved after 2 years?

[23] See, e.g., Sundquist (1985).

2. Consider a situation as in problem 1, except assume that the forcing from the aerosol cloud begins at 4 W m^{-2} but then decreases exponentially with an *e*-folding time of 2 years. Forcing = 4 W m^{-2} • exp($-t/2$ years). What is the temperature response after 2 years?
3. If the tropical SST is raised by 2°C from 299 to 301 K, and the lapse rate follows the moist adiabatic profile, estimate by how much the temperature at 300 mb will increase. Use Fig. C.1 of Appendix C.
4. A 1-km-thick layer of the atmosphere at about 55-km altitude is heated by absorption of solar radiation by ozone at a rate that would warm the layer by 10°C day^{-1}. In equilibrium this solar heating is balanced by longwave cooling. The air density in the layer is about 4×10^{-4} kg m^{-3}. The outgoing longwave radiation is about 240 W m^{-2}, on which the stratosphere at this altitude is assumed to have very little effect.

   (a) What is the broadband longwave emissivity of this stratospheric layer if it has a radiative equilibrium temperature of 280 K? The emissivity is the ratio of the actual emission from the layer to that of a black body at the same temperature, ε = (emission/σT^4). Assume local thermodynamic equilibrium so that the longwave emissivity of the layer is equal to its absorptivity.

   (b) The emissivity of the layer depends primarily on the CO_2 concentration, which was initially 300 ppmv. The CO_2 is doubled to 600 ppmv and the climate finds a new equilibrium. The albedo of the planet is assumed to remain constant during this climate change, so that the emission temperature of the planet is unchanged as a result of doubling CO_2. If the emissivity of the air increases approximately as the natural logarithm of the CO_2 concentration so that the new emissivity is 112% of the original one, what will be the new radiative equilibrium temperature of the stratosphere in the doubled CO_2 climate?
5. Suppose we can assume that the climate system has an effective heat capacity equivalent to that of 100 m of ocean, but the climate sensitivity can be either λ_R = 0.5 or 1.0 K/(W m^{-2}), and we don't know which. A climate forcing is applied that increases linearly with time as $Q = Q_t\, t$, where Q_t = 4 W m^{-2} in 50 years. If the measurement uncertainty for global mean temperature is 0.5°C, how many years will it be before we could distinguish whether the climate sensitivity is characterized by λ_R = 0.5 or 1.0 K/(W m^{-2})? That is, when would the temperature for the more sensitive climate exceed that of the less sensitive climate by 0.5°C? What are the temperature perturbations of the global climate for the two sensitivities when they differ by 0.5°C? If the climate forcing is held constant at its value when the two responses become distinguishable, what will be the equilibrium response for each sensitivity? *Hint:* Start by assuming that the time in question is significantly longer than the response time τ_R, so that the exponential transient in the solution (12.8) can be ignored. Determine a posteriori whether this assumption is justified.

Appendix A Calculation of Insolation under Current Conditions

A.1 Solar Zenith Angle

Consider a unit circle representing the Earth, as pictured in Fig. A.1. We are interested in calculating the solar zenith angle θ_s and the solar azimuth angle ξ at point X on the surface of the sphere, located at latitude ϕ. We draw a radial from the center of the sphere through point X to define the local zenith direction, Z. Another radial from the center of Earth to the sun crosses through the surface of the sphere at the subsolar point, ss. The point ss occurs at a latitude of δ, which is equal to the declination angle. At point X we draw another line to the sun. Since the sun is many Earth radii from Earth, the lines drawn to the sun from the center of the Earth and point X are parallel. The arclength on the unit sphere between X and ss is equal to the desired solar zenith angle, θ_s. We can now apply the law of cosines to the oblique spherical triangle defined by the points at the pole, P, the subsolar point, ss, and point, X, where we want the solar zenith angle. For this triangle we know the arclength of two sides and one interior angle, and we wish to know the length of the third side. The law of cosines requires that

$$\cos\theta_s = \cos(90-\phi)\cos(90-\delta) + \sin(90-\phi)\sin(90-\delta)\cos h \tag{A.1}$$

or

$$\cos\theta_s = \sin\phi\,\sin\delta + \cos\phi\cos\delta\cos h \tag{A.2}$$

The solar azimuth angle ξ, the angle between the subsolar point and due south (or equatorward) can be obtained from the law of sines for an oblique spherical triangle.

$$\frac{\sin(180-\xi)}{\sin(90-\delta)} = \frac{\sin h}{\sin\theta_s} \tag{A.3}$$

or

$$\sin\xi = \frac{\cos\delta\sin h}{\sin\theta_s} \tag{A.4}$$

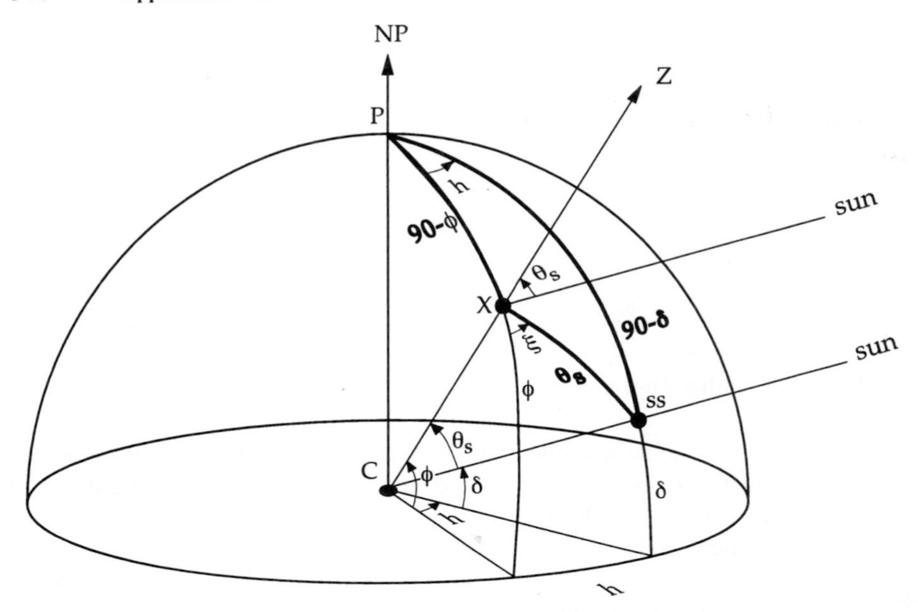

Fig. A.1 Spherical geometry for solar zenith angle calculation.

A.2 Declination Angle

The annual variation of the declination angle for current conditions can be defined to a good approximation in terms of a truncated Fourier series in the time of year. If the time of year is expressed in radians according to the following formula, where d_n is the day number, which ranges from 0 on January 1 to 364 on December 31, then

$$\theta_d = \frac{2\pi d_n}{365} \tag{A.5}$$

The declination angle is given to a good approximation by the Fourier series

$$\delta = \sum_{n=0}^{3} a_n \cos(n\theta_d) + b_n \sin(n\theta_d) \tag{A.6}$$

where the coefficients are as given in the following table.

n	a_n	b_n
0	0.006918	
1	−0.399912	0.070257
2	−0.006758	0.000907
3	−0.002697	0.001480

Because of the eccentricity of Earth's orbit, the Earth–sun distance varies according to the time of year.

A Fourier series formula for the squared ratio of the mean Earth–sun distance to the actual distance has also been derived by Spencer (1971).

$$\left(\frac{\bar{d}}{d}\right)^2 = \sum_{n=0}^{2} a_n \cos(n\theta_d) + b_n \sin(n\theta_d) \tag{A.7}$$

where the coefficients are given by

n	a_n	b_n
0	1.000110	
1	0.034221	0.001280
2	0.000719	0.000077

Appendix B | The Clausius–Clapeyron Relation

The temperature dependence of saturation pressure of water vapor over a water surface is governed by the Clausius–Clapeyron relationship.

$$\frac{de_s}{dT} = \frac{L}{T(\alpha_v - \alpha_l)} \tag{B.1}$$

In (B.1), e_s is the saturation vapor pressure above a liquid surface, L is the latent heat of vaporization, T is the temperature, and α represents the specific volume of the vapor α_v and liquid α_l forms of water. The Clausius–Clapeyron relation can be manipulated to express the fractional change of saturation vapor pressure, and thereby the specific humidity at saturation, q^*, to the fractional change of temperature.

$$\frac{\Delta q^*}{q^*} = \frac{\Delta e_s}{e_s} = \left(\frac{L}{R_v T}\right)\frac{\Delta T}{T} = r\frac{\Delta T}{T} \tag{B.2}$$

For terrestrial conditions, $T \sim 260$ K, and the factor r is approximately 20. This means that a 1% change in temperature of about 3 K will result in a 20% change in

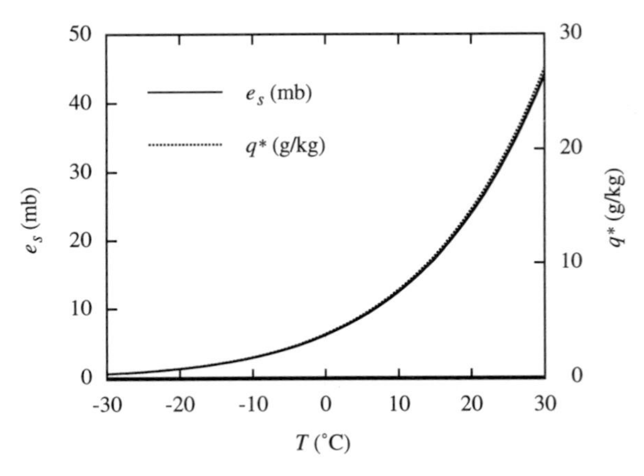

Fig. B.1 Saturation vapor pressure and specific humidity as junctions of temperature at standard pressure.

the saturation vapor pressure. If the relative humidity (the ratio of the actual humidity to the saturation humidity) remains fixed, then the actual water vapor in the atmosphere will increase by 20% for every 3 K temperature increase. This rapid logarithmic increase of saturation pressure with temperature can be seen more explicitly if we consider the approximate solution to (B.1) valid near standard pressure and temperature of 1013.25 mb and 273 K.

$$e_s \cong 6.11 \bullet \exp\left\{\frac{L}{R_v}\left(\frac{1}{273} - \frac{1}{T}\right)\right\} \tag{B.3}$$

The units of pressure in (B.3) are millibars. The exponential dependence of the saturation vapor pressure on temperature expressed by (B.3) is shown in Fig. B.1.

Appendix C | The First Law of Thermodynamics, Lapse Rate and Potential Temperature

The First Law of Thermodynamics states that energy is conserved (2.1), so that for a unit mass of gas, the applied heat dQ is equal to the sum of the change in internal energy dU and the work done dW. If we assume that the external work done by air is only that associated with volume changes, $dw = pd\alpha$, and we use the definition of the specific heat at constant volume $c_v = (dU/dT)_\upsilon$, we obtain a useful form of the First Law of Thermodynamics.

$$dQ = c_v \, dT + p \, d\alpha \tag{C.1}$$

Here p is pressure, dT is the change in temperature, and $d\alpha$ is the change in specific volume, α. Using the Ideal Gas Law

$$p\alpha = RT \tag{C.2}$$

we may write that

$$p \, d\alpha = R \, dT - \alpha \, dp \tag{C.3}$$

Using (C.3), and

$$R = c_p - c_v \tag{C.4}$$

(C.1) becomes

$$dQ = c_p \, dT - \alpha \, dp \tag{C.5}$$

or using (C.2) again,

$$dQ = c_p \, dT - \frac{RT}{p} dp \tag{C.6}$$

For an adiabatic process, $dQ = 0$, and (C.6) can be rearranged to read

$$\frac{dT}{T} - \frac{R}{c_p}\frac{dp}{p} = d \ln T - \frac{R}{c_p} d \ln p = d \ln\left(T \, p^{-R/c_p}\right) = 0 \tag{C.7}$$

so that for a parcel of gas that undergoes an adiabatic process

$$T\,p^{-R/c_p} = \text{constant} \tag{C.8}$$

If we define Θ to be the temperature at some reference pressure p_0, which is usually taken to be 1000 mb, then the potential temperature Θ is the temperature a parcel of air would have if it were brought adiabatically to the reference pressure.

$$\Theta = T\left(\frac{p_0}{p}\right)^{R/c_p} \tag{C.9}$$

C.1 Static Stability and the Adiabatic Lapse Rate

The potential temperature is useful because it remains constant as a parcel undergoes an adiabatic change of pressure. The vertical gradient of potential temperature determines the dry static stability of the atmosphere. If the potential temperature increases with height, then parcels raised adiabatically from their initial height will always be colder and thus more dense than their environment and will sink back to their original pressure. If the potential temperature decreases with height, then parcels raised up will be warmer than their environment when they reach the lower pressure and will be accelerated upward by buoyancy.

$$\begin{aligned} \frac{d\Theta}{dz} &> 0 \qquad \text{stable} \\ \frac{d\Theta}{dz} &< 0 \qquad \text{unstable} \end{aligned} \tag{C.10}$$

The rate at which temperature changes as a parcel moves up or down in the atmosphere without heating can be derived by using the hydrostatic equation

$$\alpha\,dp = -g\,dz \tag{C.11}$$

in (C.6), and setting $dQ = 0$.

$$c_p\,dT + g\,dz = 0 \tag{C.12}$$

or

$$\left.\frac{dT}{dz}\right)_{\text{adiabatic}} = -\frac{g}{c_p} = \Gamma_d \tag{C.13}$$

C.2 Moist Processes and Equivalent Potential Temperature

When moisture is present in air and an air parcel is raised adiabatically, the parcel can become supersaturated such that the water vapor condenses and latent heat is

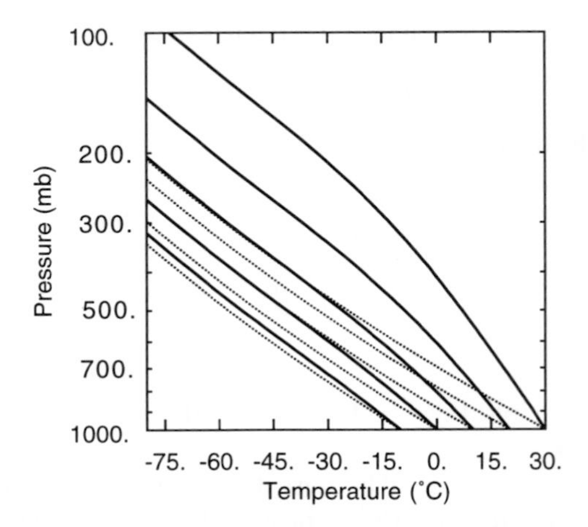

Fig. C.1 Dry (dashed) and moist (solid) adiabats for surface temperatures of –10, 0, 10, 20, and 30°C.

released. One can incorporate the latent heat release as heating in the First Law of Thermodynamics by writing the heat release in terms of the change in saturation water vapor mixing ratio, dq^*. The saturated adiabatic lapse rate can then be derived.[1]

$$\Gamma_s = \frac{\Gamma_d}{1 + (L/c_p)(dq^*/dT)} \tag{C.14}$$

The saturated adiabatic lapse rate is generally less than the dry adiabatic lapse rate, and becomes smaller as the temperature rises.

Another useful quantity is the equivalent potential temperature, which is the temperature that would be obtained by a moist air parcel if it were first raised moist adiabatically until all of its water condensed out, and then brought adiabatically back to a reference surface pressure.

$$\Theta_e = \Theta \exp\left(\frac{Lq^*}{c_p T}\right) \tag{C.15}$$

If the equivalent potential temperature decreases with height, then the air parcel is only conditionally unstable. It is unstable only if it becomes saturated. If the equivalent potential temperature increases with height, then the parcel is absolutely stable.

The important difference between dry adiabatic ascent and moist ascent can be illustrated by plotting the dry and moist adiabats for a few representative cases. The adiabats are the temperature profiles that would be experienced by parcels as they are raised upward from the surface. Examples of some dry and moist saturated adiabats are plotted in Fig. C.1. Because of the release of latent heat the temperatures of moist adiabats decrease less rapidly with height in the lower troposphere. This

[1] See, e.g., Wallace and Hobbs (1977).

difference is particularly evident at high temperatures, where the saturated parcel starts out at the surface with much more latent energy, and therefore its temperature drops less rapidly with altitude. Because the curvature of the saturated adiabats increases with temperature, when the temperature of the saturated parcel at the surface is increased, its temperature when it arrives at any layer higher in the troposphere is increased by a larger amount than the surface temperature increase. For example, parcels started from the surface at 20 and 30°C have temperatures of –8.7 and +7.2°C when they reach 500 mb. The difference of 15.9°C at 500 mb is much greater than their initial difference of 10°C, and reaches 24°C at 300 mb. This basic mechanism must explain why tropical middle troposphere temperatures increase more than the surface temperature in climate change simulations [e.g., Figs. 10.14(b) and 12.9].

Appendix D Derivation of Simple Radiative Flux Equations

To obtain the upward-directed, spectral flux density from the intensity we must integrate over all solid angles in the upper hemisphere as in (3.5). Since we assume that the intensity is independent of the azimuth angle, (3.5) becomes

$$F_\nu^{\uparrow}\left(\tau_\nu(z)\right) = 2\pi \int_0^1 \mu\, I_\nu\left(\tau_\nu(z), \mu\right) d\mu \qquad \text{(D.1)}$$

The intensity at the lower boundary, $I_\nu(0,\mu)$, must be prescribed. Most natural surfaces emit radiation isotropically with emissivity near one, so that it is reasonable to assume that the lower boundary emits like a blackbody. If we integrate (3.34) over the upper hemisphere and use (D.1), we obtain

$$F_\nu^{\uparrow}\left(\tau_\nu(z)\right) = 2\pi B_\nu\left(T(0)\right) E_3\{-\tau_\nu\} + \int_{E_3\{\tau_\nu(0)-\tau_\nu(z)\}}^{1/2} 2\pi B_\nu\left(T\left(\tau_\nu'\right)\right) dE_3\left\{\tau_\nu' - \tau_\nu(z)\right\} \qquad \text{(D.2)}$$

Here we have used the definition of the exponential integral of order n,

$$E_n\{x\} \equiv \int_1^\infty w^{-n}\, e^{(wx)}\, dw \qquad \text{(D.3)}$$

and the fact that these integrals have the following property:

$$\frac{dE_n\{x\}}{dx} = -E_{n-1}\{x\} \qquad \text{(D.4)}$$

Through similar steps we may obtain an equation for the downward-directed flux density at any level z. The main difference is a lack of a boundary flux term, since the downward flux of terrestrial radiation at the top of the atmosphere is zero.

$$F_\nu^{\downarrow}\left(\tau_\nu(z)\right) = \int_{E_3\{\tau_\nu(z)\}}^{1/2} 2\pi B_\nu\left(T\left(\tau_\nu'\right)\right) dE_3\left\{\tau_\nu(z) - \tau_\nu'\right\} \qquad \text{(D.5)}$$

The net flux of terrestrial radiation is given by the difference between the upward and downward flux:

$$F_\nu(z) = F_\nu^\uparrow - F_\nu^\downarrow \tag{D.6}$$

To obtain the energy flux needed for heating calculations, the net monochromatic flux density must be integrated over the terrestrial spectrum

$$F(z) = \int_{\nu_1}^{\nu_2} F_\nu(z)\, d\nu \tag{D.7}$$

For realistic atmospheric conditions the integration of (D.2) and (D.5) over frequency is difficult because of the rapid variation of the extinction coefficient and thereby the optical depth with frequency. Accurate solutions can be obtained by numerically integrating the equations for monochromatic flux density over the entire band of terrestrial frequencies from 4 to 200 μm, including the effect of each individual absorption line. Alternatively, schemes can be devised to represent the transmission over a range of frequencies in terms of "band models" with fewer parameters than the detailed absorption spectrum.[1] With band models that properly characterize the transmission integrated over a range of frequencies, the integrated terrestrial flux can be obtained with far fewer calculations than in a line-by-line integration.

D.1 Simplified Solutions

To understand something of how radiative transfer in the atmosphere affects climate, we may assume that a flux transmission function can be defined.

$$\mathcal{T}\{z_1, z_2\} = \frac{2}{\nu_2 - \nu_1} \int_{\nu_1}^{\nu_2} E_3\{\tau_\nu(z_2) - \tau_\nu(z_1)\}\, d\nu \tag{D.8}$$

The transmission function will always be between zero, which indicates no transmission, and one, which indicates complete transmission. The exponential integrals decrease rapidly as their argument increases. In this case the argument is the optical depth between two layers. Numeric values for $2E_3(\tau)$ and $\exp(-\tau)$ are given in Table D.1. The simple exponential represents the transmission for a parallel beam along a vertical path, while the exponential integral represents the transmission of isotropically incident radiation through a slab whose vertical optical depth is τ. Because the isotropic radiation takes all possible paths through the layer, and the vertical beam takes the shortest path, the vertical flux of isotropic radiation decreases more rapidly with optical depth than the parallel vertical beam.

To calculate accurate fluxes, the frequency interval of integration in (D.7) cannot be too large. For purposes of illustration, however, we may assume that we can

[1] Goody and Yung (1989).

Table D.1

Transmission as a Function of Vertical Optical Depth for a Parallel Beam along a Vertical Path, exp(−τ), and for Diffuse Radiation Integrated over All Angles

τ	$\exp(-\tau)$	$2\,E_3(\tau)$
0.00	1.00000	1.00000
0.01	0.99004	0.98055
0.05	0.95122	0.90983
0.10	0.90483	0.83258
0.20	0.81873	0.70389
0.50	0.60653	0.44320
0.70	0.49658	0.33212
1.00	0.36787	0.21938
1.50	0.22313	0.11347
2.00	0.13533	0.06026
2.50	0.08208	0.03259
3.00	0.04978	0.01786
3.50	0.03019	0.00989

integrate over the entire terrestrial band, so that with (3.8) and (D.8) one might integrate (D.2) and (D.5) over all frequencies to obtain

$$F^{\uparrow}(z) = \sigma T_g^4\, \mathcal{T}\{z_g, z\} + \int_{\mathcal{T}\{z_g, z\}}^{1} \sigma T(z')^4 \; d\mathcal{T}\{z', z\} \tag{D.9}$$

$$F^{\downarrow}(z) = \int_{\mathcal{T}\{z, \infty\}}^{1} \sigma T(z')^4 \; d\mathcal{T}\{z, z'\} \tag{D.10}$$

In deriving the above equations we have ignored the fact that both the transmission properties of the atmosphere and the Planck function change with frequency, so that the integral of their product is not equal to the product of their integrals. To justify ignoring the frequency dependence of transmission one could assume that the transmission functions properly represent a weighted mean of the transmission function over the band of interest.

$$\mathcal{T}\{z_1, z_2\} = \frac{\left[2/\left(\nu_2-\nu_1\right)\right] \int_{\nu_1}^{\nu_2} B_\nu(T)\; E_3\left\{\tau_\nu\left(z_2\right)-\tau_\nu\left(z_1\right)\right\} d\nu}{\left[1/\left(\nu_2-\nu_1\right)\right] \int_{\nu_1}^{\nu_2} B_\nu(T)\; d\nu} \tag{D.11}$$

Approximations like (D.11), where the transmission is assumed to be an average value independent of frequency, are called "gray absorption" approximations, because the structure of the absorption spectrum is averaged out. This is not a good approximation for terrestrial climate, because transmission in weak lines and in line wings is so important. Nonetheless, it allows the transmission function to be taken outside the frequency integral, so that the Planck function integrates to the Stefan–Boltzmann law, and yields easily interpretable forms of the flux equations used in Chapter 3.

Appendix E Symbol Definitions

Many symbols are required to represent the quantities used in physical climatology. The letters of the alphabet are fewer than the number of variables, even if both the English and Greek alphabets are fully utilized. Modern climatology crosses the boundaries between subdisciplines within earth sciences, and the traditional symbol for a quantity in one discipline may be the same as the traditional symbol for some other quantity in another discipline. In developing the symbols used in this book, an attempt was made to balance simplicity and tradition, on one hand, against the clarity of having a unique symbol for every variable, on the other. Some symbols are used to represent more than one variable in order to maintain the traditional usage. It is hoped that the meaning will be obvious from the context. A table containing the symbols used, their meaning, and the equation, section, or figure where they first appear is provided here to assist the student.

English Symbols

Symbol	Definition
a	the mean radius of Earth = 6.37×10^6 m (2.21), (6.4)
a_n	a series of expansion coefficients with index n (A.6)
a_0	semimajor axis of Earth's orbit Fig. 11.8, (11.7)
a_p	planetary absorptivity = $1 - \alpha_p$ (9.12)
A	area Fig. 2.4
A_c	total fractional area coverage by clouds (9.43)
A_b	fractional area coverage by black daisies (9.45)
A_g	fractional area coverage by ground in which daisies can grow (9.48)
A_w	fractional area coverage by white daisies (9.44)
A_ν	absorption of radiation at frequency ν (3.26)
b_e, b_Φ	regression coefficients for eccentricity and obliquity (11.17)
b_n	a series of expansion coefficients with index n (A.6)
B	a coefficient relating OLR to surface temperature (9.16), (9.21)
B_e	equilibrium Bowen ratio (4.34)
$B_\nu(T)$	Planck's blackbody emission at frequency ν and temperature T (3.7)

B_o	Bowen ratio = SH/LE (4.34)
c	heat capacity per unit area (12.3)
c^*	speed of light = 3×10^8 m s^{-1} (3.1)
c_p	specific heat of air at constant pressure (3.24), (C.1)
c_v	specific heat of air at constant volume (C.1)
c_s	specific heat of soil (4.8)
c_c	specific heat of organic matter in soil (4.8)
c_w	specific heat of water (4.5)
$\overline{C_a}$	total effective heat capacity for the atmosphere per unit area (4.4)
$\overline{C_o}$	total effective heat capacity for the ocean per unit area (4.5)
$\overline{C_{eo}}$	total effective heat capacity for the surface per unit area (4.3)
C_D	aerodynamic transfer coefficient for momentum, or drag coefficient (4.18)
C_{DE}	aerodynamic transfer coefficient for vapor (4.27)
C_{DH}	aerodynamic transfer coefficient for heat (4.26)
C_s	volumetric heat capacity of soil (4.8)
^{14}C	the isotope of carbon with atomic weight of approximately 14 Section 12.2
d	distance of Earth from the sun (2.4), (11.7)
d	total derivative prefix (3.3), (9.1)
d_a	Earth–sun distance at aphelion (11.6)
d_e	a depth scale for baroclinic disturbances (9.30)
d_n	a series of expansion coefficients with index n (A.5)
d_p	Earth–sun distance at perihelion (11.6)
ds	an increment of optical path length (3.12)
d_w	a depth of water (4.5)
D	surface condensation (dewfall plus frost) (5.1)
D_o	depth of ocean (7.13)
DJF	December, January, and February
D_T	thermal diffusivity (4.9)
D/Dt	material derivative following motion (6.4), (10.2)
e	eccentricity of an orbit (11.6)
e_s	saturation partial pressure of water vapor (9.8), (B.1)
e^x	2.71828 raised to the power x (1.6), (3.17), (7.4)
E	evapotranspiration (5.1)
E_a	energy content of the atmosphere per unit area (6.1)
E_{ao}	energy content of the atmosphere–ocean climate system (2.19)
E_{air}	component of evaporation associated with dryness of air (5.13)
E_{BB}	radiative energy emission from a blackbody (2.5)
E_{en}	component of evaporation associated with energy supply to the surface (5.10)
E_R	radiative energy emission (2.6)
E_s	energy content of the surface (4.1)

E_{sub}	mass rate of sublimation of snow	(5.15)
E_ν	emission of radiation at frequency ν	(3.26)
f	Coriolis parameter = $2\Omega \sin\phi$	(7.1)
F	energy flux	(3.12)
$F^\uparrow(z)$	upward longwave flux at altitude z	(3.33), (4.11)
$F^\downarrow(z)$	downward longwave flux at altitude z	(3.34), (4.11)
$F^\downarrow_{net}(z)$	$F^\downarrow(z) - F^\uparrow(z)$ net downward longwave flux at altitude z	(9.35)
$\vec{F}_a$	the horizontal flux of energy in the atmosphere	(6.1), (7.15)
$\vec{F}_{ao}$	horizontal energy flux in the atmosphere plus ocean	(2.19), (7.14)
F_{eo}	horizontal energy flux below the surface in earth or ocean	(4.1)
F_I	vertical flux of heat at the base of sea ice	(10.9)
$\vec{F}_o$	the horizontal flux of energy in the ocean	(7.15)
FRH	a subscript indicating a process takes place with fixed relative humidity	(9.9)
F_s	the vertical flux of energy through soil or snow	(4.6), (10.10)
F_ν	energy flux in some infinitesimal range of frequency centered on ν	(3.3)
F_ϕ	meridional flux of energy in the atmosphere–ocean climate system	(2.21)
F_∞	downward radiation flux at the top of the atmosphere	(3.17)
g	acceleration of gravity, 9.81 m s^{-2}	(1.2)
g_w	storage of water at and below the surface	(5.1)
g_{wa}	storage of water in the atmosphere	(5.3)
G	energy storage in the surface	(4.1)
h	hour angle	(2.15), (A.1)
$\hbar$	Planck's constant = 6.625×10^{-34} J s	(3.2)
h_c	water equivalent depth of soil water capacity	(5.14)
h_I	depth of sea ice	(10.11)
h_0	hour angle at sunset and sunrise ($-h_0$)	(2.16)
h_s	water equivalent depth of snow	(5.15)
h_T	penetration depth of temperature perturbations in soil	(4.10)
h_v	water equivalent depth of soil moisture below which plants transpire at less than the potential rate	(5.18)
h_w	equivalent depth of soil moisture	(5.14)
H	scale height = RT/g	(1.5), (3.18)
$\vec{i}$	unit vector in eastward direction	(7.9)
I	OLR scaled by global insolation	(9.22)
I	global ice volume	(11.18)
I_ν	intensity of radiation at frequency ν	(3.3), (3.26)
ITCZ	Intertropical Convergence Zone	
$\vec{j}$	unit vector in northward direction	(7.9)
JJA	June, July, and August	
k	Boltzmann's constant = 1.37×10^{-23} J K^{-1}	(3.7)

$\vec{k}$ vertical unit vector (7.9), (10.2)
k_{abs} absorption cross section (3.12), (11.1)
k_{ext} extinction cross section (11.1)
k_I thermal conductivity for sea ice (10.9)
k_s thermal conductivity for snow (10.9)
k_{sca} scattering cross section (11.1)
k_ν absorption cross section at frequency ν (3.27)
K_H a coefficient for horizontal heat transport (9.19), (9.26)
K_T thermal conductivity (4.6), (10.8)
L_e a horizontal mixing length scale for atmospheric disturbances (9.27)
L latent heat of vaporization for water = 2.5×10^6 J kg^{-1} (4.25)
L_0 luminosity of the sun (2.2)
LAI leaf area index Section 5.4
LE surface cooling by evaporation Fig. 2.3, (4.1)
L_f latent heat of fusion for water = 3.34×10^5 J kg^{-1} (10.12)
LP rate of latent heating of the atmosphere by precipitation (6.1)
L_R Rossby radius of deformation (9.28)
m_a molecular weight of dry air Appendix F
m_w molecular weight of water Appendix F
M angular momentum (6.16)
M_a mass mixing ratio of absorber in air (3.19), (3.25)
M_e orbital angular momentum of Earth about the sun (11.11)
MAM March, April, and May
M_r angular momentum associated with the relative zonal motion of the atmosphere (6.16)
M_s mass rate of snow melting per unit area (5.14)
M_Ω angular momentum associated with Earth's rotation (6.16)
N the buoyancy frequency (9.28)
^{18}O the isotope of oxygen with atomic weight of approximately 18 Section 8.5.4
OLR outgoing longwave radiation = $F^\uparrow(\infty)$ Section 3.11
p pressure (1.2)
p_0 reference pressure, usually 1000 mb (C.9)
p_s surface pressure (1.6), (3.19), (10.1)
P precipitation by rain and snow (5.1)
PE potential evaporation rate (5.17)
P_0 period of Earth's orbit about the sun (11.12)
P_q parameterized sources of moisture (10.4)
$\vec{P}_v$ parameterized sources of momentum (10.2)
$\hat{P}$ scattering phase function (11.4)
P_r mass rate of rainfall per unit area (5.14)
P_s mass rate of snowfall per unit area (5.15)

P_T parameterized sources of heat (10.3)
q specific humidity–mass mixing ratio of water vapor in air Fig. 4.5
q^* saturation specific humidity (4.28)
q_a specific humidity at anemometer level (4.27)
q_s specific humidity at the surface (4.27)
q_s^* saturation specific humidity value at the surface (9.37)
Q heating or energy input (2.1)
Q_{ABS} absorbed solar radiation (9.11)
Q_{abs} absorption coefficient (11.2)
Q_{ext} extinction coefficient (11.2)
Q_{sca} scattering coefficient (11.2)
r_{eff} effective particle radius Fig. 11.6
r_p radius of a planet (2.7)
r_{photo} radius of the photosphere of the sun (2.2)
R gas constant for air (1.3)
R^* universal gas constant
R_a net radiative heating of the atmosphere per unit area (6.1)
RH relative humidity = q/q^* (4.30)
Ri Richardson number (4.20)
R_s net radiative energy input at the surface (4.1), (4.11)
R_{TOA} net radiative energy input at the top of the atmosphere (2.19), (3.53), (7.14)
R_v gas constant for water vapor = 461 J K^{-1} kg^{-1} (9.9), (B.2)
s distribution function for solar radiation (9.12), (11.10)
$\tilde{s}$ distribution function for solar radiation for a circular orbit (11.8)
S salinity Fig. 7.2, 7.4
S_o solar constant at the mean Earth–sun distance (1367±3 W m^{-2}) (2.7)
S_d solar flux density in W m^{-2} at some distance d from the sun (2.4)
SH sensible cooling of the surface (4.1)
SON September, October, and November
S_{O_3} heating from solar radiation absorption by ozone (12.1)
SPCZ South Pacific Convergence Zone
$S^{\uparrow}(z)$ upward solar flux at altitude z (4.11)
$S^{\downarrow}(z)$ downward solar flux at altitude z (4.11)
T temperature (1.1)
T^* deviation of temperature from its zonal average (6.13)
T_A an atmospheric temperature (2.11), (3.39)
T_B temperature at the bottom of a layer of sea ice (10.10)
T_a temperature of air at anemometer level (4.26)
T_e emission temperature (2.8), (3.47), (9.7)
T_i local emission temperature (9.46)

T_0	a reference temperature (4.20)
T_s	surface temperature (2.12)
T_{eo}	effective temperature of the land or ocean surface for heat storage (4.3)
T_{photo}	emission temperature of the photosphere of the sun (2.2)
T_{SA}	temperature of the near-surface air (3.54)
$T_{z_{cb}}$	temperature of the air at the altitude of cloud base (3.45)
$T_{z_{ct}}$	temperature of the air at the altitude of cloud top (3.45)
t	time (2.19)
t_0	initial time (10.13)
U	internal energy (2.1)
U	wind speed (4.16)
U_E	eastward transport velocity in the oceanic Ekman layer (7.6)
U_r	a reference wind speed (4.18)
u	eastward component of velocity relative to surface (6.4)
u_E	eastward component of current velocity in the oceanic Ekman layer (7.4)
u_{earth}	eastward velocity of the surface associated with Earth's rotation (6.16)
u_ϕ	eastward velocity air will have at any altitude ϕ, if it is zero at the equator (6.16)
u_*	friction velocity (4.15)
u^*	the deviation of u from its zonal average (6.13)
v	northward component of velocity (6.4)
v^*	the deviation of v from its zonal average (6.6)
v_E	northward component of current velocity in the oceanic Ekman layer (7.5)
V	a horizontal velocity scale (9.27)
$\vec{V}$	horizontal velocity vector (7.10)
V_{eff}	variance of effective particle radius Fig. 11.7
V_E	northward transport velocity in the oceanic Ekman layer (7.6)
V_I	northward current velocity in the ocean interior (7.13)
w	vertical component of velocity (D.3), (4.22), (6.5)
w_E	vertical velocity at the base of the Ekman layer (7.9)
w_w	soil water mass per unit area (5.14)
w_s	surface snow mass per unit area (5.15)
W	a vertical velocity scale (9.33)
W	work (2.1)
x	eastward spatial coordinate and distance
x	sine of latitude (9.11)
x	a symbol for an arbitrary variable (6.12)
x_i	sine of latitude poleward of which perennial icecover exists (9.21)

y	northward spatial coordinate and distance	(7.7), (9.26)
z	vertical spatial coordinate and distance	(1.1)
z_E	Ekman depth	(7.5)
z_0	roughness height	(4.17)
z_r	a reference height	(4.19)
z_s	elevation of the surface, often zero	(3.35)
z_{cb}	altitude of cloud base	(3.45)
z_{ct}	altitude of cloud top	(3.46)

Greek Symbols

α	specific volume	(B.1), (C.2)
α_b	albedo of black daisies	(9.48)
α_g	albedo of bare ground	(9.48)
α_p	planetary albedo	(2.7)
α_s	surface albedo	(4.12)
α_w	albedo of white daisies	(9.48)
β	birth rate of daisies	(9.44)
β	meridional derivative of the Coriolis parameter	(7.11)
β_E	ratio of evapotranspiration to potential evaporation	(5.17)
γ	a heat exchange coefficient	(9.17), (10.11)
Γ	lapse rate	(1.1)
Γ_d	dry adiabatic lapse rate	(C.13)
Γ_s	saturated adiabatic lapse rate	(C.14)
∂	partial differential symbol	(1.1)
δ	γ/B ratio of horizontal transport to longwave cooling coefficients	(9.23)
δ	inverse Ekman depth	(7.4)
δ	solar declination angle	(2.15), (A.1)
$\delta^{18}O$	normalized deviation of $^{18}O/^{16}O$ fraction from normal in ‰	Section 8.5.4
Δ	a prefix to signify a difference	(3.54)
Δ	a prefix to signify a divergence operator	(2.19)
Δf	runoff or divergence of horizontal water flux at and below the surface	(5.1)
Δf_a	divergence of horizontal water flux in the atmosphere	(5.3)
ΔF_{a}	divergence of horizontal heat flux in the atmosphere	(6.1)
ΔF_{ao}	the divergence of the horizontal transport of energy by the atmosphere and ocean	(9.11)
ΔF_{eo}	divergence of the horizontal heat flux below the surface $\approx \Delta F_o$	(4.1)
ΔF_o	divergence of the horizontal heat flux in the ocean	(4.1)

$\vec{\nabla}$ vector gradient operator (7.9), (10.2)
$\vec{\nabla}\bullet$ divergence operator (10.2)
ε emissivity (2.6)
ε_ν emissivity at frequency ν (3.28)
ζ_a vertical component of absolute vorticity $= f + \zeta_r$ (7.10)
ζ_r vertical component of relative vorticity (7.10)
η horizontal transport efficiency parameter (9.51)
θ a general zenith angle (3.4), (3.13)
θ_s solar zenith angle (2.15), (A.1)
Θ potential temperature (4.20), (C.9)
Θ_e equivalent potential temperature (C.15)
Θ_v virtual potential temperature Fig. 4.5
κ von Karmann constant (4.17)
λ longitude Earth coordinate (6.4), Fig. 6.3
λ wavelength of radiation (3.1)
λ_R a gross sensitivity parameter for climate (9.2), (9.6)
Λ longitude of perihelion Fig. 11.8
μ cos θ (3.21)
ν diffusivity (7.1)
ν frequency of radiation (3.1)
ν true anomaly angle of Earth's orbit (11.7)
ξ azimuth angle between south and the position of the sun (A.3)
π pi = 3.141592654 (2.2)
ρ density of air (1.2)
ρ_a density of absorber in air (3.12)
ρ_{as} density of absorber in air at the surface (3.18)
ρ_c density of organic matter in soil (4.8)
ρ_I density of sea ice (10.12)
ρ_o mean density of seawater (7.3)
ρ_s density of soil material (4.8)
ρ_t potential density of seawater Fig. 7.1
ρ_w density of water (4.5), (5.14)
σ pressure normalized by surface pressure, p/p_s (10.1)
σ Stefan–Boltzmann constant, 5.67×10^{-8} W m^{-2} K^{-4} (2.5)
$\sigma(x)$ the standard deviation of x (11.17)
$\dot{\sigma}$ vertical velocity in the sigma coordinate (10.7)
τ optical depth (3.15)
τ a time scale (4.10)
τ_I a time scale for global ice volume (11.8)
τ_o surface stress (4.15)
τ_R response time scale (12.4)
τ_x eastward component of wind stress (7.3)
τ_y northward component of wind stress (7.3)

τ_ν	optical depth at a particular frequency ν (3.33)
$\vec{\tau}$	$= \vec{i}\,\tau_x + \vec{j}\,\tau_y$, wind stress vector (7.9)
φ	an azimuth angle Fig. 3.1, (3.4)
φ_0	a phase angle (11.17)
ϕ	latitude Earth coordinate (2.15), Fig. 6.3
Φ	geopotential height = gz (10.2)
Φ	obliquity angle (11.8), Fig. 11.8
χ	death rate of daisies (9.44)
Ψ_M	meridional mass stream function (6.9)
Ψ_Z	zonal mass stream function (6.21)
ω	angle between Earth, the sun and Earth's position at vernal equinox Fig. 11.8
ω	rate of change of pressure following an air parcel (6.5)
ω	solid angle (3.4), (11.4)
ϖ	single scattering albedo (11.3)
Ω	the rotation rate of Earth = 7.292×10^{-5} s^{-1} (6.16)

Miscellaneous Symbols

$\mathcal{T}\{z, z'\}$	broadband slab transmissivity between altitudes z and z' (3.33), (D.8)
$[x]$	the average of x around a latitude circle, the zonal average of x (6.7)
$\lvert x \rvert$	absolute value of x
$\tilde{T}$	the global average of T (9.18)
$\bar{x}$	the time average of x (6.8)
$\vec{x} \times \vec{y}$	vector cross product of $\vec{x}$ and $\vec{y}$ (10.2)
%	percent
‰	per thousand
°C	degrees Celsius temperature unit
°E	degrees east Earth coordinate
°F	degrees Fahrenheit temperature unit
°N	degrees north Earth coordinate
°W	degrees west Earth coordinate

Appendix F Système Internationale (SI) Units

In most fields of science the SI system of units is the accepted standard. The SI base units are consistent with the metric system of measurement and are given in Table F.1. In addition to the base units, derived units are used in many applications. A selection of SI-derived units frequently used in physical climatology is given in Table F.2.

Supplementary dimensionless units for angles and some derived units using these are given in Table F.3. Prefixes to indicate decimal multiples of units are given in Table F.4. Some non-SI units that are in common use or appear in older references are given in Table F.5 with their SI equivalents.

Table F.1

SI Base Units

Quantity	Name	Symbol
Length	meter	m
Mass	kilogram	kg
Time	second	s
Electric current	ampere	A
Thermodynamic temperature	kelvin	K
Amount of substance	mole	mol
Luminous intensity	candela	cd

Table F.2
Examples of SI-Derived Units

Quantity	Name	Symbol		
		In terms of SI base units	For special name	In terms of other units
Area	square meter	m^2	—	—
Volume	cubic meter	m^3	—	—
Speed, velocity	meter per second	$m\ s^{-1}$	—	—
Acceleration	meter per second squared	$m\ s^{-2}$	—	—
Divergence	per second	s^{-1}	—	—
Vorticity	per second	s^{-1}	—	—
Wavenumber	1 per meter	m^{-1}	—	—
Geopotential; dynamic height	meter squared per second squared	$m^2\ s^{-2}$	—	—
Density	kilogram per cubic meter	$kg\ m^{-3}$	—	—
Specific volume	cubic meter per kilogram	$m^3\ kg^{-1}$	—	—
Luminance	candela per square meter	$cd\ m^{-2}$	—	—
Frequency	hertz	s^{-1}	Hz	—
Force	newton	$m\ kg\ s^{-2}$	N	—
Pressure	pascal	$m^{-1}\ kg\ s^{-2}$	Pa	$N\ m^{-2}$
Energy	joule	$m^2\ kg\ s^{-2}$	J	N m
Power	watt	$m^2\ kg\ s^{-3}$	W	$J\ s^{-1}$
Electric charge	coulomb	s A	C	A s
Electric potential	volt	$m^2\ kg\ s^{-3}\ A^{-1}$	V	$W\ A^{-1}$
Capacitance	farad	$m^{-2}\ kg^{-1}\ s^4\ A^2$	F	$C\ V^{-1}$
Electric resistance	ohm	$m^2\ kg\ s^{-3}\ A^{-2}$	Ω	$V\ A^{-1}$
Conductance	siemens	$m^{-2}\ kg^{-1}\ s^3\ A^2$	S	$A\ V^{-1}$
Magnetic flux	weber	$m^2\ kg\ s^{-2}\ A^{-1}$	Wb	V s
Magnetic flux density	tesla	$kg\ s^{-2}\ A^{-1}$	T	$Wb\ m^{-2}$
Inductance	henry	$m^2\ kg\ s^{-2}\ A^{-1}$	H	$Wb\ A^{-1}$
Luminous flux	lumen	cd sr	lm	—
Illuminance	lux	m^{-2} cd sr	lx	—
Dynamic viscosity	pascal second	$m^{-1}\ kg\ s^{-1}$	—	Pa s
Moment of force	meter newton	$m^2\ kg\ s^{-2}$	—	N m
Surface tension	newton per meter	$kg\ s^{-2}$	—	$N\ m^{-1}$
Heat flux density	watt per square meter	$kg\ s^{-3}$	—	$W\ m^{-2}$
Entropy	joule per kelvin	$m^2\ kg\ s^{-2}\ K^{-1}$	—	$J\ K^{-1}$
Gas constant, universal	joule per kelvin	$m^2\ kg\ s^{-2}\ K^{-1}$	—	$J\ K^{-1}$
Specific heat capacity	joule per kilogram kelvin	$m^2\ s^{-2}\ K^{-1}$	—	$J\ kg^{-1}\ K^{-1}$
Specific energy	joule per kilogram	$m^2\ s^{-2}$	—	$J\ kg^{-1}$
Thermal conductivity	watt per meter kelvin	$m\ kg\ s^{-3}\ K^{-1}$	—	$W\ m^{-1}\ K^{-1}$
Energy density	joule per cubic meter	$m^{-1}\ kg\ s^{-2}$	—	$J\ m^{-3}$

Table F.3

SI Supplementary Units and Derived Units Formed Using Supplemental Units

Quantity	Name	Symbol
Plane angle	radian	rad
Solid angle	steradian	sr
Angular velocity	radian per second	rad s^{-1}
Angular acceleration	radian per second squared	rad s^{-2}
Radiant intensity	watt per steradian	W sr^{-1}
Radiance	watt per square meter steradian	W m^{-2} sr^{-1}

Table F.4

Prefixes for Decimal Multiples and Submultiples of SI Units

Multiple	Prefix	Symbol	Submultiple	Prefix	Symbol
10^{18}	exa	E	10^{-1}	deci	d
10^{15}	peta	P	10^{-2}	centi	c
10^{12}	tera	T	10^{-3}	milli	m
10^{9}	giga	G	10^{-6}	micro	μ
10^{6}	mega	M	10^{-9}	nano	n
10^{3}	kilo	k	10^{-12}	pico	p
10^{2}	hecto	h	10^{-15}	femto	f
10^{1}	deka	da	10^{-18}	atto	a

Table F.5

Non-SI Units Commonly Used in Current or Past Literature with Conversion Factors

	Name	Symbol	Value in SI Unit
Time	minute	min	1 min = 60 s
	hour	h	1 h = 60 min = 3600 s
	day	d	1 d = 24 h = 86,400 s
Distance	nautical mile	n mi	1 n mi = 1852 m
	knot	kt	1 kt = 1 n mi h^{-1} = (1852/3600) m s^{-1}
	ångström	Å	1 Å = 0.1 nm = 10^{-10} m
	mile (USA, statute)	mi	1 mi = 1609.3 m
	foot	ft	1 ft = 0.3048 m
Mass	pound	lb	1 lb = 0.4336 kg
	metric ton	t	1 t = 10^3 kg
Area	are	a	1 a = 1 dam^2 = 10^2 m^2
	acre	acre	1 acre = 4046.8 m^2
	hectare	ha	1 ha = 1 hm^2 = 10^4 m^2
Volume	liter	l	1 l = 1 dm^3 = 10^{-3} m^3
	gal	gal	1 gal = 3.785 × 10^{-3} m^3
Angle	degree	°	1° = (π/180) rad
	minute	'	1' = (1/60)° = (π/10,800) rad
	second	"	1" = (1/60)' = (π/648,000) rad
Force	dyne	dyn	1 dyn = 10^{-5} N
Energy	calorie	cal	1 cal = 4.186 J
	British thermal unit	Btu	1 Btu =1054.6 J
	erg	erg	1 erg = 10^{-7} J
Pressure	bar	b	1 b = 0.1 MPa = 10^5 Pa
	standard atmosphere	atm	1 atm = 101,325 Pa
Water flow	Sverdrup	Sv	1 Sv = 10^6 m^3 seawater s^{-1}

Appendix G Useful Numerical Values

Fundamental constants	
Universal gas constant (R^*)	$8.3143\ J\ K^{-1}\ mol^{-1}$
Boltzmann's constant (k)	$1.38 \times 10^{-23}\ J\ K^{-1}$
Stefan–Boltzmann constant (σ)	$5.67 \times 10^{-8}\ W\ m^{-2}\ K^{-4}$
Planck's constant ($\hbar$)	$6.63 \times 10^{-34}\ J\ s$
Speed of light (c^*)	$2.998 \times 10^{8}\ m\ s^{-1}$
Gravitational constant	$6.67 \times 10^{-11}\ Nt\ m^{2}\ kg^{-2}$
Sun	
Luminosity	$3.92 \times 10^{26}\ W$
Mass	$1.99 \times 10^{30}\ kg$
Radius	$6.96 \times 10^{8}\ m$
Earth	
Average radius (a)	$6.37 \times 10^{6}\ m$
Equatorial radius	$6.378 \times 10^{6}\ m$
Polar radius	$6.357 \times 10^{6}\ m$
Standard gravity (g)	$9.80665\ m\ s^{-2}$
Mass of Earth	$5.983 \times 10^{24}\ kg$
Mass of ocean	$1.4 \times 10^{21}\ kg$
Mass of atmosphere	$5.3 \times 10^{18}\ kg$
Mean angular rotation rate (Ω)	$7.292 \times 10^{-5}\ rad\ s^{-1}$
Solar constant (S_o)	$1367 \pm 2\ W\ m^{-2}$
Mean distance from sun (d)	$1.496 \times 10^{11}\ m$
Dry air	
Average molecular weight (m_a)	$28.97\ g\ mol^{-1}$
Gas constant (R) – R^*/m_a	$287\ J\ K^{-1}\ kg^{-1}$
Density at 0°C and 101,325 Pa	$1.293\ kg\ m^{-3}$
Specific heat at constant pressure (c_p)	$1004\ J\ K^{-1}\ kg^{-1}$
Specific heat at constant volume (c_v)	$717\ J\ K^{-1}\ kg^{-1}$

Water

Molecular weight (m_w)	18.016 g mol^{-1}
Gas constant for vapor ($R_v = R^*/m_w$)	461 J K^{-1} kg^{-1}
Density of pure water at 0°C	1000 kg m^{-3}
Density of ice at 0°C	917 kg m^{-3}
Specific heat of vapor at constant pressure	1952 J K^{-1} kg^{-1}
Specific heat of vapor at constant volume	1463 J K^{-1} kg^{-1}
Specific heat of liquid water at 0°C	4218 J K^{-1} kg^{-1}
Specific heat of ice at 0°C	2106 J K^{-1} kg^{-1}
Latent heat of vaporization at 0°C	2.5×10^6 J kg^{-1}
Latent heat of vaporization at 100°C	2.25×10^6 J kg^{-1}
Latent heat of fusion at 0°C	3.34×10^5 J kg^{-1}

Appendix H Answers to Selected Exercises

Chapter 1

3. Increases by ~670 Pa

Chapter 2

2. 233 K
4. 255 K
6. 42.8° summer, 81.3° winter at 47°N
8. 303 K conducting, 361 K nonconducting

Chapter 3

1. 30 km
5. The model convective adjustment flux is 159 W m^{-2} from surface to the lower layer, 172 W m^{-2} from lower layer to the upper layer; net surface longwave loss is 80.5 W m^{-2}.
7. 194 K, 21 km

Chapter 4

1. 270 W m^{-2}, 26 W m^{-2}
3. 5.4 W m^{-2} K^{-1} longwave; 12 W m^{-2} K^{-1} sensible flux.
5. 50.6°C dry asphalt; 33°C wet asphalt for $B_e = 0.25$.

Chapter 5

1. 69 years; 1925 years
2. (a) 0.11 mm day^{-1}; (b) 0.028 mm day^{-1}; (c) 0.78 mm day^{-1}; (d) 0.167 mm day^{-1}

Chapter 6

4. –54 m s^{-1} (easterly)
6. $w = -2 \times 10^{-4}$ m s^{-1}

Chapter 7

1. Need $S \approx 35‰ \approx$ average
2. ~13 m of sea ice
5. $\rho_t = 22.5$ becomes $\rho_t = 28.2$

Chapter 8

5. North America ~10^{19} kg, Greenland ~10^{18} kg
7. From 34 to 33‰

Chapter 9

3. The static stability will increase more than the meridional gradient.

Chapter 10

1. (a) $T_s = -2$°C, SH = 180 W m^{-2}; (b) $T_s = -23.9$°C, SH = 44 W m^{-2}; (c) $T_s = -27.6$°C, SH = 17 W m^{-2}
3. (a) 9 m; (b) 4.5 m
5. ~3 m s^{-1} in atmosphere, ~3×10^{-3} m s^{-1} in ocean

Chapter 11

2. 222 W m^{-2}, 3.7 km
4. The difference is about 110 W m^{-2} or 23%, with a larger contribution from precession (~16%) than from obliquity (~7%).

Chapter 12

1. 1°C, 4°C, 25%
4. (a) $\varepsilon = 10^{-4}$, (b) 276 K.

Glossary

absolute vorticity	The sum of the planetary vorticity, which is associated with the rotation of Earth, and relative vorticity, which is associated with the fluid motion relative to the surface of Earth.
absorption	The annihilation of a photon and an equivalent release of energy.
absorption line	A discrete frequency corresponding to an energy transition of a molecule.
aerosol	A suspension of liquid or solid matter in air.
albedo	The fraction of incoming radiation that is reflected.
aphelion	The point on the orbit of Earth that is farthest from the sun.
atmospheric boundary layer	The lowest part of the troposphere where the wind, temperature, and humidity are strongly influenced by the surface.
backscattering	Scattering toward the direction of the source of the incident beam.
biogeochemical	Having to do with the interaction between chemical and biological processes on Earth.
bioturbation	The stirring of sediment by animal life.
blackbody radiation	The electromagnetic radiation emitted by an ideal blackbody. No actual substance behaves like a true blackbody, although platinum black and other soots come close. Hypothetically, it is a body that absorbs all of the electromagnetic radiation that strikes it, neither reflecting nor transmitting any of the incident radiation.
Bowen ratio	The ratio of sensible to latent heating of the surface.
bucket model	A simple model for the soil water budget in which the surface is represented by a field of buckets of specified depth.
climate	The synthesis of weather in a particular region.

climate sensitivity — The relationship between the measure of climate forcing and the magnitude of the climate change.

cloud condensation nuclei — Aerosol particles on which molecules of vapor may condense.

conduction — Heat transport in which a medium is required to transfer heat by collisions between atoms and molecules. No mass is exchanged.

convection — Heat transfer in which mass is exchanged. A net movement of mass may occur, but more commonly parcels with different energy amounts change places, so that energy is exchanged without a net movement in mass.

convective adjustment — A method used to simulate the effect of unresolved convective motions in a climate model.

convective turbulence — Generated when warm fluid parcels are accelerated upward by their buoyancy.

Coriolis parameter — ($f = 2\Omega \sin \phi$) A measurement of twice the local vertical component of the rotation rate of a spherical planet.

cryosphere — Ice near the surface of Earth, including glaciers, snowcover, and sea ice.

declination angle — (See *season.*) The angle between the equator and the subsolar point at noon.

dynamical — Having to do with the equations of motion or variables contained therein.

eccentricity — The measure of how much the planetary orbit deviates from being perfectly circular. It controls the amount of variation of the solar flux density at the planet as it moves through its orbit during the planetary year.

eddy — Deviations of flow from the time or zonal average.

Ekman layer — A layer of transition near the surface layer where Coriolis and frictional forces are both important in the momentum balance.

Ekman spiral — An idealized mathematical description of the velocity distribution in the boundary layer of the atmosphere or ocean, within which rotation and friction jointly determine the velocity profile.

El Niño — A name given to the event when anomalously warm surface

	waters appear near the west coast of equatorial South America.
El Niño–Southern Oscillation (ENSO) phenomenon	A joint reference made to the related oceanic and atmospheric variations that accompany warm and cold events in the equatorial Pacific.
electronic excitation	A process by which photons excite the electrons in the outer shells of an atom.
emission	The creation of a photon and a corresponding loss of energy by the emitting body.
emission temperature	The temperature at which a blackbody will emit energy at a specified rate.
energy balance	A condition in which a system neither gains nor loses energy.
energy budget	The measure of energy entering and leaving a system such as Earth's climate system.
energy flux density	The energy delivered per unit time per unit area (W m^{-2}).
equilibrium time	The time required for a system to establish a new steady state after the application of a stimulus or a change in external conditions.
evaporation	A change of state from liquid to vapor.
evapo-transpiration	The sum of evaporation and transpiration.
extinction	The sum of scattering and absorption of a direct beam of radiation.
faculae	Bright regions on the sun.
faint young sun problem	The combination of a less luminous sun and a warm climate on Earth during the early history of the solar system.
feedback mechanism	A process that changes the sensitivity of the climate response.
Ferrel cell	Weaker meridional cells in midlatitudes that circulate in the direction opposite to that of the Hadley cell. Ferrel cells are thermodynamically indirect, as they transport energy from cold areas to warm areas.

field capacity — The maximum volume fraction of water that the soil can retain against gravity.

forward-scattering — Scattering in the same direction as the incident beam.

free atmosphere — The portion of the atmosphere above the planetary boundary layer in which the effect of surface friction on the air motion is weak.

Gaia — The concept of Earth as a single complex entity involving the biosphere, atmosphere, ocean, and land.

GCM — General circulation model; but sometimes means Global climate model.

general circulation — The global system of atmospheric or oceanic motions.

greenhouse effect — A condition in which the atmosphere warms the surface by being relatively transparent to solar radiation and absorbing and emitting terrestrial radiation very effectively.

Hadley cell — A meridional cell in which air rises near the equator, flows toward the pole at upper levels, and sinks in the subtropical latitudes.

halocarbon — Compounds of carbon with halogens such as chlorine, bromine, and iodine.

heat capacity — The amount of energy that is required to raise the temperature by one degree.

hour angle — The longitude of the subsolar point relative to its position at noon.

hydrologic cycle — The movement of water among the reservoirs of ocean, atmosphere, and land.

hydrostatic balance — A condition in which the gravity force that pulls atmospheric molecules toward the center of the planet is equal to the pressure-gradient force that pushes them out into space.

infiltration — The transfer of surface water to the soil.

instrumental record — The past history of climate as directly measured using instruments, such as the thermometer and barometer.

internal energy — Energy associated with the temperature of the atmosphere.

Intertropical Convergence Zone (ITCZ)	The axis, or a portion thereof, of the broad tradewind current of the tropics where the northerly and southerly tradewinds meet, either in narrow bands or in the broader convergence zones over Indonesia, South America, and Africa.
irradiance	Radiant energy flux density.
isothermal	Of equal or constant temperature, with respect to either space or time.
isotropic scattering	Scattering that is equally probable in all directions.
Jovian planets	The planets of the solar system with physical characteristics similar to Jupiter, which include Jupiter, Saturn, Uranus, and Neptune.
K-T boundary	The time period marking the end of the Cretaceous and the beginning of the Tertiary epochs about 65 million years ago.
Lambert–Bouguet–Beer law of extinction	Absorption is linear in the intensity of radiation and the absorber amount.
lapse rate	Rate of decline of temperature with height in the atmosphere, defined by $\Gamma = -\partial T/\partial z$.
leaf area index (LAI)	The ratio of the area of the horizontal projection of the top sides of all the leaves in the canopy to the surface area.
line broadening	An increase in the range of frequencies that can be absorbed or emitted during a particular molecular energy transition. The broadening mechanisms are natural, pressure (collision), and Doppler broadening.
lithosphere	The outer, solid portion of Earth, also known as Earth's crust.
Little Ice Age	A period of expansion of mountain glaciers from about 1350 to 1800 in the Alps, Norway, Iceland, Alaska, and probably elsewhere.
longwave radiation	Thermal radiation emitted by Earth which has wavelengths between about 4 and 200 μm.
luminosity	The rate at which energy is released from the sun by radiation.
Maunder minimum	The period between 1645–1715 when sunspots were very few in number.

meridional	North–south, along a line of longitude.
meridional wind	The wind, or wind component, along the local north–south meridian, usually defined positive towards the north.
mesosphere	The atmospheric layer extending from the top of the stratosphere to the upper temperature minimum (the mesopause).
monsoon	A name for a wind system that changes in speed and direction with season.
nuclear fusion	The process whereby lighter elements combine to form heavier ones, releasing energy in the process.
numerical integration	The process of solving the equations of a numerical model on a computer.
numerical model	A model expressed in mathematical formulas and solved approximately on a computer.
numerical simulation	The sequence of states of the climate system as represented by a computer model, or the statistics of such a sequence.
obliquity	The angle between the axis of rotation of a planet and the normal to the plane of its orbit.
oceanic mixed layer	The top 20–200 m of water in contact with the atmosphere in which water properties are almost independent of depth because of rapid turbulent mixing.
outgassing	The process of releasing gases from the interior of a planet.
paleoclimatic record	Climate variables that are indirectly measured using physical, biological, and chemical information contained on land, in lake and ocean sediments, and in ice sheets.
parameterization	A process by which the effects of sub-grid-scale phenomena are specified from the knowledge of the variables at the grid scale of a climate model.
photodissociation	A process by which an energetic photon breaks the bond that holds together the atoms of a molecule.
photoionization	A process by which a photon removes electrons from the outer shells surrounding the nucleus, producing ionized atoms and free electrons.
photolysis	Chemical decomposition due to the interaction of a photon with a molecule.

photosphere	The region of the sun from which most of its energy emission is released to space.
Planck's law of blackbody radiation	An expression for the variation of monochromatic radiation emissive power of a blackbody as a function of wavelength and temperature.
planetary albedo	The fraction of incoming solar energy reflected back to space.
porosity of soil	The volumetric fraction of the soil that can be occupied by air or water.
potential density	The density that sea water with a particular salinity and temperature would have at zero water pressure, or the density at surface air pressure.
potential energy	Energy associated with the gravitational potential of air some distance above the surface.
potential evaporation	The rate of evaporation that would occur if the surface was wet.
precession parameter	($e \sin \Lambda$) The critical parameter that describes the influence of eccentricity and longitude of perihelion on northern summer insolation.
precipitation	Any or all forms of water particles, whether liquid or solid, that fall from the atmosphere and reach the surface.
primitive equations	A simplified form of the equations of motion used by most global climate models.
radiation	Energy transport in which no medium is required and no mass is exchanged. Pure radiant energy moves at the speed of light.
radiative cloud forcing	Change of the radiative energy flux caused by the presence of liquid water and ice in the atmosphere.
relative humidity	The dimensionless ratio of the actual water vapor mixing ratio of the air to the saturation mixing ratio.
Richardson number	A scaling parameter measuring the ratio of buoyancy to inertial forces in a fluid.
rotational energy	The energy associated with rotation (e.g., of a molecule).
salinity	Number of grams of dissolved salts in a kilogram of sea water.

scattering	An interaction between an object and radiation in which the radiation changes direction without a change in energy.
season	Expressed in terms of the declination angle of the sun, which is the latitude of the point on the surface of the Earth directly under the sun at noon.
shadow area	The area that a body sweeps out of a beam of parallel energy flux. (See Figs. 2.1 and 2.4.)
shortwave radiation	Solar radiation which has wavelengths between about 0.2 and 4 μm.
sigma coordinate	$\sigma = (p/p_s)$. A vertical coordinate that is equal to pressure normalized by surface pressure.
sink	A point, line, or area where energy or mass is removed from a system.
soil water zone	The region that extends downward to the depth penetrated by the roots of the vegetation.
solar constant	The energy flux density of the solar emission at a particular distance. Also known as total solar irradiance.
solar zenith angle	The angle between the local normal to Earth's surface and a line between a point on Earth's surface and the sun. It depends on latitude, season, and time of day.
source	A point, line, or area where mass or energy is added to a system, either instantaneously or continuously.
South Pacific convergence zone (SPCZ)	The band of convection rooted near Indonesia and extending southeastward over the South Pacific Ocean that is most prominent in southern summer.
specific humidity	The dimensionless ratio of the mass of water vapor to the total mass of dry air, often given in units of grams of water vapor per kilogram of dry air.
SST	Sea surface temperature.
stationary eddy fluxes	Fluxes of heat, mass, or momentum associated with stationary waves.
stationary planetary waves	East–west variations of the time-average state of the atmosphere.
storm track	The path followed by a center of low atmospheric pressure. Also the axis along which such midlatitude systems frequently travel.

stratosphere The atmospheric layer above the tropopause and below the mesosphere where the temperature generally increases with height.

sub-grid-scale phenomena Phenomena that occur at scales smaller than the grid resolution of a climate model.

sublimation The direct conversion of snow and ice to water vapor, without an intermediate liquid phase.

subsolar point The point on Earth's surface that falls on a line between the center of Earth and the center of the sun.

sunspots Dark spots seen on the sun's photosphere.

surface albedo The fraction of the downward solar flux density that is reflected by the surface.

surface energy budget The energy flux per unit area passing through the surface boundary of the atmosphere, measured in watts per square meter, and apportioned between radiative, sensible, and latent fluxes, storage in the surface, and horizontal energy transport below the surface.

surface layer The thin layer of air adjacent to Earth's surface, where surface friction dominates the momentum balance and vertical fluxes are almost independent of height above the surface.

T-Tauri solar wind The intensified solar wind associated with a young star.

temperature inversion A region of negative lapse rate where the temperature increases with altitude.

terrestrial planets The planets of the solar system whose physical characteristics most resemble Earth, which include Mercury, Venus, Mars, and Earth.

thermocline A layer of thermally stratified water about 1 km deep between the warmer surface layer of the ocean and the colder, deeper layer, both of which are of almost uniform temperature.

thermosphere The atmospheric layer extending from the top of the mesosphere to outer limits of the atmosphere in which temperature increases with height.

transient eddy fluxes Fluxes associated with rapidly developing weather disturbances, especially in midlatitudes.

translational energy The energy associated with the movement of molecules or atoms through space (kinetic energy, temperature).

transmission — An interaction between an object and radiation in which the radiation passes the object without absorption or reflection.

transpiration — The release of moisture through the surface of leaves or other parts of plants.

tropopause — The altitude where the positive temperature lapse rate of the troposphere changes to the weak or negative lapse rate of the stratosphere.

troposphere — The atmospheric layer from Earth's surface to the tropopause; that is, the lowest 10–20 km of the atmosphere where the temperature generally decreases with altitude.

true anomaly — (v) The angle formed at the center of the sun between Earth and perihelion.

turnover time — The time required for the fluxes through a reservoir to completely replace the content of the reservoir.

upwelling, downwelling — Upward or downward mean vertical motion in the ocean.

vibrational energy — The energy associated with rapid variations of the interatomic distances within molecules.

vorticity — The curl of the velocity vector and a measure of the local rotation of a fluid.

Walker circulation — The largest circulation cell oriented along the equator with rising in the Indonesian region and sinking to the east and west.

Wien's law of displacement — The frequency of maximum emission is proportional to temperature.

Younger Dryas — A cold event returning Europe to nearly glacial conditions about 11,000 years ago.

zonal — East–west, along a line of latitude.

zonal wind — The wind, or wind component, along the local parallel of latitude, usually defined positive toward the east.

References

Allen, C. W., 1973: *Astrophysical Quantities,* 3rd ed. Athlone Press, London, 291 pp.

Alvarez, L. W., 1987: Mass extinctions caused by large bolide impacts. *Physics Today,* **40,** 24–33.

Andreae, M. O., 1990: Ocean-atmosphere interactions in the global biogeochemical sulfur cycle. *Mar. Chem.,* **30,** 1–29.

Andreae, M. O., R. J. Ferek, F. Bermond, K. P. Byrd, R. T. Engstrom, S. Hardin, P. D. Houmere, F. LeMarrec, H. Raemdonck, and R. B. Chatfield, 1985: Dimethyl sulfide in the marine atmosphere. *J. Geophys. Res.,* **90,** 12,891–12,900.

Angell, J. K., 1988: Variations and trends in tropospheric and stratospheric global temperatures: 1958–1987. *J. Clim.,* **1,** 1296–1313.

Arakawa, A., and W. H. Schubert, 1974: Interaction of a cumulus cloud ensemble with the large-scale environment: Part I. *J. Atmos. Sci.,* **31,** 674–701.

Arrhenius, S., 1896: On the influence of carbonic acid in the air upon the temperature of the ground. *Phil. Mag.,* **41,** 237–276.

Arya, S. P., 1988: *Introduction to Micrometeorology.* Academic Press, Orlando, FL, 307 pp.

Ash, M. E., I. I. Shapiro, and W. B. Smith, 1967: Astronomical constants and planetary ephemerides deduced from radar and optical observations. *Astro. J.,* **72,** 338–350.

Baliunas, S., and R. Jastrow, 1990: Evidence for long-term brightness changes of solar-type stars. *Nature,* **348,** 520–522.

Bambach, R. K., C. R. Scotese, and A. M. Ziegler, 1981: Before Pangea: The geographies of the Paleozoic world. In *Climates Past and Present,* B. J. Skinner, ed., Kaufmann, Los Altos, CA, pp. 46–60.

Barnett, T. P., 1988: Global sea level change. In *NCPO, Climate Variations over the Past Century and the Greenhouse Effect.* National Climate Program Office/NOAA, Rockville, MD.

Barnola, J. M., D. Raynaud, Y. S. Korotovitch, and C. Lorius, 1987: Vostok ice core provides 160,000 year record of atmospheric CO_2. *Nature,* **329,** 408–414.

Barron, E. J., and W. H. Peterson, 1990: Mid-Cretaceous ocean circulation: Results from model sensitivity studies. *Paleoceanography,* **5,** 319–337.

Barron, E. J., and W. M. Washington, 1985: Warm cretaceous climates: High atmospheric CO_2 as a plausible mechanism. In *The Carbon Cycle and Atmospheric CO_2: Natural Variations Archean to Present,* E. T. Sundquist and W. S. Broecker, eds., Vol. 32, American Geophysical Union, pp. 546–553.

Baumgartner, A., and E. Reichel, 1975: *The World Water Balance: Mean Annual Global, Continental and Maritime Precipitation and Run-Off.* Elsevier Scientific Publishers, Amsterdam, 179 pp., 31 maps.

Berger, A. L., 1978: Long-term variations of daily insolation and Quaternary climate changes. *J. Atmos. Sci.,* **35,** 2362–2367.

Berner, R. A., 1993: Paleozoic atmospheric CO_2: Importance of solar radiation and plant evolution. *Science,* **261,** 68–70.

Blackmon, M. L., 1977: An observational study of the Northern Hemisphere winter time circulation. *J. Atmos. Sci.,* **34,** 1040–1053.

Bond, G., H. Heinrich, W. S. Broecker, L. Labeyrie, J. McManus, J. Andrews, S. Huon, R. Jantschik, S. Clasen, C. Simet, K. Tedesco, M. Klas, G. Bonani, and S. Ivy, 1992: Evidence for massive discharges of icebergs into the North Atlantic Ocean during the last glacial period. *Nature,* **360,** 245–249.

Boyle, E. A., and L. Keigwin, 1987: North Atlantic thermohaline circulation during the past 20,000 years linked to high-latitude surface temperature. *Nature,* **330,** 35–40.

Bras, R. L., 1990: *Hydrology.* Addison-Wesley, Reading, MA, 643 pp.

Broccoli, A. J., and S. Manabe, 1987: The influence of continental ice, atmospheric CO_2, and land albedo on the climate of the last glacial maximum. *Clim. Dyn.,* **1,** 87–99.

Broecker, W. S., and G. H. Denton, 1990: What drives glacial cycles? *Sci. Am.,* **262,** 48–56.

Broecker, W. S., G. Bond, M. Klas, E. Clark, and J. McManus, 1992: Origin of the northern Atlantic's Heirich events. *Clim. Dyn.,* **6,** 265–273.

Broecker, W. S., M. Andreae, W. Wolfli, H. Oescher, G. Bonani, J. Kennett, and D. Peteet, 1988: The chronology of the last deglaciation: Implications to the cause of the Younger Dryas event. *Paleoceanography,* **3,** 1–19.

Brown, R. A., 1991: *Fluid Mechanics of the Atmosphere.* Academic Press, San Diego, 489 pp.

Brutsaert, W., 1982: *Evaporation into the Atmosphere.* Reidel, Dordrecht, Netherlands, 299 pp.

Bryan, K., 1991: Poleward heat transport in the ocean: A review of a hierarchy of models of increasing resolution. *Tellus,* **43A,** 104–115.

Bryan, K., S. Manabe, and M. J. Spelman, 1988: Interhemispheric asymmetry in the transient response of a coupled ocean-atmosphere model to a CO_2 forcing. *J. Phys. Ocean.,* **18,** 851–867.

Bryden, H. L., and M. M. Hall, 1980: Heat transport by currents across 25°N latitude in the Atlantic Ocean. *Science,* **207,** 884–886.

Budyko, M. I., Ed., 1963: *Atlas Teplovogo Balansa.* Gidrometeorologicheskoe, Izdatelszo, Moscow, 69 pp.

Budyko, M. I., 1969: The effects of solar radiation on the climate of the earth. *Tellus,* **21,** 611–619.

Cerling, T. E., 1991: Carbon dioxide in the atmosphere: Evidence from Cenozoic and Mesozoic paleosols. *Am. J. Sci.,* **291,** 377–400.

Cess, R. D., G. L. Potter, J. P. Blanchet, G. J. Boer, A. D. Del Genio, M. Déqué, V. Dymnikov, V. Galin, W. L. Gates, S. J. Ghan, J. T. Kiehl, A. A. Lacis, H. Le Treut, Z. -X. Li, X. -Z. Liang, B. J. McAvaney, V. P. Meleshko, J. F. B. Mitchell, J. -J. Morcrette, D. A. Randall, L. Rikus, E. Roeckner, J. F. Royer, U. Schlese, D. A. Sheinin, A. Slingo, A. P. Sokolov, K. E. Taylor, W. M. Washington, R. T. Wetherald, I. Yaga, and M.-H. Zhang, 1990: Intercomparison and interpretation of climate feedback processes in seventeen atmospheric general circulation models. *J. Geophys. Res.,* **95,** 16,601–16,615.

Chamberlain, J. W., and D. M. Hunten, 1987: *Theory of Planetary Atmospheres,* 2nd ed., Academic Press, San Diego, 481 pp.

Charlson, R. J., 1988: Have the concentrations of tropopheric aerosol particles changed? In *Physical, Chemical, and Earth Sciences Research Reports.* F. S. Rowland and I.S.A. Isaksen, eds., Wiley-Interscience, Chichester, UK, pp. 79–90.

Charlson, R. J., J. E. Lovelock, M. O. Andreae, and S. G. Warren, 1987: Atmospheric sulphur: Geophysiology and climate. *Nature,* **326,** 655–661.

Charlson, R. J., J. Langner, H. Rodhe, C. B. Leovy, and S. G. Warren, 1991: Perturbation of the Northern Hemisphere radiation balance by backscattering from anthropogenic sulfate aerosols. *Tellus,* **43AB,** 152–163.

Charlson, R. J., S. E. Schwartz, J. M. Hales, R. D. Cess, J. A. Coakley, J. E. Hansen, and D. J. Hofmann, 1992: Climate forcing by anthropogenic aerosols. *Science,* **255,** 423–430.

Charney, J. G., 1975: Dynamics of deserts and drought in the Sahel. *Quart. J. Roy. Met. Soc.,* **101,** 193–202.

Charney, J. G., R. Fjortoft, and J. von Neumann, 1950: Numerical integration of the barotropic vorticity equation. *Tellus,* **2,** 237–254.

CLIMAP Project Members, 1976: The surface of the ice age earth. *Science,* **191,** 1131–1137.

CLIMAP Project Members, 1981: *Seasonal Reconstruction of the Earth's Surface at the Last Glacial Maximum.* Geological Society of America, maps, microfiche.

Coakley, J. A. J., R. L. Bernstein, and P. A. Durkee, 1987: Effect of ship track effluents on cloud reflectivity. *Science,* **237,** 1020 pp.

COHMAP Members, 1988: Climate changes of the last 18,000 years. *Science,* **241,** 1043–1052.

Colin, C., C. Henin, P. Hisard, and C. Oudot, 1971: Le courant de Cromwell dans le Pacifique central de février. *Cah. ORSTOM, Ser. Oceanogr.,* **9,** 167–186.

Crawford, K. C., and H. R. Hudson, 1973: The diurnal wind variation in the lowest 1500 feet in central Oklahoma, June 1966–May 1967. *J. Appl. Meteorol.,* **12,** 127–132.

Croll, J., 1890: *Climate and Time in their Geological Relations: A Theory of Secular Changes of the Earth's Climate,* 2nd ed. Appleton, New York, 577 pp.

Crowley, T. J., 1983: The geologic record of climate change. *Rev. Geophys. Sp. Phys.,* **21,** 828–877.

Crowley, T. J., and G. R. North, 1991: *Paleoclimatology.* Oxford University Press, Oxford, 339 pp.

Crutcher, H. L., and J. M. Meserve, 1970: *Selected Level Heights, Temperatures, and Dew Points for the Northern Hemisphere.* NAVAIR 50-1C-52, U.S. Naval Weather Service, Washington DC.

de Bergh, C., 1993: The D/H ratio and the evolution of water in the terrestrial planets. *Orig. Life. Evol. Biol.,* **23,** 11–21.

Delworth, T. L., and S. Manabe, 1988: The influence of potential evaporation in the variabilities of simulated soil wetness and climate. *J. Clim.,* **1,** 523–547.

Denman, K. L., and M. Miyake, 1973: Upper-layer modification at Ocean Station *Papa:* Observations and simulation. *J. Phys. Ocean.,* **3,** 185–196.

Denton, G. H., and W. Karlén, 1973: Holocene climatic changes, their pattern, and possible cause. *Quat. Res.,* **3,** 155–205.

de Vaucouleurs, G., 1964: Geometric and photometric parameters of the terrestrial plants. *Icarus,* **3,** 187–235.

Dickinson, R. E., 1983: Land-surface processes and climate: Surface albedos and energy balance. *Adv. Geophys.,* **25,** 305–353.

Dickinson, R. E., 1984: Modeling evapotranspiration for three-dimensional global climate models. *Geophys. Monogr.* J. E. Hansen and T. Takahashi, Eds.; American Geophysical Union, **29,** 58–72.

Dickinson, R. E., 1991: Global change and terrestrial hydrology: A review. *Tellus,* **43AB,** 176–181.

Dickinson, R. E., and A. Henderson-Sellers, 1988: Modeling tropical deforestation: A study of GCM land-surface parameterizations. *Quart. J. Roy. Meteorol. Soc.,* **114,** 439–462.

Dietrich, G., K. Kalle, W. Kraus, and G. Siedler, 1980: *General Oceanography: An Introduction.* Wiley-Interscience, New York, 626 pp.

Dole, S. H., 1970: *Habitable Planets for Man,* 2nd ed., Elsevier Scientific Publishers, Dordrecht, Netherlands, 158 pp.

Dreidonks, A. G. M., and H. Tennekes, 1984: Entrainment effects in the well-mixed atmospheric boundary layer. *Bound. Layer Meteorol.,* **30,** 75–105.

Duvick, D. N., and T. J. Blasing, 1981: A dendroclimatic reconstruction of annual precipitation amounts in Iowa since 1680. *Water Resources Res.,* **17,** 1183–1189.

Eagleman, J. R., 1976: *The Visualization of Climate.* Lexington Books, Lexington, MA, 227 pp.

Eddy, J. A., 1976: The maunder minimum. *Science,* **192,** 1189–1202.

Fein, J. S., and P. L. Stephens, 1987: *Monsoons.* Wiley, New York, 632 pp.

Folland, N. P., T. N. Palmer, and D. E. Parker, 1986: Sahel rainfall and worldwide sea surface temperatures: 1901–1985. *Nature,* **312,** 602–607.

Foukal, P., and J. Lean, 1990: An empirical model of total irradiance variations between 1884 and 1988. *Science,* **247,** 556–558.

Fouquart, Y., J. C. Buriez, M. Herman, and R. S. Kandel, 1990: The influence of clouds on radiation: A climate-modelling perspective. *Rev. Geophys.,* **28,** 145–166.

Friedli, H., H. Lotscher, H. Oeschger, U. Seigenthaler, and B. Stauffer, 1986: Ice core record of the $^{13}C/^{12}C$ record of atmospheric CO_2 in the past two centuries. *Nature,* **324,** 237–238.

Gill, A. E., 1982: *Atmosphere–Ocean Dynamics.* Academic Press, Orlando, FL, 662 pp.

Gleissberg, W., 1944: A table of secular variations of the solar cycle. *Terrest. Magnet. Atmos. Elect.,* **49,** 243–244.

Goody, R. M., and J. C. G. Walker, 1972: *Atmospheres.* Prentice-Hall, Englewood Cliffs, NJ, 150 pp.

Goody, R. M., and Y. L. Yung, 1989: *Atmospheric Radiation.* Oxford University Press, New York, 519 pp.

Griffith, K. T., S. K. Cox, and R. G. Knollenberg, 1980: Infrared radiative properties of tropical cirrus clouds inferred from aircraft measurements. *J. Atmos. Sci.,* **37,** 1077–1087.

Hansen, J., and S. Lebedeff, 1988: Global surface air temperatures: Update through 1987. *Geophys. Res. Lett.,* **15,** 323–326.

Hansen, J., A. Lacis, R. Ruedy, and M. Sato, 1992: Potential climate impact of the Mount Pinatubo eruption. *Geophys. Res. Lett.,* **19,** 215–218.

Hansen, J., A. Lacis, D. Rind, G. Russell, P. Stone, I. Fung, R. Ruedy, and J. Lerner, 1984: Climate sensitivity: Analysis of feedback mechanisms. In *Climate Processes and Climate Sensitivity,* J. E. Hansen and T. Takahashi, eds., Vol. 29, American Geophysical Union, pp. 130–163.

Hansen, J. E., I. Fung, A. Lacis, D. Rind, S. Lebedeff, R. Ruedy, G. Russell, and P. Stone, 1988: Global climate changes as forecast by the Goddard Institute of Space Studies three-dimensional model. *J. Geophys. Res.,* **93,** 9341–9364.

Harrison, E. F., P. Minnis, B. R. Barkstrom, V. Ramanathan, R. C. Cess, and G. G. Gibson, 1990: Seasonal variation of cloud radiative forcing derived from the Earth Radiation Budget Experiment. *J. Geophys. Res.,* **95,** 18,687–18,703.

Harshvardhan, D. A. Randall, T. G. Corsetti, and D. A. Dazlich, 1989: Earth radiation budget and cloudiness simulations with a general circulation model. *J. Atmos. Sci.,* **46,** 1922–1942.

Hartmann, D. L., 1993: Radiative effects of clouds on Earth's climate. In *Aerosol–Cloud–Climate Interactions,* P. V. Hobbs, ed., Academic Press, San Diego, pp. 151–170.

Hartmann, D. L., and D. A. Short, 1979: On the role of zonal asymmetries in climate change. *J. Atmos. Sci.,* **36,** 519–528.

Hartmann, D. L., and M. L. Michelsen, 1993: Large-scale effects on the regulation of tropical sea surface temperature. *J. Clim.,* **6,** 2049–2062.

Hartmann, D. L., M. E. Ockert-Bell, and M. L. Michelsen, 1992: The effect of cloud type on Earth's energy balance: Global analysis. *J. Climate,* **5,** 1281–1304.

Hays, J. D., J. Imbrie, and N. J. Shackleton, 1976: Variations in the Earth's orbit: Pacemaker of the ice ages. *Science,* **194,** 1121–1132.

Hecht, A. D., ed., 1985: *Paleoclimate Analysis and Modeling.* Wiley, New York, 445 pp.

Held, I. M., 1978: The vertical scale of an unstable baroclinic wave and its importance for eddy heat flux parameterizations. *J. Atmos. Sci.,* **35,** 572–576.

Held, I. M., 1982: Climate models and the astronomical theory of ice ages. *Icarus,* **50,** 449–461.

Held, I. M., and M. J. Suarez, 1974: Simple albedo feedback models of the icecaps. *Tellus,* **26,** 613–629.

Henderson-Sellers, A., and V. Gornitz, 1984: Possible climatic impacts of land cover transformations with particular emphasis on tropical deforestation. *Clim. Change,* **6,** 231–257.

Herzberg, G., and L. Herzberg, 1957: Constants of polyatomic molecules. In *American Institute of Physics Handbook,* D. E. Gray, ed., McGraw-Hill, New York, pp. 7-145–7-161.

Hobbs, P. V., ed., 1993: *Aerosol–Cloud–Climate Interactions.* Academic Press, San Diego, 237 pp.

Hobbs, P. V., and M. P. McCormick, eds., 1988: *Aerosols and Climate.* A. Deepak Publishing, Hampton, VA, 486 pp.

Holton, J. R., 1992: *An Introduction to Dynamic Meteorology,* 3rd ed. Academic Press, San Diego, 511 pp.

Houghton, J. T., G. J. Jenkins, and J. J. Ephraums, eds., 1990: *Climate Change: The IPCC Scientific Assessment.* Cambridge University Press, Cambridge, UK, 365 pp.

Houze, R. A., Jr., 1993: *Cloud Dynamics.* Academic Press, San Diego, 576 pp.

Hsiung, J., 1985: Estimates of global, oceanic meridional heat transport. *J. Phys. Ocean.,* **15,** 1405–1413.

Hunt, G. E., 1973: Radiative properties of terrestrial clouds at visible and infrared thermal window wavelengths. *Quart. J. Roy. Meteorol. Soc.,* **99,** 346–369.

Imbrie, J., and K. P. Imbrie, 1979: *Ice Ages: Solving the Mystery.* Enslow Publishers, Short Hills, NJ, 224 pp.

Imbrie, J., and J. Z. Imbrie, 1980: Modeling the climatic response to orbital variations. *Science,* **207,** 943–953.

IPCC Working Group I, 1990: Policymakers summary. In *Climate Change: The IPCC Scientific Assessment,* J. T. Houghton, G. J. Jenkins, and J. J. Ephraums, eds., Cambridge University Press, Cambridge, UK, vii–xxxiii.

Jaeger, L., 1976: Monthly precipitation maps for the entire Earth. *Ber. Dtsch. Wetterdienstes,* **18** (139), 38 pp.

Jouzel, J., C. Lorius, J. R. Petit, C. Genthon, N. I. Barkov, V. M. Kotlyakov, and V. M. Petrov, 1987: Vostok ice core: A continuous isotope temperature record over the last climatic cycle (160,000 years). *Nature,* **329,** 403–408.

Karl, T. R., G. Kukla, and J. Gavin, 1984: Decreasing diurnal temperature range in the U.S. and Canada, 1941–1980. *J. Clim. Appl. Meteorol.,* **23,** 1489–1504.

Karl, T. R., H. F. Diaz, and G. Kukla, 1988: Urbanization: Its detection and effect in the United States climate record. *J. Clim.,* **1,** 1099–1123.

Kasting, J. F., and D. H. Grinspoon, 1991: The faint young sun problem. In *The Sun in Time,* C. P. Sonett, M. S. Giampapa, and M. S. Matthews, eds., University of Arizona Press, Tucson, pp. 447–462.

Keeling, C. D., R. B. Bacastow, A. F. Carter, S. C. Piper, T. P. Whorf, M. Heimann, W. G. Mook, and H. Roeloffzen, 1989: A three-dimensional model of atmospheric CO_2 transport based on observed winds. 1. Analysis of observational data. *Geophys. Monogr.* D. H. Peterson, ed.; American Geophysical Union, **55,** 165–236.

Kelly, P. M., and T. M. L. Wigley, 1992: Solar cycle length, greenhouse forcing, and global climate. *Nature,* **360,** 328–330.

Kiehl, J. T., and B. P. Briegleb, 1993: The relative roles of sulfate aerosols and greenhouse gases in climate forcing. *Science,* **260,** 311–314.

King, M. D., 1993: Radiative properties of clouds. In *Aerosol–Cloud–Climate Interactions.* P. V. Hobbs, ed., Academic Press, San Diego, pp. 123–149.

Korzun, V. I., A. A. Sokolov, M. I. Budyko, K. P. Voskresensky, G. P. Kalinin, A. A. Konoplyantsev, E. S. Korotkevich, P. S. Kuzin, and M. I. Lvovich, eds., 1978: *World Water Balance and Water Resources of the Earth.* USSR National Committee for the International Hydrological Decade, UNESCO Press, Paris, 663 pp.

Kukla, G., 1979: Climatic role of snow covers. In *Sea Level, Ice, and Climatic Change,* I. Allison, ed., IAHS-AISH Publ. No. 131, International Association of Hydrological Sciences, Oxfordshire, UK, pp. 79–107.

Kukla, G., and D. Robinson, 1980: Annual cycle of surface albedo. *Mon. Wea. Rev.,* **108,** 56–68.

Kuo, H. L., 1974: Further studies of the parameterization of the influence of cumulus convection on large-scale flow. *J. Atmos. Sci.,* **31,** 1232–1240.

Kutzbach, J. E., and F. A. Street-Perrott, 1985: Milankovitch forcing of fluctuation in the level of tropical lakes from 18 to 0 kyr BP. *Nature,* **317,** 130–134.

Kutzbach, J. E., and P. J. Guetter, 1986: The influence of changing orbital parameters and surface boundary conditions on climate simulations for the past 18,000 years. *J. Atmos. Sci.,* **43,** 1726–1759.

Lachenbruch, A., and B. Marshall, 1986: Changing climate: Geothermal evidence from permafrost in the Alaskan Arctic. *Science,* **234,** 689–696.

Lacis, A., J. Hansen, and M. Sato, 1992: Climate forcing by stratospheric aerosols. *Geophys. Res. Lett.,* **19,** 1607–1610.

LaMarche, V. C., 1974: Paleoclimatic inferences from long tree-ring records. *Science,* **183,** 1043–1048.

Lamb, H. H., 1969: Climatic fluctuations. In *World Survey of Climatology, Vol. 2, General Climatology,* H. Flohn, ed., Elsevier Scientific Publishers, Dordrecht, Netherlands, pp. 173–249.

Landsberg, H., 1985: Historic weather data and early meteorological observations. In *Paleoclimate Analysis and Modeling,* A. D. Hecht, ed. Wiley, New York, pp. 27–70.

Langner, J., and H. Rodhe, 1991: A global three-dimensional model of the tropospheric sulfur cycle. *J. Atmos. Chem.,* **13,** 225–263.

Lean, J., E. O. Hurlburt, A. Skumanich, and O. White, 1992: Estimating the sun's radiative output during the Maunder Minimum. *J. Geophys. Res,* **19,** 1591–1594.

Ledger, D. C., 1969: The dry season flow characteristics of West African rivers. In *Environment and Land Use in Africa,* M. F. Thomas and G. W. Whittington, eds. Methuen, London, pp. 83–102.

Lee, R. B. III, M. A. Gibson, N. Shivakumar, R. Wilson, H. L. Kyle, and A. T. Mecherikunnel, 1991: Solar irradiance measurements: Minimum through maximum solar activity. *Metrologia,* **28,** 265–268.

Legrand, M. R., R. J. Delmas, and R. J. Charlson, 1988: Climate forcing implications from Vostok ice-core sulphate data. *Nature,* **334,** 418–420.

Lettau, H. H., and B. Davidson, 1957: *Exploring the Atmosphere's First Mile.* Pergamon Press, Oxford, 578 pp.

Levitus, S., 1982: *Climatological Atlas of the World Ocean.* NOAA Professional Paper 13, Rockville, MD, 173 pp.

Lewis, J. S., and R. G. Prinn, 1984: *Planets and their Atmospheres: Origin and Evolution.* Academic Press, Orlando, FL, 470 pp.

Liou, K.-N., 1980: *An Introduction to Atmospheric Radiation.* Academic Press, New York, 392 pp.

Lorenz, E. N., 1967: *The Nature and Theory of the General Circulation of the Atmosphere.* World Meteorological Organization, Geneva, 161 pp.

Lorius, C., J. Jouzel, C. Ritz, L. Merlivat, N. I. Barkov, Y. S. Korotkevich, and V. N. Kotlyakov, 1985: A 150,000 year climatic record from Antarctic ice. *Nature,* **316,** 591–596.

Lovelock, J. E., 1979: *Gaia: A New Look at Life on Earth.* Oxford University Press, 157 pp.

Mahrt, L., R. C. Heald, D. H. Lenschow, B. B. Stankov, and I. Troen, 1979: An observational study of the structure of the nocturnal boundary layer. *Bound. Layer. Meteor.,* **17,** 247–264.

Manabe, S., and A. J. Broccoli, 1985: The influence of continental ice sheets on the climate of an ice age. *J. Geophys. Res.,* **90,** 2167–2190.

Manabe, S., and R. J. Stouffer, 1980: Sensitivity of a global climate model to an increase in the CO_2 concentration in the atmosphere. *J. Geophys. Res.,* **85,** 5529–5554.

Manabe, S., and R. J. Stouffer, 1988: Two stable equilibria of a coupled ocean–atmosphere model. *J. Clim.,* **1,** 841–866.

Manabe, S., and R. F. Strickler, 1964: Thermal equilibrium in the atmosphere with a convective adjustment. *J. Atmos. Sci.,* **21,** 361–385.

Manabe, S., and R. T. Wetherald, 1967: Thermal equilibrium of the atmosphere with a given distribution of relative humidity. *J. Atmos. Sci.,* **24,** 241–259.

Manabe, S., J. Smagorinsky, and R. F. Strickler, 1965: Simulated climatology of a general circulation model with a hydrologic cycle. *Mon. Wea. Rev.,* **93,** 769–798.

Manabe, S., J. L. Holloway, and H. M. Stone, 1970: Tropical circulation in a time-integration of a global model of the atmosphere. *J. Atmos. Sci.,* **27,** 580–613.

Manabe, S., K. Bryan, and M. Spelman, 1990: Transient response of a global ocean–atmosphere model to a doubling of atmospheric carbon dioxide. *J. Phys. Ocean.,* **20,** 722–749.

Manabe, S., R. J. Stouffer, M. J. Spelman, and K. Bryan, 1991: Transient response of a coupled ocean–atmosphere model to gradual changes in atmospheric CO_2. Part I: Annual mean response. *J. Climate,* **4,** 785–818.

Manley, G., 1974: Central England temperatures: Monthly means 1659–1973. *Quart. J. Roy. Meteorol. Soc.,* **100,** 389–405.

Marland, G., 1989: Fossil fuels CO_2 emissions: Three countries account for 50% in 1986. Carbon Dioxide Information Analysis Center Communications, Oak Ridge, TN, Winter, pp. 1–2.

Martinson, D. G., N. G. Pisias, J. D. Hays, J. Imbrie, T. C. Moore, and N. J. Shackleton, 1987: Age dating and the orbital theory of the ice ages: Development of a high-resolution 0 to 300,000-year chronostratigraphy. *Quat. Res.,* **27,** 1–29.

Maykut, G. A., and N. Untersteiner, 1971: Some results from a time-dependent thermodynamic model of sea ice. *J. Geophys. Res.,* **76,** 1550–1575.

McCartney, E. J., 1983: *Absorption and Emission by Atmospheric Gases: The Physical Processes.* Wiley, New York, 320 pp.

McCormick, M. P., P.-H. Wang, and L. R. Poole, 1993: Stratospheric aerosols and clouds. In *Aerosol–Cloud–Climate Interactions,* P. V. Hobbs, ed. Academic Press, San Diego, pp. 205–219.
McFarlane, N. A., G. J. Boer, J. P. Blanchet, and M. Lazare, 1992: The Canadian Climate Centre second-generation circulation model and its equilibrium climate. *J. Clim.,* **5,** 1013–1044.
McGinnies, W. G., B. J. Goldman, and P. Paylore, eds., 1968: *Deserts of the World: An Appraisal of Research into Their Physical and Biological Environments.* University of Arizona Press, Tucson, 788 pp.
Meehl, G. A., and W. M. Washington, 1988: A comparison of soil moisture sensitivity in two global climate models. *J. Atmos. Sci.,* **45,** 1476–1492.
Mellor, G. L., and T. Yamada, 1974: A hierarchy of turbulence closure models for the planetary boundary layer. *J. Atmos. Sci.,* **31,** 1791–1806.
Mercer, J. H., 1978: West Antarctic ice sheet and CO_2 greenhouse effect: A threat of disaster. *Nature,* **271,** 321–325.
Milankovitch, M., 1941: *Canon of Insolation and the Ice Age Problem.* U.S. Department of Commerce, Israel Program for Scientific Translations, TT 67-51410, 484 pp.
Mirinova, Z. F., 1973: Albedo of the Earth's surface and clouds. In *Radiation Characteristics of the Atmosphere and the Earth's Surface,* I. A. Kondratev, ed. NASA, Springfield, VA, 580 pp.
Mitchell, J. F. B., and D. A. Warrilow, 1987: Summer dryness in northern midlatitudes due to increased CO_2. *Nature,* **330,** 238–240.
Mitchell, J. F. B., S. Manabe, V. Meleshko, and T. Tokioka, 1990: Equilibrium climate change and its implications for the future. In *Climate Change: The IPCC Scientific Assessment,* J. T. Houghton, G. J. Jenkins, and J. J. Ephraums, eds. Cambridge University Press, Cambridge, UK, pp. 131–170.
Mitchell, J. M., Jr., 1963: On the worldwide pattern of secular temperature change. In *Changes of Climate,* Arid Zone Research XX, UNESCO, Paris, pp. 161–181.
Monin, A. S., and A. M. Obukhov, 1954: Osnovnye zakonomernosti turbulentnogo peremeshi-vaniia v prizemnom sloe atmosfery. (Basic laws of turbulent mixing in the atmosphere near the ground.) *Akademiia Nauk SSSR, Geolfizicheskii Institut, Trudy,* **24,** 163–187.
Munk, W. H., 1950: On the wind-driven ocean circulation. *J. Meteorol.,* **7,** 79–93.
Nace, R. L., 1964: Water of the world. *Natur. Hist.,* **73,** no. 1.
National Research Council, 1975: *Understanding Climate Change: A Program for Action.* National Academy of Sciences, Washington, D.C., 239 pp.
Neftel, A., E. Moor, H. Oeschger, and B. Stauffer, 1985: Evidence from polar ice cores for the increase in atmospheric CO_2 in the past two centuries. *Nature,* **315,** 45–47.
Niven, L., 1970: *Ringworld.* Ballantine Books, New York, 342 pp.
Nobre, C., P. J. Sellers, and J. Shukla, 1991: Amazonian deforestation and regional climate change. *J. Clim.,* **4,** 957–988.
North, G. R., 1975: Theory of energy balance climate models. *J. Atmos. Sci.,* **32,** 2033–2043.
Oberhuber, J. M., 1988: *The Budget of Heat, Buoyancy and Turbulent Kinetic Energy at the Surface of the Global Ocean.* Max Planck Institut für Meteorologie, March, report no. 15 19 pp., maps.
Oort, A. H., 1971: The observed annual cycle in the meridional transport of atmospheric energy. *J. Atmos. Sci.,* **28,** 325–339.
Oort, A. H., 1983: *Global Atmospheric Circulation Statistics, 1958–1973.* NOAA Prof. Pap., 14, U.S. Government Printing Office, Washington DC, 180 pp.
Oort, A. H., and T. H. Vonder Haar, 1976: On the observed annual cycle in the ocean–atmosphere heat balance over the Northern Hemisphere. *J. Phys. Oceanogr.,* **6,** 781–880.
Pedlosky, J., 1987: *Geophysical Fluid Dynamics,* 2nd ed. Springer-Verlag, New York, 710 pp.
Peixóto, J. P., and A. H. Oort, 1984: Physics of climate. *Rev. Mod. Phys.,* **56,** 365–429.
Peixóto, J. P., and A. H. Oort, 1992: *Physics of Climate.* American Institute of Physics, New York, 520 pp.
Penman, H. L., 1948: Natural evaporation from open water, bare soil and grass. *Proc. Roy. Soc. London,* **193,** 120–145.

Peterson, J. T., and C. E. Junge, 1971: Sources of particulate matter in the atmosphere. In *Man's Impact on Climate,* W. H. Matthews, W. W. Kellog, and G. D. Robinson, eds. Massachusetts Institute of Technology, Cambridge, MA, pp. 310–320.

Philander, G. S., 1990: *El Niño, La Niña, and the Southern Oscillation.* Academic Press, San Diego, 293 pp.

Phillips, N. A., 1956: The general circulation of the atmosphere: A numerical experiment. *Quart. J. Royal. Meteorol. Soc.,* **82,** 123–164.

Pickard, G. L., and W. J. Emery, 1990: *Descriptive Physical Oceanography: An Introduction.* Fifth Edition, Pergamon Press, Oxford, 320 pp.

Prell, W. L., and J. E. Kutzbach, 1987: Monsoon variability over the past 150,000 years. *J. Geophys. Res.,* **92,** 8411–8425.

Ramanathan, V., and J. A. Coakley, Jr., 1978: Climate modeling through radiative–convective models. *Rev. Geophys. Space Phys.,* **16,** 465–489.

Ramanathan, V., R. J. Cicerone, H. B. Singh, and J. T. Kiehl, 1985: Trace gas trends and their potential role in climate change. *J. Geophys. Res.,* **90,** 5547–5566.

Ramanathan, V., B. R. Barkstrom, and E. F. Harrison, 1989: Climate and the Earth's radiation budget. *Physics Today,* **42,** 22–32.

Ramaswamy, V., M. D. Schwarzkopf, and K. P. Shine, 1992: Radiative forcing of climate from halocarbon-induced global stratospheric ozone loss. *Nature,* **355,** 810–812.

Rampino, M. R., and S. Self, 1984: Sulphur-rich volcanic eruptions and stratospheric aerosols. *Nature,* **310,** 677–679.

Rampino, M. R., S. Self, and R. B. Strothers, 1988: Volcanic winters. *Ann. Rev. Earth Plan. Sci.,* **16,** 73–99.

Randall, D., R. D. Cess, J. P. Blanchet, G. J. Boer, D. A. Dazlich, A. D. Del Genio, M. Deque, V. Dymnikov, V. Galin, S. J. Ghan, A. A. Lacis, H. Le Treut, H., Z.-X. Li, X. -Z. Liang, B. J. McAvaney, V. P. Meleshko, J. F. B. Mitchell, J.-J. Morcrette, G. L. Potter, L. Rikus, E. Roeckner, J. F. Royer, U. Schlese, D. A. Sheinin, J. Slingo, A. P. Sokolov, K. E. Taylor, W. M. Washington, R. T. Wetherald, I. Yagai, and M.-H. Zhang, 1992: Intercomparison and interpretation of surface energy fluxes in atmospheric general circulation models. *J. Geophys. Res.,* **97,** 3711–3724.

Rasool, S. I., 1984: On dynamics of deserts and climate. In *The Global Climate,* J. T. Houghton, ed. Cambridge University Press, Cambridge, UK, pp. 189–204.

Raval, A., and V. Ramanathan, 1989: Observational determination of the greenhouse effect. *Nature,* **342,** 758–761.

Raymo, M. E., W. F. Ruddiman, N. J. Shackleton, and D. W. Oppo, 1990: Evolution of Atlantic-Pacific $\delta\,^{13}C$ gradients over the last 2.5 m. y. *Earth Planet Sci. Lett.,* **97,** 353–368.

Richardson, L. F., 1922: *Weather Prediction by Numercial Process.* Dover Publications, New York, 236 pp.

Rind, D., and D. Peteet, 1985: Terrestrial conditions at the last glacial maximum and CLIMAP sea-surface temperature estimates: Are they consistent? *Quat. Res.,* **24,** 1–22.

Ripley, E. A., and R. E. Redmann, 1976: Grassland. In *Vegetation and the Atmosphere,* J. L. Monteith, ed. Vol. 2, Academic Press, San Diego, pp. 349–398.

Rossow, W. B., and R. A. Schiffer, 1991: ISCCP cloud data products. *Bull. Am. Meteorol. Soc.,* **72,** 2–20.

Rotty, R. M., and G. Marland, 1986: Production of CO_2 from fossil fuel burning by fuel type, 1860–1982. Carbon Dioxide Information Center, Oak Ridge National Laboratory, report NDP-006, 20 pp.

Ryaboshapko, A. G., 1983: The atmospheric sulfur cycle. In *The Global Biogeochemical Sulfur Cycle,* M. V. Ivanov and G. R. Freney, eds. Wiley, New York, pp. 203–296.

Sarachik, E. S., 1978: Tropical sea-surface temperature: An interactive one-dimensional atmosphere ocean model. *Dyn. Atmos. Ocean,* **2,** 455–469.

Sarmiento, J. L., and J. R. Toggweiler, 1984: A new model for the role of the oceans in determining atmospheric PCO_2. *Nature,* **308,** 621–624.

Schlesinger, M. E., ed., 1988: *Physically Based Modelling and Simulation of Climate and Climatic Change.* Kluwer Academic Publishers, Dordrecht, Netherlands, 1084 pp.

Schlesinger, M. E., and J. F. B. Mitchell, 1987: Climate model simulations of the equilibrium climatic response to increased carbon dioxide. *Rev. Geophys.*, **25,** 760–798.

Schlesinger, M. E., and N. Ramankutty, 1992: Implications for global warning of intercycle solar irradiance variations. *Nature,* **360,** 330–333.

Schneider, S. H., 1972: Cloudiness as a global climatic feedback mechanism: The effects on the radiation balance and surface temperature of variation in cloudiness. *J. Atmos. Sci.*, **29,** 1413–1422.

Schneider, S. H., 1989: *Global Warming: Are We Entering the Greenhouse Century*? Sierra Books, Pasadena, CA, 317 pp.

Sellers, W. D., 1965: *Physical Climatology.* University of Chicago Press, Chicago, IL, 272 pp.

Sellers, W. D., 1969: A climate model based on the energy balance of the earth–atmosphere system. *J. Appl. Meteorol.*, **8,** 392–400.

Sellers, P. J., P. J. Mintz, Y. C. Sud, and A. Dalcher, 1986: A simple biosphere model (SiB) for use within general circulation models. *J. Atmos. Sci.*, **43,** 505–531.

Semtner, A. J., 1976: A model for the thermodynamic growth of sea ice in numerical investigations of climate. *J. Phys. Ocean.*, **6,** 379–389.

Semtner, A. J., 1984: On modeling the seasonal cycle of sea ice in studies of climatic change. *Clim. Change,* **6,** 27–37.

Semtner, A. J, and R. M. Chervin, 1992: Ocean general circulation from a global eddy resolving model. *J. Geophys. Res.*, **97,** 5493–5550

Shackleton, N. J., and N. D. Opdyke, 1973: Oxygen istotope and paleomagnetic stratigraphy of equatorial Pacific core V28-238: Oxygen isotope temperatures and ice volumes on a 10^5 and 10^6 year scale. *Quat. Res.*, **3,** 39–55.

Shapiro, I. I., 1967: Resonance rotation of Venus. *Science,* **157,** 423–425.

Shea, D. J., 1986: *Climatological Atlas: 1950–1979. Surface Air Temperature, Precipitation, Sea-Level Pressure, and Sea-Surface Temperature.* NCAR Technical Note, NCAR/TN-269+STR, Boulder, CO, 210 pp.

Shimanouchi, T., 1967a: *Tables of Molecular Vibrational Frequencies, Part 1.* National Bureau of Standards, NSRDS-NBS6, March, U.S. Government Printing Office, Washington, DC.

Shimanouchi, T., 1967b: *Tables of Molecular Vibrational Frequencies, Part 2.* National Bureau of Standards, NSRDS-NBS11, October, U.S. Government Printing Office, Washington, DC.

Shimanouchi, T., 1968: *Tables of Molecular Vibrational Frequencies, Part 3.* National Bureau of Standards, NSRDS-NBS17, March, U.S. Government Printing Office, Washington, DC.

Shukla, J., C. Nobre, and P. J. Sellers, 1990: Amazon deforestation and climate change. *Science,* **247,** 1322–1325.

Slingo, A., 1990: Sensitivity of the Earth's radiation budget to changes in low clouds. *Nature,* **343,** 49–51.

Slingo, A., and H. M. Schrecker, 1982: On the shortwave radiative properties of stratiform water clouds. *Quart. J. Roy. Meteorol. Soc.*, **108,** 407–426.

Slingo, A., S. Nicholls, and J. Schmetz, 1982: Aircraft observations of marine stratocumulus during JASIN. *Quart. J. Roy. Meteorol. Soc.*, **108,** 833–856.

Smagorinsky, J., 1963: General circulation experiments with the primitive equations. I: The basic experiment. *Mon. Wea. Rev.*, **91,** 99–164.

Smagorinsky, J., 1974: Global atmospheric modeling and numerical simulation of climate. In *Weather and Climate Modification.* W. N. Hess, ed. Wiley, New York, pp. 633–686.

SMIC (Report of the Study of Man's Impact on Climate), 1971: *Inadvertent Climate Modification.* Massachusetts Institute of Technology, Cambridge, MA, 308 pp.

Somerville, R., and L. A. Remer, 1984: Cloud optical thickness feedbacks and the CO_2 climate problems. *J. Geophys. Res.*, **89,** 9668–9672.

Sonett, C. P., M. S. Giampapa, and M. S. Matthews, eds., 1991: *The Sun in Time.* University of Arizona Press, Tuscon, 990 pp.

Spencer, J. W., 1971: Fourier series representation of the position of the sun. *Search,* **2,** 172 pp.

Spiro, P. A., D. J. Jacob, and J. A. Logan, 1992: Global inventory of sulfur emissions with a $1° \times 1°$ resolution. *J. Geophys. Res.,* **97,** 6023–6036.

Stephens, G. L., 1978: Radiation profiles in extended water clouds. Part II: Parameterization schemes. *J. Atmos. Sci.,* **35,** 2123–2132.

Stommel, H., 1948: The westward intensification of wind-driven ocean currents. *Trans. Amer. Geophys. Union,* **99,** 202–206.

Stommel, H., 1965: *The Gulf Stream: A Physical and Dynamical Description.* University of California Press, Berkeley, 248 pp.

Stommel, H., and E. Stommel, 1983: *Volcano Weather.* Seven Seas Press, Newport, RI, 177 pp.

Stouffer, R. J., S. Manabe, and K. Bryan, 1989: Interhemispheric asymmetry in climate response to a gradual increase of atmospheric carbon dioxide. *Nature,* **342,** 660–662.

Street-Parrott, F. A., and S. P. Harrison, 1984: Temporal variations in lake levels since 30,000 yr BP: An index of the global hydrological cycle. In *Climate Processes and Climate Sensitivity,* J. E. Hansen and T. Takhashi, eds., Vol. 29, *American Geophysical Union,* pp. 118–129.

Stuiver, M., and P. D. Quay, 1981: Atmospheric ^{14}C changes resulting from fossil fuel CO_2 release and cosmic-ray flux variability. *Earth Plan. Sci. Lett.,* **53,** 349–362.

Stuiver, M., T. F. Braziunas, B. Becker, and B. Kromer, 1991: Climatic, solar, oceanic and geomagnetic influences on late-glacial and Holocene atmospheric $^{14}C/^{12}C$ change. *Quat. Res.,* **35,** 1–24.

Stull, R. B., 1988: *Introduction to Boundary-Layer Meteorology.* Kluwer Academic Publishers, Dordrecht, Netherlands, 666 pp.

Suarez, M. J., A. Arakawa, and D. A. Randall, 1983: The parameterization of the planetary boundary layer in the UCLA general circulation model: Formulation and results. *Mon. Wea. Rev.,* **111,** 2224–2243.

Sundquist, E. T., 1985: Geological perspectives on carbon dioxide and the carbon cycle. In *The Carbon Cycle and Atmosphere CO_2: Natural Variations Archean to Present,* E. T. Sundquist and W. S. Broecker, eds., Vol. 32, America Geophysical Union, pp. 5–59.

Sverdrup, H. U., M. W. Johnson, and R. H. Fleming, 1942: *The Oceans: Their Physics, Chemistry and General Biology.* Prentice-Hall, Englewood Cliffs, NJ, 1087 pp.

Taljaard, J. J., H. Van Loon, H. L. Crutcher, and R. L. Jenne, 1969: *Climate of the Upper Air: Southern Hemisphere,* Vol. 1, *Temperature, Dew Point, and Heights at Selected Pressure Levels.* NAVAIR 50-1C-55, U.S. Naval Weather Service, Washington, DC.

Tanner, C. B., 1960: Energy balance approach to evapotranspiration from crops. *Soil Sci. Soc. Am. Proc.,* **24,** 1–9.

Tans, P. P., I. Y. Fung, and T. Takahashi, 1990: Observational constraints on the global atmospheric carbon dioxide budget. *Science,* **247,** 1431–1438.

Thornthwaite, C. W., 1948: An approach toward a rational classification of climate. *Am. Geogr. Rev.,* **38,** 55–94.

Trenberth, K. E., ed., 1992: *Climate System Modeling.* Cambridge University Press, Cambridge, UK, 788 pp.

Trenberth, K. E., W. G. Large, and J. G. Olson, 1990: The mean annual cycle in global ocean wind stress. *J. Phys. Ocean.,* **20,** 1742–1760.

Turekian, K. K., 1968: *Oceans.* Prentice-Hall, Englewood Cliffs, NJ, 120 pp.

Twomey, S., 1977: *Atmospheric Aerosols.* Elsevier Scientific Publishing Company, Amsterdam, 302 pp.

Twomey, S. A., M. Piepgrass, and T. L. Wolfe, 1984: An assessment of the impact of pollution on global cloud albedo. *Tellus,* **36B,** 356–366.

U.S. Standard Atmosphere Supplements, 1966: Environmental Science Services Administration, NASA, U.S. Air Force, U.S. Government Printing Office, Washington, DC, 289 pp.

Untersteiner, N., 1984: The cryosphere. In *The Global Climate,* J. T. Houghton, ed. Cambridge University Press, Cambridge, UK, pp. 121–140.

Untersteiner, N., ed., 1986: *The Geophysics of Sea Ice*. Plenum Press, New York, 1196 pp.

Valley, S. L., ed., 1965: *Handbook of Geophysics and Space Environments*. McGraw-Hill, New York, 683 pp.

Van der Hammen, T., T. A. Wijimstra, and W. M. Zagwijn, 1971: The floral record of the late Cenozoic of Europe. In *The Late Cenozoic Glacial Ages*, K. Turekian, ed. Yale University Press, New Haven, CT, pp. 391–424.

Vehrencamp, J. E., 1953: Experimental investigation of heat transfer at an air–surface interface. *Trans. Am. Geophys. Union*, **34**, 22–30.

Verniani, F., 1966: The total mass of the Earth's atmosphere. *J. Geophys. Res.*, **71**, 385–391.

Vonder Haar, T. H., and A. H. Oort, 1973: New estimate of annual poleward energy transport by Northern Hemisphere oceans. *J. Phys. Ocean.*, **2**, 169–172.

Walker, J. C. G., 1977: *Evolution of the Atmosphere*. Macmillan, New York, 318 pp.

Wallace, J. M., and P. V. Hobbs, 1977: *Atmospheric Science: An Introductory Survey*. Academic Press, San Diego, 464 pp.

Warneck, P., 1988: *Chemistry of the Natural Atmosphere*. Academic Press, Orlando, FL, 757 pp.

Warren, S. G., C. J. Hahn, J. London, R. M. Chervin, and R. J. Jenne, 1986: *Global Distribution of Total Cover and Cloud Type Amounts Over Land*. National Center for Atmospheric Research, TN-273+STR.

Warren, S. G., C. J. Hahn, J. London, R. M. Chervin, and R. J. Jenne, 1988: *Global Distribution of Total Cloud Cover and Cloud Type Amounts Over the Ocean*. National Center for Atmospheric Research, TN-317+STR.

Warrick, R., and J. Oerlemans, 1990: Sea level rise. In *Climate Change: The IPCC Scientific Assessment*, J. T. Houghton, G. L. Jenkins, and J. J. Ephraums, eds. Cambridge University Press, Cambridge, UK, pp. 256–281.

Washington, W. M., and G. A. Meehl, 1989: Climate sensitivity due to increased CO_2: Experiments with a coupled atmosphere and ocean general circulation model. *Clim. Dyn.*, **4**, 1–38.

Washington, W. M., and C. L. Parkinson, 1986: *An Introduction to Three-Dimensional Climate Modeling*. University Science Books, Mill Valley, CA, 422 pp.

Watson, A. J., and J. E. Lovelock, 1983: Biological homeostasis of the global environment: The parable of Daisyworld. *Tellus*, **35B**, 284–289.

Watson, R. T., H. Rodhe, H., Oeschger, and U. Siegenthaler, 1990: Greenhouse gases and aerosols. In *Climate Change: The IPCC Scientific Assessment*, J. T. Houghton, G. J. Jenkins, and J. J. Ephraums, eds. Cambridge University Press, Cambridge, UK, pp. 1–40.

Watson, R. T., L. G. Meira Filho, E. Sanhueza, and A. Janetos, 1992: Greenhouse gases: Sources and sinks. *Climate Change 1992*, J. T. Houghton, B. A. Callander, and S. K. Varney, eds. Cambridge University Press, Cambridge, UK, pp. 25–46.

Webster, P. J., 1983: Large-scale structure of the tropical atmosphere. In *Large-Scale Dynamical Processes in the Atmosphere*, B. J. Hoskins and R. P. Pearce, eds. Academic Press, New York, pp. 235–275.

Wetherald, R. T., and S. Manabe, 1986: An investigation of cloud cover change in response to thermal forcing. *Clim. Change*, **8**, 5–23.

Wetherald, R. T., and S. Manabe, 1988: Cloud feedback processes in a general circulation model. *J. Atmos. Sci.*, **45**, 1397–1415.

Wigley, T. M. L., and T. P. Barnett, 1990: Detection of the greenhouse effect in the observations. In *Climate Change: The IPCC Scientific Assessment*, J. T. Houghton, G. J. Jenkins, and J. J. Ephraums, eds. Cambridge University Press, Cambridge, UK, pp. 239–255.

Williamson, S. J., 1973: *Fundamentals of Air Pollution*. Addison-Wesley, Reading, MA, 472 pp.

Worthington, L. V., 1954: Three detailed cross sections of the Gulf Stream. *Tellus*, **6**, 116–123.

Yapp, C. J., and H. Poths, 1992: Ancient atmospheric CO_2 pressures inferred from natural geothites. *Nature*, **355**, 342–343.

Zdunkowski, W. G., and W. K. Crandall, 1971: Radiative transfer of infrared radiation in model clouds. *Tellus*, **23**, 517–527.

Index

International Geophysics Series

EDITED BY

RENATA DMOWSKA
Division of Applied Sciences
Harvard University
Cambridge, Massachusetts

JAMES R. HOLTON
Department of Atmospheric Sciences
University of Washington
Seattle, Washington

Volume 1 BENO GUTENBERG. Physics of the Earth's Interior. 1959*

Volume 2 JOSEPH W. CHAMBERLAIN. Physics of the Aurora and Airglow. 1961*

Volume 3 S. K. RUNCORN (ed.). Continental Drift. 1962*

Volume 4 C. E. JUNGE. Air Chemistry and Radioactivity. 1963*

Volume 5 ROBERT G. FLEAGLE AND JOOST A. BUSINGER. An Introduction to Atmospheric Physics. 1963*

Volume 6 L. DUFOUR AND R. DEFAY. Thermodynamics of Clouds. 1963*

Volume 7 H. U. ROLL. Physics of the Marine Atmosphere. 1965*

Volume 8 RICHARD A. CRAIG. The Upper Atmosphere: Meteorology and Physics. 1965*

Volume 9 WILLIS L. WEBB. Structure of the Stratosphere and Mesosphere. 1966*

Volume 10 MICHELE CAPUTO. The Gravity Field of the Earth from Classical and Modern Methods. 1967*

Volume 11 S. MATSUSHITA AND WALLACE H. CAMPBELL (eds.). Physics of Geomagnetic Phenomena. (In two volumes.) 1967*

Volume 12 K. YA. KONDRATYEV. Radiation in the Atmosphere. 1969*

*Out of print.

Volume 13 E. PALMEN AND C. W. NEWTON. Atmospheric Circulation Systems: Their Structure and Physical Interpretation. 1969

Volume 14 HENRY RISHBETH AND OWEN K. GARRIOTT. Introduction to Ionospheric Physics. 1969*

Volume 15 C. S. RAMAGE. Monsoon Meteorology. 1971*

Volume 16 JAMES R. HOLTON. An Introduction to Dynamic Meteorology. 1972*

Volume 17 K. C. YEH AND C. H. LIU. Theory of Ionospheric Waves. 1972*

Volume 18 M. I. BUDYKO. Climate and Life. 1974*

Volume 19 MELVIN E. STERN. Ocean Circulation Physics. 1975

Volume 20 J. S. JACOBS. The Earth's Core. 1975*

Volume 21 DAVID H. MILLER. Water at the Surface of the Earth: An Introduction to Ecosystem Hydrodynamics. 1977

Volume 22 JOSEPH W. CHAMBERLAIN. Theory of Planetary Atmospheres: An Introduction to Their Physics and Chemistry. 1978*

Volume 23 JAMES R. HOLTON. An Introduction to Dynamic Meteorology, Second Edition. 1979*

Volume 24 ARNETT S. DENNIS. Weather Modification by Cloud Seeding. 1980

Volume 25 ROBERT G. FLEAGLE AND JOOST A. BUSINGER. An Introduction to Atmospheric Physics, Second Edition. 1980

Volume 26 KUO-NAN LIOU. An Introduction to Atmospheric Radiation. 1980

Volume 27 DAVID H. MILLER. Energy at the Surface of the Earth: An Introduction to the Energetics of Ecosystems. 1981

Volume 28 HELMUT E. LANDSBERG. The Urban Climate. 1981

Volume 29 M. I. BUDYKO. The Earth's Climate: Past and Future. 1982

Volume 30 ADRIAN E. GILL. Atmosphere to Ocean Dynamics. 1982

Volume 31 PAOLO LANZANO. Deformations of an Elastic Earth. 1982*

Volume 32 RONALD T. MERRILL AND MICHAEL W. MCELHINNY. The Earth's Magnetic Field: Its History, Origin, and Planetary Perspective. 1983

Volume 33 JOHN S. LEWIS AND RONALD G. PRINN. Planets and Their Atmospheres: Origin and Evolution. 1983

Volume 34 ROLF MEISSNER. The Continental Crust: A Geophysical Approach. 1986

Volume 35 M. U. SAGITOV, B. BODRI, V. S. NAZARNEKO, AND KH. G. TADZHIDINOV. Lunar Gravimetry. 1986

Volume 36 JOSEPH W. CHAMBERLAIN AND DONALD M. HUNTEN. Theory of Planetary Atmospheres: An Introduction to Their Physics and Chemistry, Second Edition. 1987

Volume 37 J. A. JACOBS. The Earth's Core, Second Edition. 1987

Volume 38 J. R. APEL. Principles of Ocean Physics. 1987

Volume 39 MARTIN A. UTMAN. The Lightning Discharge. 1987

Volume 40 DAVID G. ANDREWS, JAMES R. HOLTON, AND CONWAY B. LEOVY. Middle Atmosphere Dynamics. 1987

Volume 41 PETER WARNECK. Chemistry of the Natural Atmosphere. 1988

Volume 42 S. PAL ARYA. Introduction to Micrometeorology. 1988

Volume 43 MICHAEL C. KELLEY. The Earth's Ionosphere. 1989

Volume 44 WILLIAM R. COTTON AND RICHARD A. ANTHES. Storm and Cloud Dynamics. 1989

Volume 45 WILLIAM MENKE. Geophysical Data Analysis: Discrete Inverse Theory, Revised Edition. 1989

Volume 46 S. GEORGE PHILANDER. El Nino, La Nina, and the Southern Oscillation. 1990

Volume 47 ROBERT A. BROWN. Fluid Mechanics of the Atmosphere. 1991

Volume 48 JAMES R. HOLTON. An Introduction to Dynamic Meteorology, Third Edition. 1992

Volume 49 ALEXANDER A. KAUFMAN. Geophysical Field Theory and Method. Part A: Gravitational, Electric, and Magnetic Fields. 1992. Part B: Electromagnetic Fields I. 1994. Part C: Electromagnetic Fields II. 1994.

Volume 50 SAMUEL S. BUTCHER, GORDON H. ORIANS, ROBERT J. CHARLSON, AND GORDON V. WOLFE. Global Biogeochemical Cycles. 1992

Volume 51 BRIAN EVANS AND TENG-FONG WONG. Fault Mechanics and Transport Properties in Rock. 1992

Volume 52 ROBERT E. HUFFMAN. Atmospheric Ultraviolet Remote Sensing. 1992

Volume 53 ROBERT A. HOUZE, JR. Cloud Dynamics. 1993

Volume 54 PETER V. HOBBS. Aerosol-Cloud-Climate Interactions. 1993

Volume 55 S. J. GIBOWICZ AND A. KIJCO. An Introduction to Mining Seismology. 1994